PHYSICAL AND CHEMICAL BASES OF BIOLOGICAL INFORMATION TRANSFER

International Colloquim on Physical and Chemical Information Transfer in Regulation of Reproduction and Aging, 1st, Varna, Bulgaria, 1974.

PHYSICAL AND CHEMICAL BASES OF BIOLOGICAL INFORMATION TRANSFER

Edited by

Julia G. Vassileva-Popova

Central Biophysics Laboratory
Bulgarian Academy of Sciences
Sofia, Bulgaria

QH508
A1
I49
1975

PLENUM PRESS • NEW YORK AND LONDON

Library of Congress Cataloging in Publication Data

International Colloquium on Physical and Chemical Information Transfer in Regulation of Reproduction and Aging.
    Physical and chemical bases of biological information transfer.

    Includes bibliographical references and index.
    1. Biological control systems—Congresses. 2. Information theory in biology—Congresses. 3. Reproduction—Congresses. 4. Aging—Congresses. I. Vassileva-Popova, Julia G., 1932-   II. Title.
QH508.I49 1975   574.1'8   75-30849
ISBN 0-306-30862-2

Proceedings of the First International Colloquium on Physical and Chemical Information Transfer in Regulation of Reproduction and Aging, October 7-14, 1974, Varna, Bulgaria

© 1975 Plenum Press, New York
A Division of Plenum Publishing Corporation
227 West 17th Street, New York, N.Y. 10011

United Kingdom edition published by Plenum Press, London
A Division of Plenum Publishing Company, Ltd.
Davis House (4th Floor), 8 Scrubs Lane, Harlesden, London, NW10 6SE, England

All rights reserved

No part of this book may be reproduced, stored in a retrieval system, or transmitted, in any form or by any means, electronic, mechanical, photocopying, microfilming, recording, or otherwise, without written permission from the publisher

Printed in the United States of America

Preface

This volume encompasses the Proceedings of the First International Colloquium on Physical and Chemical Information Transfer and Regulation of Reproduction and Ageing (Colloquium '74-PCITRRA), held at the F. Joliot-Curie International House of Scientists, Varna, Bulgaria, October 7-14. This international event was made possible by the financial support afforded by the International Union of Biochemistry through Doctor P.N. Campbell, the International Union of Pure and Applied Biophysics through Doctor R.D. Keynes, UNESCO, and the Bulgarian Academy of Sciences.

The characteristic and stimulating feature of the Colloquium'74-PCITRRA was the interdisciplinary audience (physicists, biologists, physicians, chemists, mathematicians, engineers) and the orientation to information transfer and biological recognition in ageing. The attention was focused on receptor transformation, especially that in ageing.

Of special interest were the mathematical models and the theoretical basis of bio-information transfer, the age changes of the receptor recognition, the chemical information transfer in development, the hormone-hormone interaction envisaged in their biophysical and biochemical aspects.

The Colloquium '74 provided a very frank, critical and stimulating atmosphere for work and discussion on the latest results and concepts in topics of interest. The active role of the chairmen, Doctors H.R. Arnstein, H.-J. Bartels, M. Blecher, J.A. Edwardson, M.R.P. Hall, E.V. Jensen, B.B. Saxena, J.R. Tata, and Ya.M. Varshavsky, was a great contribution in this respect and we would like to express our gratitude to them.

We would also like to pay tribute to our correspondent consultants, Doctors F. Lipmann, B. Chance, V.A. Engelhart, and V.T. Ivanov, for their support and useful advice.

The Organizing Committee is indebted to Mr. C. Milushev for the technical organization of this publication and the correction of part of the English wording of some of the papers.

We extend our thanks to the members of the Organizing Committee, the individual participants at the Colloquium '74-PCITRRA for their cooperation, and to Plenum Publishing Corporation Limited for their interest and for the realization of this book.

Since the photo-offset mode of production was adopted to ensure the rapidity of the publication, the Editor has of necessity transferred the entire responsibility for the scientific contents, formation, and style of the various manuscripts to the individual authors. Only in rare cases could corrections be made. For technical reasons and space limitations, the discussions will not be published. Some of the papers have not been submitted because of previous publication engagements.

The successful end of the Colloquium '74-PCITRRA resulted in the formation of an international advisory body which will act as a Colloquia Committee. Its members are Professor H.R.V. Arnstein, Professor M.R.P. Hall, Professor E.V. Jensen, Professor B.B. Saxena, Dr. J.R. Tata, and Dr. J.G. Vassileva-Popova, chairman. The decision was taken that the Second Colloquium - PCITRRA will be held in 1977.

Sofia, March 1975   J.G. Vassileva-Popova

Contents

Mathematical Approach and Models of Regulatory Mechanisms

A New Mathematical Approach of Hormonal
    Regulatory Mechanisms during Growth   3
  F. Collot

The Allosteric Model of Monod, Wyman and
    Changeux and the Phenomenon of Rising
    B/F-Curves in Hormone -
    Antibody Reactions 13
  H.-J. Bartels and R.D. Hesch

Oxytocin Effect of the Depolarized Rat Uterus:
    A Mathematical Approach Using
    System Identification 21
  O. Wanner and V. Pliška

Method for Measuring the Development
    of Control Systems in Time 35
  S.B. Stefanov, T.K. Yanev, and E.L. Chakarov

Analytical Investigation of the Oscillatory
    Phenomenon in Hormone Regulation 43
  J.G. Vassileva-Popova, A. Alexandrov,
  and M. Kotarov

Substrate Concentration and Its Effect on the
    Application of the Law of Mass Action -
    A Brownian Model 53
  T. Yanev and J.G. Vassileva-Popova

Information Transfer in Ageing: An Allosteric
    Model of Hormonal Regulation 63
  J.G. Vassileva-Popova and T.K. Yanev

Physical Mechanisms of Information Transfer

The Conformational Changes of Biopolymers Retaining Their Initial Structure: A Necessary Stage of Intracellular Information Transfer 99
  Ya. M. Varshavsky

Flexoelectric Model for Active Transport 111
  A.G. Petrov

On the Role of the Conformation Changes at the Active Site for Allosteric Interactions . . . 127
  V. Kavardjikov and I. Zlatanov

A Possible Transfer of Physical Information Through the Conversion of Mechanical Energy into Electrical: A Piezoelectric Effect in Proteins 137
  J.G. Vassileva-Popova, N.V. Vassilev, and R. Iliev

Protein Hormones and Their Receptor Interactions

Effect of PGF-2alpha on the Synthesis of Placental Proteins and HPL 147
  O. Genbačev, M. Krainčanic, V. Sulović, G. Kostić, and M. Aleksić

Bioenergetic Effect of the Gonadotropic Hormones in Vivo 157
  S. Milanov, D. Panayotov, P. Kolarov, M. Protich, and A. Maleeva

Bioenergetic Effect of Thyreostimulating Hormone in Vitro 163
  S. Milanov, A. Maleeva, and N. Visheva

Sex-Dependent Prolactin Pattern during Development 169
  J.G. Vassileva-Popova and J.R. Tata

The Correlation between the Binding of the Human Chorionic Gonadotropin to Luteal Cells and Plasma Membrane of Luteal Cells 177
  S. Papaionannou and D. Gospodarowicz

CONTENTS

Particulate and Solubilized Hepatic
    Receptors for Glucagon, Insulin,
    Secretin, and Vasoactive Intestinal
    Polypeptide 201
  M. Blecher

Protein Hormones Binding to Blood and
    Sperm Cells 203
  J.G. Vassileva-Popova and S. Shulman

Small Molecules, Steroids, and Receptor Interactions

Receptor Transformation - A Key Step in
    Estrogen Action 211
  E.V. Jensen, S. Mohla, T.A. Gorell,
  and E.R. DeSombre

Testicular Feminization Syndrome: A Model
    of Chemical Information Non-Transfer 229
  R.C. Northcutt and D.O. Toft

Effect of the Synthetic Estrogen
    Dioxydiphenylhexane on the Content
    of Nucleic Acids and Proteins during
    Development 237
  G. Uzunov

Possibility for a Direct Hormone - Hormone
    Interaction 251
  J.G. Vassileva-Popova and D. Dimitrov

Cyclic AMP, Enzyme, and Hormone Networks in Bio-Development

Growth and Developmental Hormones as
    Chemical Messengers 261
  J.R. Tata

Age-Dependent Sensitivity and Specific
    Isoenzyme Inhibition of Glucose-6-Phosphate
    Dehydrogenase by Dehydroepiandrosterone . . . 279
  M.S. Setchenska, E.M. Russanov, and
  J.G. Vassileva-Popova

The Biosynthesis of Haemoglobin, Enzymes, and
    Nucleic Acids during Differentiation and
    Development of Erythroid Bone Marrow Cells . . 285
  H.R.V. Arnstein

Maturation Dependence of Adenylate Cyclase
   in Red Blood Cells 291
   D. Maretzki, M. Setchenska, and
   A.G. Tsamaloukas

Calcium and Cyclic AMP Effects of Rabbit
   Epididymal Spermatozoa 297
   B.T. Storey

Ontogenines as Carriers of Information during
   the Early Embryonic Development 307
   G.K. Roussev

New Developments in Analytical Techniques and Apparatuses

Spectrophotometric Equipment with Flash-Light
   Excitation 313
   N. Koralov, T. Todorov, G. Kostov,
   A. Shosheva, V. Ivanova, E. Fessenko,
   N. Orlov, and V. Kulakov

A Rapid-Scanning Spectrophotometer 319
   P. Shishmanov, N. Koralov, G. Georgiev,
   J.G. Vassileva-Popova, S. Kovachev, N. Uzunov,
   T. Todorov, V. Neitchev, A. Mogilevsky,
   V. Slavnyi, and M. Georgiev

Equipment for Data Registration and Processing
   in Fast Spectroscopy 331
   A. Stoimenov and M. Kotarov

ECG Telemetric Device and System for Population
   Attendance 335
   G. Georgiev, A. Venkov, Vl. Prokopov,
   S. Kovachev, N. Hadjivanova, R. Mateeva,
   and G. Kostov

Modern Instrumentation in Protein Chemistry 341
   W. Zainzinger

Physical and Chemical Transfer in Reproductive Processes

Hypothalamic - Pituitary Function and Ageing . . . 349
   M.R.P. Hall

Adenohypophysis and the Male Gonadal Axis
    Following Partial Body γ-Irradiation 361
  G.S. Gupta

The Testicular Function Efficiency Determined
    by the Spermatogenic Activity Test (SAT)
    in Diabetic and Alcoholic Patients 371
  P. Kolarov, D. Panayotov, M. Protich,
  D. Strashimirov, and S. Milanov

New Protein - Glycolipoprotein Complex
    Formation in Diluted Semen 377
  G. Kichev

Changes in EEG and in the Activity of the Whole
    Cervical Vagus upon Application of
    Sex Hormones 385
  A. Varbanova, P. Doneshka, and
  J.G. Vassileva-Popova

Epinephrine Reception by Sperm Cells 391
  P. Tsocheva-Mitseva, K. Veleva, and
  J.G. Vassileva-Popova

        Physical and Chemical Information
        Transfer in cAMP and Enzyme Network

Control of Size and Shape in the Rodent Thymus . . 401
  D. Bellamy

The DNA Polymerase and the cAMP Content of
    Human Tonsillar Lymphocytes 409
  M. Staub, A. Faragó, and F. Antoni

Insulin and Adrenalin Effect on Lung
    Triglyceride Lipases 419
  N.B. Hadjivanova and G.A. Georgiev

Magnetic Pumping of Blood in the
    Vascular System 427
  V.K. Sud and R.K. Mishra

Biomagnetic Effect of the Constant Magnetic
    Field Action on Water and Physiological
    Activity 441
  M.S. Markov, S.I. Todorov, and M.R. Ratcheva

List of Participants 451

Index 457

Mathematical Approach and Models of Regulatory Mechanisms

A NEW MATHEMATICAL APPROACH OF HORMONAL REGULATORY MECHANISMS DURING GROWTH

F. Collot

Société Internationale de Biologie Mathématique

92 - Antony, France

The growth of a living body depends upon two main parameters.

The first one looks as tied by mathematical law of logarithmic growing. At every time, a living aggregate occupying a determined volume grows proportionally to the number of its cells.

$$dw = W.dt$$

To apply the infinitesimal calculus to the problem it is necessary to define w as an unit of living matter which, at each moment, receives an equal quantity of energy (incident-energy). This unit is called "bioton". (Ref. Revue de Bio.Mathematique, 1962 No 1).

The second one, is a global of regulation which sums up the different hormonal effects, due to the hormonal simultaneous or successive actions which slow down and stop the growth process.

A new mathematical model is built, which is based on an energetic balance. At every time, the growing system gets an energetic flow φ_a: the accepted flow, which is the difference between the incident energetic flow φ_i and the reflected flow φ_r which is part of the incident flow send back.

Hence a differential relation is drawn:

$$\frac{dw}{dt} = \varphi_i - \varphi_r = \varphi_a \qquad (1)$$

Integrating this relation leads to express φ_i and φ_r in terms of w.

φ_i - looks proportional to the number of "biotons": (approximately to the number of its cells).

$$\varphi_i = a \cdot w \qquad (2)$$

(a being an appropriate coefficient)

with regard to φ_r some difficulties arise and lead to introduce two concepts:

The first one is "energetic efficiency" (denoted R). We have in effect:

$$R = \frac{\varphi_i - \varphi_a}{\varphi_i} = \frac{\varphi_r}{\varphi_i}$$

If we could know the value of R, we could conclude the value of φ_r

$$\varphi_r = R\,\varphi_i = Raw \qquad (3)$$

Secondly, that of complexity which we already tried to define by mathematics (Revue de Bio-Mathématique No.4, 1969) as follows:
We stated that: any structure, whatever be its physical, chemical, biological, sociological nature, can be split into a certain number (n) of entities called nodes and by a certain number l of liaisons between these nodes. In order to keep the coherence in such a structure, we need to have the following relation:

$$l \geq n-1$$

The minimum number lmin is, therefore, equal to $n-1$. We shall call hyper-liaison h the number of liaisons over the minimum number of liaisons:

$$h = l - lmin$$

In such a structure, we shall call complexity X the total number of distinct combinations we can get by having the position of each individual node and that of the hyper-liaisons h vary.

A combinatory analysis gives an answer to each particular case, but it is possible to establish some general formulas:

1) As a first step, we shall try to define an energetic

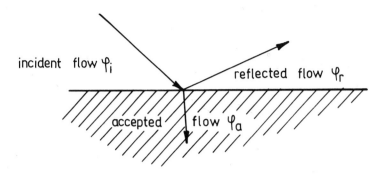

Figure 1

Diagram of energetic flows in an organism

efficiency similar to that of Carnot $R = \frac{T - T'}{T}$ but by using complexity X.

For this purpose, we shall consider that any structure immersed in the surrounding world is able to give a limited number of convenient answers r to a certain number of "situations constituting problems" among the very large number of "situations constituting problems".

We shall admit that there exists a genuine specifity of the couple: adequate answer - determined problem: in a certain way similar to the specificity of the antigen / anti-body reaction.

As a first approach, we shall state that number X, or any number proportional to it, representing the number of various aspects, the structure can show theoretically or effectively, also represents the number of specific answers r to the surrounding world. This structure may allow:

$$r = b.X$$

Let us consider a cycle of biological transformations such as that of growth, we shall call X the complexity of the incident structures S (i.e. of the ingesta) and X' the complexity of the final structure S' (i.e. the body itself), the adequate answers will be r and r'.

Now here the incident structure is absorbed by the final structure and the number of answers which can be attributed to S' only are r' - r.

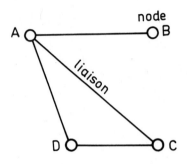

Figure 2

In this case we have:
  n(nodes) = 4
  l(liaisons) = 4
  lmin (minimum number of liaisons) = n - 1 = 3
  h (hyper-liaisons) = l - -min = 4 - 3 = 1
  H (maximum number of hyper-liaisons)=
    = 1/2 (n-1)(n-2) = 3

The complexity X is the product of 2 factors:
  - The first named mutability μ is here the number of possible permutations:
    $\mu = 4!$

  - The second factor is named liability λ and we have in the general case:

  $\lambda = \dfrac{H!}{(H-h)!\ h!}$

  Here $\lambda = \dfrac{3!}{(3-1)!\ 1!} = 3$

  and $X = \mu.\lambda = 3 \times 4!$

The "efficiency" of these very answers relative to the total number of answers which can be obtained at the end of the cycle is as follows:

$\mathcal{R} = \dfrac{r'-r}{r} = \dfrac{b_{x'} - b_x}{b_{x'}} = \dfrac{x' - x}{x'}$

Let us take an example in the sociological field and suppose that we try to determine the efficiency of the activities of the group of individuals composing nation. The incident structures S will be represented by import: raw material or machineries which are by themselves supposed to answer to a certain number of situations. The final structures would be represented by the whole of the industrial structures which is called industrial potentialities of a nation. The answers proper to this nation will be obviously equal to those gathered at the end of the cycle minus those which were already brought by import.

If you assume that to execute one answer a structure will give a quantity of energy q we have (if R is the cannot's efficiency)

$$R = \frac{q_i - q_r}{q_i} = \frac{r'q_1 - rq_1}{r'q_1} = \Re$$

$$R = \Re = \frac{\varphi_i - \varphi_r}{\varphi_i} = \frac{x' - x}{x'} = 1 - \frac{x}{x'} \tag{4}$$

Such a result can also be obtained by two other different means:

1) On one hand, by using Carnot's theorem on a thermic machine efficiency and the notion of generalized entropy which we have developed and explained, in the: (Collot, Cassé and Ricard, Revue de Bio-Mathématique, No.3, 1973).

2) On the other hand, by using the concept of generalized efficiency. The latter concept is an abstract system as "an instantaneous energetical hyper-volume (denoted Ω) as well for the incident structure, as for the reflected structure. This hyper-volume is the product of various types of energy (kinetical, calorifical (=m.cT), formal ($1/\theta$), structural ($1/X$) and so on), which are to be changed into another form of energy (mechanical work, electrical and chemical energy, psychical energy and so on).

From this point of view we shall assume that the aptitude of a certain quantity of energy to be transformed into an another form is all the greater as the incident energy is brought through a simpler structuration (simplicity being defined as the inverse number of the complexity: $\sigma = 1/X$). For example: at highest temperatures: matter is changed into plasma.

Let Ω_i be the energetical hyper-volume of the incident structure.

Let Ω_a be the energetical hyper-volume of the accepted structure.

Let Ω_r be the energetical hyper-volume of the reflected structure.

Then the generalized efficiency is defined as follows:

$$\mathcal{R} = \frac{\Omega_i - \Omega_a}{\Omega_i} = \frac{\Omega_r}{\Omega_i}$$

and finally

$$\Omega_r = \mathcal{R}\, \Omega_i$$

In the particular case of the thermodynamical efficiency where all the Ω parameters are unchanged except temperature, the generalized efficiency assumes the Carnot - efficiency shape.

$$\mathcal{R} = \frac{T - T'}{T} = R$$

In the case of the organic growth, we obtain:

$$R = \frac{X' - X}{X'}$$

Now, in order to reach integration it is necessary to express X' in relation to w (X is effectively constant).

We show that the complexity X' of a body does not depend on the number of its hyper-liaisons.

We shall obtain the result by raising several simple hypotheses on the way the number of hyper-liaisons h may vary when a body grows as a dependent variable of W.

1) <u>First hypothesis</u> - h is optimum (its value h_{op} maximizes the complexity X). $h_{op} = H/2$ (H being the maximum number of possible hyper-liaisons)

- Or - h is proportional to that optimum. Under these condition the body complexity X' becomes as follows:

$$X' = e^{\beta w^2} \qquad (5)$$

(if h is optimum, $\beta = \log 2 \# 0,7$)

It is possible to show that in a living body the complexity X' is a function of number w of biotons considered as nodes. We have

$$X' = \frac{H!}{h!(H-h)!}$$

If we call L the minimum number of possible liaisons between the w nodes of living structure, we have:

$$L = C_w^2 = \frac{w(w-1)}{2!} = 1/2\, w(w-1)$$

and

$$H = L - l_{min} = 1/2\, w(w-1) - (w-1) = 1/2(w-1)(w-2)$$

If h = H/2 we have:

$$X' = \frac{H!}{H/2!\, H/2!} = \frac{1/2\,(w-1)(w-2)!}{1/4\,(w-1)(w-2)!\quad 1/4(w-1)(w-2)!}$$

It is possible to write too:

$$X' = e^{\log X'}$$

and so to applicate the Stirling's formula to log X' since w is a great number. We have:

$$X' \# e^{\beta w^2}$$

*   *   *

By combining (1), (2), (3), (4) and (5) we obtain

$$\frac{dw}{dt} = \varphi_i - \varphi_r = aw - Raw = aw(1-R) = aw \cdot \frac{X}{X'}$$

and since the complexity X of the incident structures can be considered as constant (X = c) we finally obtain:

$$\frac{dw}{dt} = aw\,\frac{c}{X'} = awce^{\beta w^2} = \alpha w e^{\beta w^2} \tag{6}$$

(stating that a.c = α)

Equation (6) can be integrated by consecutive developments and give the following implicit equation.

$$t = \frac{1}{\alpha}\left[\log w + \frac{\beta w^2}{2} + \frac{\beta^2 w^4}{4.2!} + \frac{\beta^3 w^6}{6.3!} \cdots\right]$$

The resulting curve is a sigmoide one with an inflexion

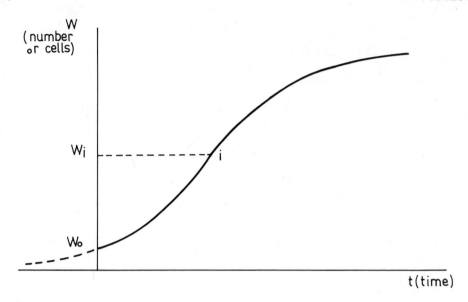

Figure 3

Representative curve of equation (6)

point for $w_i = 1/\sqrt{2\beta}$. If $w_o > \frac{1}{\sqrt{2\beta}}$ the inflexion point disappears.

2) <u>Other hypotheses</u>

If we consider other hypotheses such as stating that the number of hyper-liaisons increases proportionally to the number of cells w or to a power n comprises between 0 and 2, we get a similar equation.

* * *

This pattern allows an interpretation of evil growth which is of an expotential type. We only have to assume that the number of hyper-liaisons remains unvariable: $h = k$

Thus: $\frac{dw}{dt} = awe^{-k} = gw \Longrightarrow w = e^{gt}$

As a summary, we hope to have succeeded in bringing to light the importance of the organization concept which we could define as being liability λ or as being its logarithm (the λ liability gives the number of variances of the structure exclusively due to the shifting of the hyper-liaisons h).

This organization concept would represent the functional set of the hormonal or regulatory connections from which the whole growth depends.

THE ALLOSTERIC MODEL OF MONOD, WYMAN AND CHANGEUX AND THE PHENOMENON OF RISING B/F-CURVES IN HORMONE - ANTIBODY REACTIONS

H.-J. Bartels and

R. D. Hesch

Mathematical Inst. d. Univ. Göttingen

34 Göttingen, BRD

The question of conformational change of antibody molecules when combined with antigen molecules is not new. So, in 1965 GROSSBERG and coworkers have investigated the effect of hapten on the chymotryptic proteolysis of antibody and they found, that the presence of the specific hapten reduced the rate of antibody proteolysis. This result justifies the assumption that when a hapten is bound to an antibody the configuration of the antibody changes to another form less readily accessible for chymotrypsin /1/. At Göttingen in 1967 the kinetics of several antigen antibody reactions have been studied in more detail by FROESE with the so called temperature-jump method. Unfortunately only in a few experiments there were indications for a second relaxation effect and these effects were only slightly marked. On the basis of these kinetic experiments, it has been impossible to say whether or not a conformational change accompanies the antibody-hapten reaction /2/.

The recently observed phenomenon of increasing B/F-curves in radioimmunochemical determination methods geves rise to a renewed discussion of the above mentioned questions. Such increasing B/F-curves have been found in several radioimmunochemical assay systems, namely for adrenocorticotropic hormone ACTH (Matsukura et al. 1971 /3/), triiodothyronine (T_3), human calcitonin (HCT), parathormone (PTH), human placental lactogen (HPL), human thyroid stimulating hormone (TSH), human chorionic gonadotropin (HCG) (Weintraub et al. 1972) and gastrin. The occurrence of increasing B/F-curves seems not to be an isolated phenomenon but rather the manifestation of a general principle.

Our empirically found equation for a first simplified description of rising B/F-curves in the case of T_3 /4/ starts from an equilibrium

$$2H + AB \rightleftharpoons H_2AB$$
$$K[h]^2[ab] = [h_2ab]$$
$$KF^2([AB] - 1/2B) = 1/2B$$

This scheme totally neglects the complexes HAB. One calculates from the last equation:

$$\frac{B}{2 \cdot [AB]} = \frac{KF^2}{1 + KF^2}$$

and this agrees with the classical Hill equation

$$Y = \frac{K \cdot p^n}{1 + K \cdot p^n} \qquad (n \approx 2,6)$$

for the description of the binding of oxygen to hemoglobin (p: O_2 - pressure, Y: saturation function).

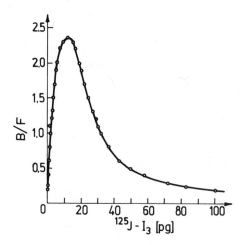

Fig. 1

Figure 1 shows a B/F-curve computed after this simplified scheme. It seems justified to discus the possibility that a well marked homotropic cooperative effect accompanies the binding of the two hormone molecules to the antibody and that this effect allows in first approximation the neglect of the complex HAB. If one assumes that the cooperative effect is mediated by some kind of molecular transition (allosteric transition) of the antibody molecules one can apply well known models to our situation: the allosteric model of MONOD-CHANGEUX-WYMAN (1965) or the sequential model of KOSHLAND and cowerkers (1966). The allosteric model postulates the existence of a preequilibrium between two conformations (the socalled R- and T-states) of the allosteric protein, each of which has a different affinity for the ligand.

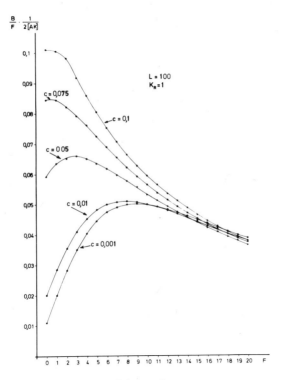

Fig. 2

Figure 2 shows theoretical B/F-curves as a function of free hormone which result from the allosteric model of MONOD. C is the quotient of the two affinity constants, L indicates the equilibrium constant between the two conformations in absence of ligands and K_R is the affinity constant for the binding of the ligand in the R-state.

In other words, it seems not justified to say that there are no reasonable explanations for increasing B/F-curves now /5/ . But on the other hand, one should not be satisfied if a model can, from plausible assumptions, explain the data observed. Instead, one should try to think of new and different experiments whose outcome can be predicted by the model so that the new observation can serve to support the model. Some experiments were undertaken in an attempt at verifying some of the conclusions derived from this model for hormone-antibody reactions.

1. Since IgG-antibody molecules show structure similarities the occurence of increasing B/F-curves has to be the normal behaviour and not an isolated phenomenon. This is veryfied in several radioimmunochemical assay systems.

2. Increasing B/F-curves have independently to be obtained using different separation methods. Indeed, increasing B/F-curves have been observed, when ammonium sulphate precipitation, electrophoresis, ethanol precipitation, double antibody precipitation and adsorption methods were used to separate antibody bound from free hormone.

3. In analogy to classical results for the binding of oxygen to hemoglobin the postulated conformational changes of the antibody molecules will show a specific temperature- and pH-value-dependence. This we have proved in the case of T_3 (compare figure 3).

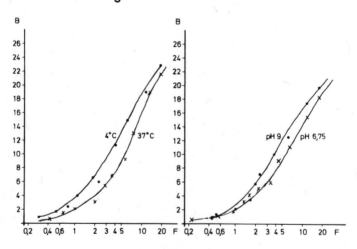

Fig. 3

4. For univalent antibody-fragments the so called Fab-fragments no increasing B/F-curve has to be expected. Recently this conclusion has been veryfied by WEINTRAUB and coworkers (1972) for HCG-antibodies.

5. The dependence of the maximal B/F-ratio on the antibody concentration can easily be calculated and has to give a linear relation. This mentioned dependence qualitatively agrees with the experimental results of MATSUKURA /3/.

6. A well marked heterogeneity of the respective antibody population can superpose totally the cooperative binding behaviour. This has been proved in the following way: Mixing three different specific antisera against T_3 we enlarged the heterogeneity of the antiserum and we obtained a changed binding behaviour.

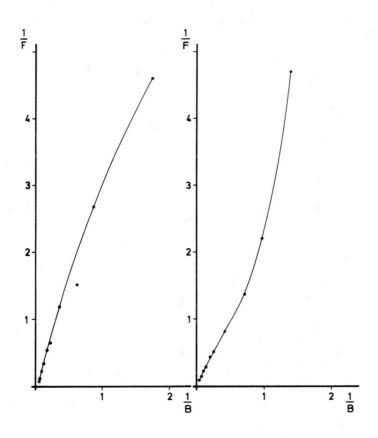

Fig. 4

Figure 4 shows on the left the binding behaviour of one of the unmixed antisera only, on the right one can see the changed behaviour of the mixture of three antisera. Here it must be pointed out, that a plot of 1/F versus 1/B showing a curvature in the direction of the 1/B - axis is equivalent to the existence of an increasing part of the B/F-curve /4/.

The experimental results of point 6 allow a further important conclusion: Since most antigens have different binding sites against specific antibodies one might reason that networks cause the increasing B/F-curves (cf. theory of L.PAULING). The totally changed binding behaviour of the mixture of the three antisera seems to exclude such an explanation of rising B/F-curves.

The above results are of practical importance for the problem of optimal sensitivity and precision in radioimmunochemical procedures. For more details compare referance /6/. Let me conclude this paper with one remark cocerning the question of measurements of other

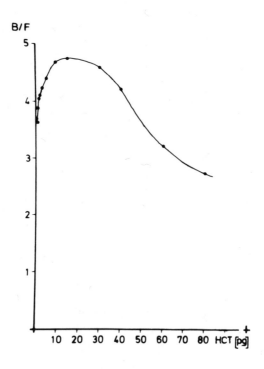

Fig. 5

antigens than ACTH on the increasing part of the B/F-curve: The results of MATSUKURA predict a greater sensitivity of such an assay system. For these purposes a high antibody concentration has to be used. Figure 5 shows a human calcitonin standard curve, 1 pg calcitonin can significantly be distinguished from zero.

REFERENCES

1. Grossberg, A.L., Markus, G. and Pressman, D. (1965), Proc. Nat. Acad. Sci. USA 54, 942-945.

2. Froese, A. (1968), Immunochemistry 5, 253-264.

3. Matsukura, S., West, C.D., Ichikawa, Y., Jubiz, W., Harada, G. and Tyler, F.H. (1971), J. Lab. Clin. Med. 77, 490-500.

4. Bartels, H.-J., Hesch, R.D. and Hüfner, M. (1972), Z. Klin. Chem. Klin. Biochem. 10, 351-354.

5. Odell, W.D. and W. H. Daughaday, Principles of Competitive Protein-Binding Assays, Lippincott Co. Philadelphia (1971) p.22.

6. Bartels, H.-J. and Hesch, R.D. (1973), Z. Klin. Chem. Klin. Biochem. 11, 311-318.

OXYTOCIN EFFECT ON THE DEPOLARIZED RAT UTERUS:

A MATHEMATICAL APPROACH USING SYSTEM IDENTIFICATION

Oskar Wanner[+] and Vladimir Pliška[*]

[*] Institute of Molecular Biology and Biophysics

[+] Institute of Measurement and Control

Swiss Federal Institute of Technology

8049 Zürich, Switzerland

INTRODUCTION

It is becoming increasingly recognized that "classical" pharmacological receptor theory, operating with relatively few fundamental concept such as receptor binding, stimulus generation, and stimulus-effect coupling is not fully adequate in describing the behaviour of actual experimental systems /1/. Nor have the recent advances in molecular endocrinology, however important in themselves, contributed to this particular problem since they have mainly been concerned with simplified biological systems at the cellular or subcellular level.

We have encountered such a situation in some studies of the action of oxytocin and some of its analogues on the depolarized rat uterus /2/. It was the aim of these studies to elucidate the contribution of hormone metabolism to the intensity and time course of the response. In a qualitative way, this aim was achieved /3/. However, attempts at a more quantitative treatment revealed that the system was more complex than we had assumed. We therefore decided to examine it by <u>modelling techniques</u>; although this approach has not yielded definitive results we believe that our experience with it is of sufficient interest to be reported here.

MODEL OF THE OXYTOCIN EFFECT

Models can be used to state a <u>hypothesis</u> in a form which allows it to be confirmed or rejected. In order to do so we need a clear definition of the system being studied, a mathematical description of its essential features and suitable experiments to test this model.

The system under investigation was defined as the K^+-depolarized rat uterus plus the organ bath /4/. The uteri were taken from rats in natural oestrus, suspended in the organ bath and attached to an isometric transducer. For a period of about one hour the muscle was incubated in a Mg^{++}-containing van Dyke-Hastings medium according to Munsick /5/ at 30°C. In order to suppress the rhythmic contractions of the uterus, it was then depolarized in a high-potassium Ringer solution /4/ and kept in this medium. Under these conditions the analysis of the response to oxytocin was simplified to a great extent and the scatter between different uteri caused mostly by variations in their hormonal states was considerably decreased.

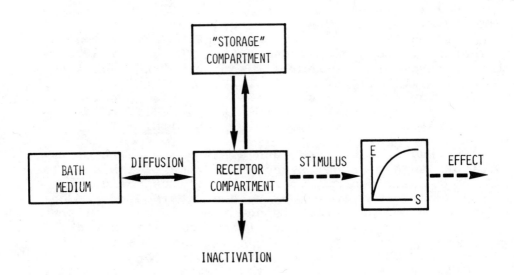

Fig.1: Conceptual model of the oxytocin effect on the rat uterus.

The conceptual model of the system is shown in Figure 1. The hormone is distributed between three compartments (as defined by Jacquez /6/). These are the bath compartment, the volume of which is large compared to the other compartments, and the receptor /7/ and "storage" compartments representing a free and bound form of the hormone, respectively, in the vicinity of the hormone receptors. The term "storage" is preferred to "residual" used in an earlier discussion of a three-compartment model /7/ since binding species for neurohypophysial hormones have been found in several target tissues, including the uterus /8/. It is assumed that the free hormone in the receptor compartment is in equilibrium with the hormone-receptor complex. The stimulus eventually inducing the muscle contraction is proportional to the fraction of receptors occupied by the hormone. The concentration of the hormone in the receptor compartment is determined by transport from the bath compartment, by inactivation processes within the receptor compartment and by reversible attachment of the hormone to various binding sites. This last process (constituting the "storage" compartment) introduces a simple nonlinearity which might account for effects not explicable by linear relations.

The dynamic part of this model is described mathematically by three simultaneous differential equations:

$$\dot{C}_0 = -K_0 C_0 \qquad\qquad\qquad + K_0 C_1 \qquad (3)$$

$$\dot{C}_1 = +K_1 C_0 - /K_1 + W_{12}(P-C_2) + \varkappa_1 / C_1 + W_{21} C_2 \qquad (2)$$

$$\dot{C}_2 = \qquad\qquad\qquad W_{12}(P-C_2)C_1 - W_{21} C_2 \qquad (3)$$

The variable C_0 stands for the hormone concentration in the bath compartment, C_1 for the concentration of free hormone in the receptor compartment and C_2 for the concentration in the "storage" compartment. The coefficients K_0 and K_1 are the ratios of the rate constants of transport and the volumes of the compartments; $\varkappa_1$ is the rate constant of inactivation; P is the concentration of binding sites; W_{12} and W_{21} are rate constants of the reversible binding. Equations (2) and (3) are nonlinear but on the assumption that P is always much greater than C_2 they can be converted to a linearized form

$$\dot{C}_1 = K_1 C_0 - (K_1 + W_{12} P + \varkappa_1) C_1 + W_{21} C_2 \qquad (4)$$

$$\dot{C}_2 = \qquad\qquad\qquad W_{12} P C_1 - W_{21} C_2 \qquad (5)$$

The effect was assumed to be a hyperbolic function of the hormone concentration in the receptor compartment:

$$E = C_1/(C_1^* + C_1) \qquad (6)$$

where C_1^* stands for the concentration eliciting one-half of the maximal effect.

To test the model, experimental data provided by earlier investigations /2,3/ were used. In these experiments, a given amount of oxitocin was added to or washed out of the organ bath, the isometric tension of the muskle was measured and the time course of contraction and relaxation ("time response") was recorded. This experiment was simulated with our model: The hormone concentration in the bath compartment, i.e. the variable C_0 was changed according to the experiment, the transient of the effect was computed and fitted to the experimental data. This optimization was carried out on a digital computer using a nonlinear least squares refinement procedure /9/.

For some experiments the fit was good, for others the model did not work at all. This was due to large variations among the experimental time responses: The reproducibility of the experiments seemed to be poor, thus the verification of the model was impossible. Since this problem was crucial, it will be discussed in a more detail in the next section.

EXAMINATION OF EXPERIMENTAL DATA

There are indeed mathematical methods to deal with the uncertainty of measurements, but in order to apply them properly we must know whether we are dealing with unpredictable influences, i.e. "noise", whether the variation is due to differences among various uteri or whether it is caused by alterations in the environment or in the system itself, such as tachyphylaxis.

To clear up this point, a new series of experiments was carrid out in which the behaviour of the system was observed under stable conditions for long periods of time. Instead of measuring the sequential responses to varying doses or different peptides, we recorded strictly timed series of contraction-relaxation cycles, each being initiated by the same dose of oxytocin. A typical recording is shown in Figure 2; it clearly demonstrates alterations in the response of the system.

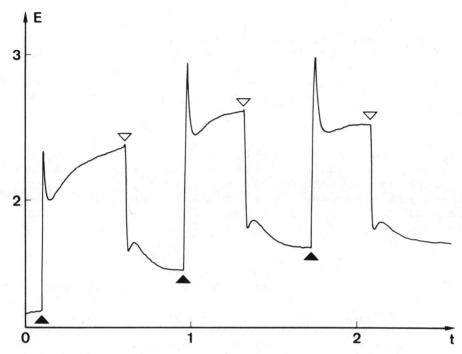

Fig. 2: Response of the rat uterus to periodic excitation by oxytocin: Isometric tension E (ponds) vs. time (hours). Increase of the hormone concentration in the bath from 0 to 2×10^{-10} M indicated by full triangles, wash-out by outline triangles.

Empirical characteristics of the time response such as peak and steady state tension were followed along the recording in order to identify the sources of these alterations. Base-line tension, for instance, increased for hours, sometimes by more than 100 %, and then dropped again slightly. Since regression analysis revealed no correlation to any other characteristic value, it was concluded that the system's response was superimposed on the base-line tension. Similar considerations using other characteristics suggested the existence of nonlinear and time-dependent relations within the system. Without giving a detailed explanation, it should be stressed that these phenomena are observed in every single experiment and therefore must be considered in any model.

To determine the extent of unpredictable influences, parallel experiments using the two uterine horns from a

single animal were carried out. Sampled values of the
two recordings were subtracted from each other and the
variance of the differences was computed. It revealed
a share of noise in the response of usually less than
5%. If this exceeded a certain threshold, the behaviour
of the system was considered "irregular" and the results
were rejected. This was the case in three curves out of
thirty.

The <u>variations between individual uteri</u> were examined. All showed the same behaviour qualitatively, i.e.
their responses could be described in terms of the same
set of phenomena which could be observed, with varying
weight, all along the recording. Any model with time-
dependent parameters which would reproduce these phenomena adequately must be able to fit all the experimental
time responses.

If our model of the oxytocin effect is to meet this
requirement, the set of its possible transients must include the set of all the "regular" time responses. A dimensionless representation of the two sets readily revealed that this condition was not fulfilled. We therefore
re-examined the assumptions and simplifications on wich
our model was based. Even if the nonlinear differential
equations (2) and (3) were used in the model in place
of equations (4) and (5) the fit was not appreciably im-
proved in spite of extensive parameter variations. Other
assumptions, such as the description of the concentration-effect function by a static relation or the ideal
step function used to model the changes of the hormone concentration in the bath, were also tested but proved not to be responsible for the inadequacy of the model. Our conclusion was that the <u>a priori</u> information
on which the model was based was either incomplete or
(partly) wrong.

To summarize, suitable collected experimental data
revealed that there was no qualitative difference in the
responses of various uteri. The responses were almost
free of noise, but varied with time. Finally it was shown
from qualitative considerations that the suggested model was inconsistent with the experimental data.

SYSTEM IDENTIFICATION

It was not known if the inconsistency of the model
was mainly due to nonlinear or time-dependent relations
or compartments of the system which were not considered

in the model. This made us search for the missing information by an identification technique.

On the basis of the input and output signal, i.e. of the hormone concentration in the bath and the isometric tension, the identification procedure determines the _essential_ _dynamic_ _components_ of the system. If no _a_ _priori_ information about the arrangement of those components is available, we are confined to _black_ _box_ _identification_. This approach describes the dynamic properties of the system by an appropriate mathematical expression, the transfer function, but eludes the problem of structure. The idea of identification is to subject system and black box to the same input, to compare the two output signals and to adjust the transfer function of the black box until the output of system and black box is equal. Then according to theory the two transfer functions are identical /10/.

A TR-10 analog computer was used to determine the transfer function, i.e. its order and the values of its coefficients. The procedure is to find the order according to the number of processes which can be distinguished from the time response of the system. With the program implemented on the computer, the numerical values of the coefficients are determined by iteration. The method uses the fact that the sensitivity of the transient to a given coefficient varies along the transient. The more the rate constants of the processes in the system differ, the more widely the zones of maximal sensitivity to different coefficients are separated from each other so that the coefficients can be determined independently. Once the best possible fit is obtained, it must be decided whether the order of the function should be increased or whether the fit is adequate. The whole procedure is a double iteration process.

An interesting question is what information can be obtained by the identification. As was mentioned above, the purpose of any identification is to determine the essential dynamic components of the system. As "essential" we designate those elementary processes which determine the form of the output signal and therefore can be picked up by the identification. Their time constants are of the same order of magnitude as the duration of one contraction-relaxation cycle, T. Very slow processes with time constants much greater than T, such as alterations of the properties of the system, are not picked up by the identification. The consequence of this is that the coefficients of those processes which are included in the transfer function seem to be time-dependent, because they have to compensate for changes in the response of the system which are caused by the slow processes. On the other hand it is also difficult to obtain results for processes with time costants much less than T. This is due to practical limitations imposed by the nonideal form of the experimental input signal and by the response time of the measuring devices. Input signal and experimental setup must be selected so as to keep their time constants below the range of the time constants of those processes in which we are interested.

Using a recording with five contraction-relaxation cycles and small input signal steps allowing linearization, the following transfer function $G(s)$ in which s is the Laplace variable and a_i, b_i are the fitted coefficients, was obtained:

$$G(s) = \frac{a_2 s^2 + a_1 s + a_0}{b_6 s^6 + \ldots + b_1 s + b_0} \qquad (7)$$

This Laplace transform, in the form provided by a black box identification, indicates that a model of sixth order is necessary to describe completely the dynamic behaviour of the system. However, six first-order elements can be connected in an infinite number of ways to a model whose transfer function is equal to equation (7). This clearly demonstrates that black box identification can not provide a single model but only a <u>class of models</u>. It is therefore pointless to discuss the significance or numerical values of the coefficients a_i and b_i as long as the problem of structure remains unsolved.

MODEL BUILDING

The central problem of model building is the problem of structure and our <u>a priori</u> knowledge was insufficient for its solution. Examination of the experimental data and the identification results provided additional information on the complexity of the system but not on its structure. In order to determine the structure, it was necessary to combine the different approaches: To postulate a structure using the results of the identification, to interpret it in terms of our knowledge of the system and to test it by experiment.

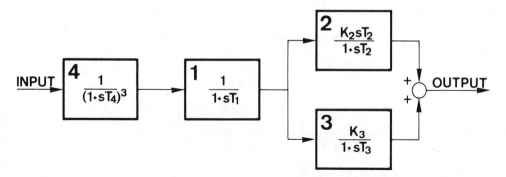

Fig. 3: Structure I. The block diagram displays the components of the system and their connections in a symbolic form. The dynamics of each component are described by its transfer function, with the Laplace variable s, the gains K and the time constants T. For detailed information see /11/.

The block diagram in Figure 3 shows one possible model of the class of models described by equation (7). This particular stucture was selected for practical reasons: First, it could be seen from the shape of the time response that structure I would be suitable to fit the time response; and second, the parameters of the parallel branches could be determined more or less indepentdently. This is important because the slowest processes, which mainly determine the shape of the response, are located in the parallel branches.

Each block in Figure 3 is described by its transfer function and represents a dynamic process. An attempt was made to ascribe a physical significance to the single blocks. Block 3 was assumed to be the receptor compartment; block 1 was assumed to be an additional diffusion barrier between bath an receptor compartment, whereas block 4 seemed to account for the nonideal form of the experimental input signal. So far, it has not been possible to find any real equivalent for block 2. This additional element, first detected by the identification, is responsible for the so-called "fade phenomenon" /2/ and is very important for a good fit because its time constant T_2 is greater than all others except T_3, and its absence is the main reason for the failure of the model based only on a priori knowledge.

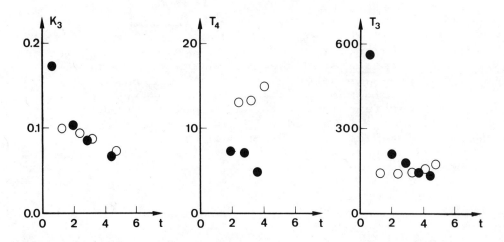

Fig. 4: Numerical values of the parameters of blocks 3 and 4 of structure I: Gain K_3 (ponds/mU), time constant T_3 (sec) and time constant T_4 (sec) vs. time (hours). The full circles are the parameter values of contraction transients, the outline circles correspond to relaxation.

The numerical values of the parameters of the model were computed from the coefficients a_i and b_i of equation (7). Since a structure was selected, they now possess physical meaning and could be given an interpretation. Some typical parameters are shown in Figure 4. The continuous decrease of the gain K_3 might be due to a slow loss of the muscle's ability to contract, e.g. due to the irreversible depletion of energy substrates, changes in the contractile apparatus, bacterial contamination etc. Block 4 had been assumed to express the deviation of the experimental input signal from an ideal step function. If this assumption were correct, then the difference between the contraction and relaxation values of T_4 indicated that the wash-out process was slower than the concentration increase of the hormone upon its addition to the organ bath. In order to test the assumption we added the hormone gradually and showed that the differences between contraction and relaxation values of T_4 disappeared.

The "drift" of the time constants, i.e. the slow decrease of the contraction and increase of the relaxation values, remained unexplained until an attempt was made to link it to the alteration of the base-line tension. When this phenomenon was taken into account by an additional element, the drift disappeared from the time constants - a typical example of a slow process which influences the parameters of the model when neglected.

The model was tested on several previously recorded series of contraction-relaxation cycles and in all cases the results were satisfactory. This means that all the information obtained from our standard experiment was now included in the model and that for its verification <u>additional</u> <u>experiments</u> had to be designed.

One such experiment examined the effect of a stepped positive input signal by recording a cumulative dose-effect curve. An interesting question was how the output signals of the two independent branches were changed when saturation was reached. The result showed that the gains K_2 and K_3 became smaller as saturation was approached; however, their ratio remained constant for the whole range of the nonlinear concentration-effect function described by eq. (6). This meant that the signals from both parallel branches were subject to the same nonlinearity. Since this nonlinearity is most probably caused by a saturation of the hormone receptors, we concluded that the receptor compartment and the fade generating element were connected in series and not in parallel. With this additional information we set up a <u>further</u> <u>model</u> <u>structure</u>, whose block diagram is shown in Figure 5. Its important components are the two feedback loops and the nonlinearity between them.

The parallel structure was replaced by the positive feedback loop because we did not know any physical phenomenon related to the hormone compartmentation which could have explained the two parallel branches.

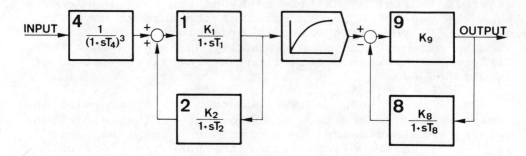

Fig. 5: Structure II. For symbols see legend to Fig. 3.

Furthermore, this positive feedback loop provides a link to the originally suggested model of the oxytocin effect, since equations (4) and (5) also form a positive feedback loop if they are expressed in the form of a block diagram. Therefore block 1 was interpreted as receptor compartment, block 2 as "storage" compartment and the nonlinearity was ascribed to the nonlinear effect function introduced by equation (6). Block 4 was still related to the nonideal form of the input signal, whereas an interpretation for blocks 8 and 9, which correspond to block 2 of structure I, cannot be given as yet. Since, however, the model is nonlinear, it was possible to prove that the fade generating blocks 8 and 9 are located behind the nonlinear element, and we tend to believe that they might represent a feedback mechanism operating in the contraction-relaxation system of the muscle.

So far we have found no evidence incompatible with structure II; however, the experimental verification of the hypotheses is still in progress.

SUMMARY AND DISCUSSION

The model based on the a priori knowledge of the system has turned out to be inconsistent with the experimental data. It was not known whether elements of the suggested model had to be changed or whether additional elements were needed to improve the model. Therefore a black box identification was carried out in order to obtain a mathematical description of the experimental time response. The identification showed that the response could be fitted with a linearized model, but it also revealed that the order of the suggested model had been too low: An additional linear element, the "fade generating element", had to be added to the model. In structure I we postulated that this element was in parallel with the receptor compartment. However, an experiment designed to test this hypothesis showed that the two elements were connected in series and not in parallel, which resulted in model structure II.

Although this structure is consistent with the available a priori information and experimental data, it should not be considered as final, but rather as a further step in an iterative procedure. The iterative approach is necessary as the system is complex and only one variable, the output, was measured. The iteration could be speeded up by measuring additional variables of the system.

ACKNOWLEDGEMENTS

This work was generously supported by the Swiss National Science Foundation grants 3.0780.73 and 3.424.70.

REFERENCES

1. J. Rudinger, V. Pliška & I. Krejčí, Rec. Progr. Horm. Res. $\underline{28}$, 131-165 (1972).

2. J. Furrer & J. Rudinger, Experientia $\underline{28}$, 742 (1972).

3. G. Jutz, V. Pliška, O. Keller & J. Rudinger, Experientia $\underline{30}$, 696 (1974).

4. H.O. Schild, Brit. J. Pharmacol. $\underline{36}$, 329-349 (1969).

5. R. A. Munsick, Endocrinology, $\underline{66}$, 451-457 (1960).

6. J.A. Jacquez, "Compartmental Analysis in Biology and Medicine", Elsevier Publishing Company, Amsterdam, 1972.

7. V. Pliška, Il Farmaco, Ed. Sci. 23, 623-641 (1968).

8. M. Ginsburg & K. Jayasena, J. Physiol. 197, 53-63 and 65-76 (1968).

9. J. Lang & Müller, Computer Phys. Commun. 2, 79-86 (1971).

10. P. Eykhoff, "System Identification", J. Wiley & Sons, New York, 1974.

11. J.J. DiStefano, A.R. Stubberud, I.J. Williams, "Feedback and Control Systems", Schaum Publishing Company, New York, 1967.

12. W.D.M. Paton, Proc. Roy. Soc. London B 154, 21-69 (1961).

METHOD FOR MEASURING THE DEVELOPMENT OF CONTROL SYSTEMS IN TIME

S.B. Stefanov
Institute of Poliomyelitis and Virous Encephalites, USSR Acadamy of Sciences, Moscow, USSR

T.K. Yanev
Central Laboratory of Biophysics, Bulgarian Academy of Sciences, 1113 Sofia, Bulgaria

E.L. Chakarov
Institute of Oncology, Bulgarian Academy of Sciences, Sofia, Bulgaria

The changes observed in complex control systems (CS) is due to regulating factors (RF) exceeding the action of the random outside effects. The system retains its complexity, but in the general case both the separate links and their relations and interactions change. The state of the system is described by a determining set of signs a_j and by a function $F(A_j)$ specific for the system. If a comparative analysis of objects with the same function F, but with different values of a_j is carried out, then the type of function F is of no importance and to each state O_j corresponds isomorphically an image-vector R_j in the multidimensional space $M_{(j)}$ of the sign a_j. Such is the case of the ontogenetic development of biological objects of the same species in an interval of time where the type of function F remains the same (e.g. the function of metabolism). Even in this case, however, the number of the signs a , resp. of the dimensionality in the space $M_{(j)}$, is usually considerable. On account of this the researchers tend to reduce this number, even when it leads to a compromise with the accuracy of the analysis. One possibility for such a compromise is the accuracy of the scale used for measuring the differences between the values of a_j in the different states (moments of time).

In some previous works (1 - 4) a method was suggested having an evaluation scale of three degrees: +, -,

0 depending on the quantities compared: differ more, less, or have no reliable difference. This seemingly rough scale may be made accurate enough in our case of CS provided the smallest time intervals, for which a_j are compared, are small enough. However, such a scale gives the possibility of reducing the multidimensional space M to a bidimensional one through a suitable transformation described in some works (2 - 5). Stefanov et al. (5) have suggested an iteration method with which the analysis in the bidimensional space of the degree of similarity between the objects investigated is carried out without any loss of imformation (with the already assumed accuracy of the threedegree scale). To this advantage we should add also the nondimensionality of the evaluations of the scale permitting the comparison of signs with different dimensionality.

In this case every momentary state of the CS may be considered as an object characterized by the real values of the set of signs a_j, while the various momentary states may be compared between them as different objects. The reduction of the multidimensional space M to a bidimentional one brings foreward all the well-known advantages of a plane image. A procedure for the application of this approach has already been developed (1 - 5), and here its method of application will be described.

Every momentary state of the CS is called an "object" and an ordinal number is assigned to it. Every sign of the object also gets an ordinal number. This is a code number (see the working card) under which the value of the sign and its confidential interval are registered. Each object has a card filled in bearing its ordinal number. It contains a matrix whose horizontal lines refer to the various objects, and the columns refer to the different signs. In the card there is place for carrying out arithmetic operations comprised in the algorithm. In this way the number of cards corresponds to the number of objects (states of the CS), the number of columns to that of the signs and the number of lines to that of the objects.

The card is filled in by putting one of the three degrees of the scale: +, -, or 0 in each cell according to the presence or absence of reliable difference between the values of the respective signs of the objects compared. Evidently, in each card the line corresponding to the object for which the card is filled in will contain only naughts because the object does not differ from itself by any of its signs.

DEVELOPMENT OF CONTROL SYSTEMS

Then, the following arithmetic operations are carried out in each card:

$\Sigma(+) : \Sigma(+,-,0) = CR_+$ (coefficient of the positive relations)

$\Sigma(-) : \Sigma(+,-,0) = CR_-$ (coefficient of the negative relations)

$[\Sigma(+) + \Sigma(-)] : \Sigma(+,-,0) = CD$ (coefficient of differentiation)

$[\Sigma(+) - \Sigma(-)] : \Sigma(+,-,0) = CP$ (coefficient of predominance).

The coefficients of relations CR_+ and CR_- are considered as coordinates in a bidimensional space: "field of relations". They determine a vector whose apex is an analogue (image) of each object of the multidimensional space $M_{(i)}$ in the bidimensional space of the "field of relations". The latter represents an isoceles right-angled triangle (Fig.1). The intercepts of its hypotenuse

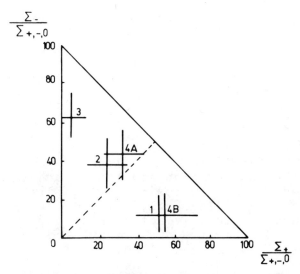

Figure 1

A system of liver mitochondria from female rats, regulated by the sex cycle and measured by a set of 6 morphometric and 8 polarographic signs. Sequence of the cycle phases: №№1, 2, 3, 4A, 4B.

with the coordinate axes represent 100% positive, resp. 100% negative, differences of a given object in comparison with the others. In this way the position of each point in the field of relations is "a system characteristics" for the given object, because it represents its "place" among the others evaluated by the three-degree scale. The confidential limits of the coordinates of this "place" are determined by the tables of the binominal distribution (there is a difference "+", or there is not, resp. there is a difference "-", or there is not).

The system of points-analogues in the field of relations is not isomorphous to the system of objects, as far as it represents an image of "many in one", since it is possible different objects to have the same coordinates in the field of relations. However, this manner of imaging has the significant advantage that it is convenient for easy graphic operations and good visuality. With a suitable iteration algorithm (5) this non-isomorphism may be overcome, so that only similar objects (in the sense of the three-degree scale of evaluation of their similarity) are in spatial proximity in the field of relations.

The most important properties of the field of relations will be discussed hereafter. In the case of the cyclic development of the CS the points-analogues of its states will shift from the initial position in the field of relations along an almost closed curve (it will be completely closed in the ideal case when there is an exact cyclic development where the initial and the final positions coincide). Fig.1 shows the field of relations for mitochondria controlled by the sex cycle (3). It is clear that the disposition meets the expectations: the initial and the final state are close, but do not coincide, while the two intermediary states are located one near the other and are approximately equidistant from the initial and physiologically antipode state No.3. The reason for such an image is not to bear out the result expected, but to give it a qualitative and obective appearance, which has been achieved.

In a previous study (14) we considered an example of the noncyclic development in time which is reflected in the field of relations via a progressive trajectory.

Since for the whole set of objects investigated the number of positive and negative relations is equal (property of the relations matrices), then the points-analogues will be distributed on both sides of the bisector of the field of relations, and the more all the objects of the set considered are similar, the nearer to the beginning of the field of relations they are and vice versa. For this reason the action of a given regulating factor (RF) may be evaluated by the dispersion of the points-analogues in the field of relations: the greater the dispersion and the more markedly directed to the hypotenuse of the field of relations, the stronger is the regulating action of the RF, and vice versa. Irrespective of that whether the cycle is opened or closed, the curve-image of the process controlled intercepts the bisector (the points-analogues are situated on both sides of it).

As noted above, the imaging through the points-analogues in the field of relations is not isomorphous: the different objects are described by different points-analogues, yet it is possible that different objects correspond to identic points-analogues. In this case one or more repetitions of the procedure for the determination of the coefficients of the relations CR_+ and CR_- by the iteration method described above (each time the repetitions comprise only objects grouped in one region of proximity) provide the possibility of eliminating the objects "foreign" to the region of proximity in the field of relations. The criterion for proximity is chosen according to the methods of the theory of pattern recognition, or it is done intuitively. The iterations of the procedure are carried on until a region is obtained situated on the bisector. If the region is homogenous, then it is situated in the imediate proximity of the beginning of the coordinate system (if it is composed only of unrecognizable objects, then it is right at the beginning). If the region is composed of several subregions, they are situated in the mutual proximity with the bisector (when it is composed only of objects unrecognizable inside the region, they are situated exactly on one point of the bisector). The position of the bisector determins the number of the sub-regions, but the number of objects in each region is the same.

In the cases when a verification is required to establish whether one or several objects belong to a given region, because of doubts about their being "foreign" to that region, it is convenient to use the coefficients CP and CD. Evidently, two objects are identic only if the CP and the CD are identic (they are calculated only for the two objects compared). The large value of the CP indicates a great difference between the two objects, while the large value of the CD shows that this difference is shifted considerably in the direction of positive or negative differences.

Thus, through the points-analogues in the field of relations it seems possible to express by a two-component numerical estimation (resp. to express graphically): a) the momentary state of the CS at a given moment of time; b) the relative "position" of a given state in connection with the others in the field of relations; c) the strength of the action of a given regulating factor.

The method of the morphokinetic synthesis is suitable for application only when the CS state is described by a large number of signs which renders irrelevent the researcher's intuitive estimation.

This method has been applied in a number of problems in biophysics, biochemistry, age morphology, microbiology and cytophotometry (6-14).

The method has a simple procedure and does not require any special mathematical qualification.

WORKING CARD

	Sign	1	2	3	...	m	CD	CP
	Value							
Object No.1	Confidential limits						...	...
Object No.1		0	0	0	...	0		
Object No.2		+	−	0		−		
Object No.3		−	+	+		+		
. . .		. .	.	.	. . .			
Object No. n		+	0	−		−		

$\Sigma(+) = \ldots, \quad \Sigma(-) = \ldots, \quad \Sigma(+,-,0) = \ldots, CR_+ = \ldots, CR_- = \ldots$

REFERENCES

1. Stefanov S.B., Proc. IVth Biophys. Cong. Moscow, 1972.
2. Stefanov S.B., Depon. VINITY №1185-74, 6/V, Moscow, 1974 (in Russian)
3. Stefanov S.B., Biofizika, 19, №5, 1974 (in Russian)
4. Stefanov S.B., Proc. IVth All-union Meet. Automation Image Analysis, Nauka, Moscow, 1974 (in Russian)
5. Stefanov S.B., Yanev T.K., Chakarov E.L., Meditsinska tekhnika, Sofia, 1974 (in Bulgarian)
6. Stefanov S.B., Liul'kina E.I., Palilova A.N., Atrashonok N.V., Proc. IXth All-union Conf. Electron Microscopy, Tbilissy, 1973 (in Russian)
7. Stefanov S.B., Preprint, Pushchino, 1974 (in Russian)
8. Palilova A.N., Liul'kina E.I., Atrashonok N.V., Stefanov S.B., Bull. Acad. Sci. Bieloruss. SSR, №2, 1974 (in Russian).
9. Stefanov S.B., Motlokh N.N., Alexeeva L.V., in: New Methods in Age Physiology, №3, Pedagogika, Moscow, 1974 (in Russian).
10. Lapina Z.V., Stefanov S.B., in: New Methods in Age Physiology №3, Pedagogika, Moscow, 1974 (in Russian).
11. Rylkin S.S., Stefanov S.B., Petrikevich S.B., Bull. Acad. Sci. USSR, Biological Series, Moscow, (in press; in Russian).
12. Stefanov S.B., Nossova L.S., in: Collec. papers of the Instit. Bacteriol., Gor'ky (in press; in Russian).
13. Semenova L.K., Antipov E.E., Stefanov S.B., Collec. papers Pedagogical Inst., Gor'ky, (in press; In Russian).
14. Stefanov S.B., Yanev T.K., Chakarov E.L., (in press; in Russian).

ANALYTICAL INVESTIGATION OF THE OSSCILLATORY PHENOMENON IN HORMONE REGULATION

Julia G. Vassileva-Popova, A. Alexandrov and M. Kotarov

Central Laboratory of Biophysics, Bulgarian Academy of Sciences, 1113 Sofia, Bulgaria

The field of chronobiology /1-6/, the biological clocks /7/ and the rhytmicity in the biological functions /8/ became a stimulating part of research. In this respect the work done on the circadian rhythm in the hormonal regulation /9, 10/ provides a good support for an interest in this field. Some of the works are oriented to the study of ageing in the aspect of circadian rhythms /11/.

The object of the present study is to introduce the hormonal regulation as an oscillatory periodic process and to present its possible analysis. The physical meaning of oscillation is assumed as an application to this kind of bio-regulation.

As a formal description of the phenomenon investigated from a biological point of view it is known that in a definite location the central signal (CS) is formed. The CS incorporated in the central regulatory protein (CRP), otherwise the protein hormone (PH) propagates in bio-space transferring the information to the target receptors (TR) specialized in recognition, registration and interpretation of the CS. The CS, resp. PH, change their quantity in portions perhaps as a "quantum" process in time and cause cascade reactions as a specific answer at the target level.

In this respect the present study describes the results of:

- a possible analytical representation of the laboratory data for a more occurate characterization via the Fourie exponsion and

- the comparative analysis of the periodic changes of the hormonal concentration and its reflection on the periodic changes of the hormonal receptors.

METHODS AND DISCUSSION

It has been shown that a large class of periodic functions satisfying some general mathematical conditions may be expanded into a series of hormonic functions, i.e. of sines and cosines with frequencies divisible to the basic frequency of the function $\omega = \frac{2\pi}{T}$. The functions obtained experimentally, which express the periodic processes in physics, always satisfy the condition to be represented in a Fourie series. For this reason we shall not dwell upon the mathematical conditions where the expansion in a Fourie series is possible /12, 13, 14/.

If we have a periodic function $F(t) = F(t + T)$ which is expandable into a Fourie series, then in the most general case this series consists of a infinite number of terms from which in practice way is accepted only the first terms of the series.

$$F(t) = \frac{1}{2} a_0 + a_1 \cos \omega t + b_1 \sin \omega t + a_2 \cos 2\omega t + b_2 \sin 2\omega t + \ldots + \ldots ,$$

or

$$F(t) = \frac{1}{2} a_0 + \sum_{n=1}^{\infty} a_n \cos n\omega t + \sum_{n=1}^{\infty} b_n \sin n\omega t ,$$

where $\omega = \frac{2\pi}{T}$.

Besides in this form the expansion in the Fourie series may take the form:

$$F(t) = \frac{1}{2} a_0 + \sum_{n=1}^{\infty} A_n \sin(n\omega t + \varphi_n)$$

The comparison of the two forms of representation gives:

$$A_n \sin(n\omega t + \varphi_n) = a_n \cos n\omega t + b_n \sin n\omega t,$$

from where

$$A_n \cos \varphi_n = b_n, \quad A_n \sin \varphi_n = a_n \quad \text{or}$$

$$\text{tg } \varphi_n = \frac{a_n}{b_n} \quad \text{and} \quad A_n^2 = a_n^2 + b_n^2$$

The parameters investigated in the present study are:

1. The quantity (Q) of the central regulatory protein (in our case LH) per unit blood serum in female sex for a period (T) of 28 days.

2. The capacity of saturation (C) of the target receptors with LH for the same period of time.

The first parameter Q represents an average value of the empirical data obtained via radio-immunoassay /15,16/ and our results. The description of this parameter is given and analysed as $f(x)$ function. The second parameter C includes our results obtained by means of radio-ligand receptor assay based on the principles accepted /17-19/. With some approximation the second parameter is analysed as $\varphi(x)$ function. For both functions /$f(x)$ and $\varphi(x)$/T = 28 days.

Attention is paid mainly to two classes of problems:

1. A numerical valuation of the amplitudes (A) and the phase shifts between $f(x)$ and $\varphi(x)$,

2. A comparative analysis of sets of functions $\varphi(x)$ and $f(x)$ obtained in various conditions.

The solution of these problems requires functions $f(x)$ and $\varphi(x)$ to be represented as a minimal number of parameter which would facilitate the comparative analysis conserving the necessary accuracy of the description.

The curves describing Q and C are of the type given in Fig. 1.

Let us introduce the Fourie expansion of the functions $f(x)$ and $\varphi(x)$. On the basis of experimental data

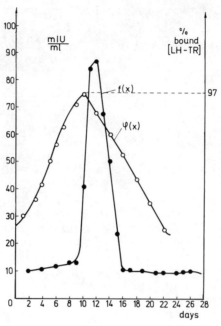

Fig.1

fed in a computer the following values for function f(x) were obtained:

f(x) = 18.03 -
- 17.52 cos x + 12.77 sin x +
+ 9.48 cos2x - 13.39 sin2x -
- 1.12 cos3x + 11.18 sin3x -
- 7.17 cos4x - 8.10 sin4x +
+ ...

and for function φ(x):

φ(x) = 44.05 -
- 24.38 cos x - 7.20 sin x +
+ 2.77 cos2x - 0.37 sin2x -
- 2.04 cos3x + 0.31 sin3x +
+ ...

The Fourie expansion obtained for the functions f(x) and φ(x) shows that the periodical process is expanded in a spectrum. Its components are harmonics of which the periodical process is constructed. The spectrum is discrete and is composed of the separate parts of the harmonics forming an arithmetical series x, 2x, 3x, ... Very often x is assumed to be the basic frequency, and 2x, 3x, ... as its harmonic ones.

The data from the Fourie expansion of the functions $f(x)$ and $\varphi(x)$ allow to be represented as:

$$f(x) = 18.03 +$$
$$+ 21.68 \sin(x - 54°) +$$
$$+ 16.40 \sin(2x - 35°) +$$
$$+ 11.23 \sin(3x - 5°50') +$$
$$+ \ldots$$

$$\varphi(x) = 44.05 +$$
$$+ 25.38 \sin(x - 106°30') +$$
$$+ 2.80 \sin(2x - 82°10') +$$
$$+ \ldots$$

In our case the steep slope of curve $f(x)$ was obtained mainly under the influence of the amplitude of the basic harmonic. The amplitudes of the other harmonics were considerably smaller.

The free term of function $\varphi(x)$ is approximately 2.5 times larger than the free term of function $f(x)$. They express the mean values of the respective functions for the period T = 28 days.

The phase shift between $f(x)$ and $\varphi(x)$ suggests that the peak of the hormonal concentration appeare because of receptor saturation. The decrease of the hormonal concentration appears usually after the realization of the maximum hormonal action at the TR. The relatively not so quick decline of the hormonal level after the maximum hormonal action (as C of TR) is possibly due to a comparatively slow classical feedback control. Probably this is the purpose of the existance at the TR of a quick and programmed mechanism for the quantitative control of the hormonal response.

The useful information can be obtained by the comparison of function $f(x)$ with function $\varphi(x)$. It could be assmed that the function $f(x)$ is a primary oscillation in relation to function $\varphi(x)$. Because of the complexity of the system investigated the existence of different levels of oscillations should also be taken into consideration.

The second part of the present study deals with the approximated experimental data. On the basis of the analysis of the Q of LH - $f(x)$ and the C of TR for LH - $\varphi(x)$, two additional behaviours of the TR were shown. The first one represents the value of C of TR in the aged gruop - $\varphi_1(x)$. In the second case the TR were isolated from patients treated with clomiphene - $\varphi_2(x)$ *
(Fig. 2).

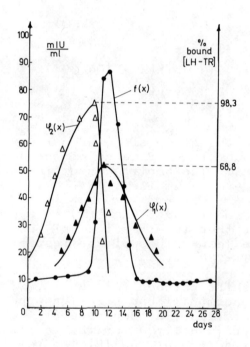

Fig.2

The Fourie expansion of the functions $\varphi_1(x)$ and $\varphi_2(x)$ is:

---

* Because of difficulties in obtaining complete results for C of TR in $\varphi(x)$ and especially in $\varphi_1(x)$ and $\varphi_2(x)$ a mathematical approximation of these curve was made.

$$\varphi_1(x) = 19.66 +$$
$$+ 32.38 \sin(x - 50') +$$
$$+ 16.91 \sin(2x - 17°) +$$
$$+ \ldots$$

$$\varphi_2(x) = 18.97 +$$
$$+ 25.38 \sin(x - 82°50') +$$
$$+ 10.62 \sin(2x - 68°40') +$$
$$+ \ldots$$

Another observation is that C of TR for functions $\varphi(x)$ and $\varphi_2(x)$ is that their values are relatively close, while the value of function $\varphi_1(x)$ differs considerably from them.

The comparison of the representative functions $\varphi(x)$, $\varphi_1(x)$ and $\varphi_2(x)$ indicates the possibility of a precise enough for the moment analysis. This analytical approach suggests a possibility for a programmed computer recognition and diagnosis of the series of different functions.

GENERAL DISCUSSION

On the basis of analysis made and the data from yet unfinished experiments it is possible to have the following comments: in the previous part of the present study we discuss the analytical results from the correspondance between the hormonal signal and receptors recognition where: $\varphi(x)$ depends on $f(x)$.

Besides this the theoritical possibilities for a different correspondance between the hormonal signal and the response at the receptor level should be introduced as an idealized combinations (Fig. 3).

The basic correlation between the Q of PH and the C of TR does not exclude the opportunity the receptor recognition (receptors at the local and central levels) to be dependent only in one hand on the basic hormone as a signal, and on the other on the releasing hormone (RH) as an amplifaer. With the assumption of the dual control of RH (of the PH release and the mediation of the hormone receptor recognition) the picture becomes too complicated for the kind of analysis indicated above.

Another point is dealing with the possible superpositions of the different oscillatory levels. Here the primary PH oscillation and the corresponding secondary oscillation at the receptor level are indicated. The primary oscillation of PH and the secondary oscillation of TR are assumed for a more accurate analysis, but this assumption is quite relative because it is more likely that PH is a secondary oscillation instead of the primary RH oscillation. In this case the RH should have a periodicity corresponding to that of the PH release. The complexity of such a system as hormonal regulation does not allow an easy simplification for the mathematical approach.

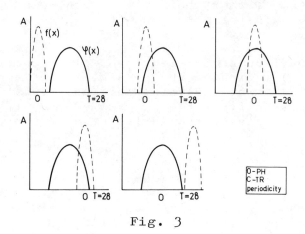

Fig. 3

The question raised in the present study should suggest a dynamic accumulation of the different oscillation levels:

RH oscillation ⟶ PH oscillation ⟶ TR oscillation ⟶ cellular enzyme activity oscillation /20, 21/.

The possibility of studying the hormonal regulation on the basis of the oscillatory mechanism and the wavelength process as a temporary dissipative matter should be also of theoretical interest.

Coming back to the analytical background of the present study the following conclusions should be drawn:

- the oscillatory mechanism is a possible way for the preservation of programmed balance of the hormonal regulation;

- the useful information is received via an analytical approach: in this case the Fourie expansion;

- the correspondance of the rhythms between hormonal signalization and target reception allow a comparative analysis;

- the present study should be a kind of preparation for a programmed computer analysis and diagnosis of a series of laboratory results.

Some of the theoretical approximations should be the basis for further laboratory and mathematical examinations.

REFERENCES

1. Luce, G., in:Biological Rhythms in Psychiatry and Medicine, U.S.Public Health Service Publication No.2088, Chevy Chase, Md., 1972.

2. Halberg, F., Ann.Rev.Physiol., 31(969)675.

3. Reinberg, A., & Halberg, F., Ann.Rev.Pharmacol., 2(1971)465.

4. Bünning, E., in: The Physiological Clock., Springer-Verlag, Berlin 1974.

5. Kleitman, N., in: Sleep and Wakefulness, University of Chicago Press. Chicago, Ill.1965.

6. Mills,J., Phys.Rev., 46(1966)71.

7. Winfree, A. in: Temporal Aspects of Therapeutics (Urquardt,J., Yates, F., eds.) Plenum, New York 1973.

8. Halberg, F., Annals of the New York Academy of Sciences 231(1974)108.

9. Rashevsky, N., Bull.Math.Biophys., 30(1968)735.

10. Rashevsky, N., Bull.Math.Biophys., 34(1972)65.

11. Yunis, E.J., Fernandes, G., Nelson, W. and Halberg,F., in Chronobiology (Scheving, L.E., Halberg, F. and Pauly, J.E., eds.) Igaku shoin LTD., Tokyo, 1974.

12. Korn, G.A. and Korn, T.M., in: Mathematical Handbook for Scientists and Engineers, McGrow-Hill Book Co., New York 1968.

13. Feynman, R.P., Leighton, R.B. and Sands, M., in: The Feynman Lectures on Physics, v.I and 2, Addison-Wesley Publ.Co., London 1963.

14. Davidov, A.S. in: Quantum Mechanics, Nauka, Moscow 1973 (in Russian).

15. Speroff, F., Wiele, V., Amer.J.Obstet.Gynec. 107 (1970)IIII.

16. Saxena, B.B., Rathnam, P. and Römmler, A., Endocr. Experimentalis 7(1973)19.

17. Haour, F. and Saxena, B.B., J.Biol.Chem., 249(1974) 2195.

18. Tsurahara, T., Van Hall, E.V., Dufau, M.L. and Catt, K.J., Endocrinology 91(1972)463.

19. Gospodarowicz, D. (in press, 1974).

20. Dynnik, V.V., Sel'kov, E.E. and Semashko, L.R., Studia biophysica, 41(1973)193.

21. Gerisch, G. and Hess, B., Proc.Nat.Acad.Sci. USA 71(1974)2118.

SUBSTRATE CONCENTRATION AND ITS EFFECT ON THE APPLICATION OF THE LAW OF MASS ACTION - A BROWNIAN MODEL

T. Yanev and Julia Vassileva-Popova

Central Laboratory of Biophysics, Bulgarian Academy of Sciences, 1113 Sofia, Bulgaria

The law of mass action has been postulated for an ideal solution with the following assumptions:
 a) space isotropy of the physical properties of the solution;
 b) indipendence of the free transition of the interacting molecules from their concentration.

Despite the verified applicability of this model, there is evidence that it is not adequate in some real conditions. For instance, when modelling the cytochrome complexes of mitochondria Klingernberg established a significant discrepancy with the law of mass action /1/. Probably in this case the first requirement of the model was distrubed on account of the linear structure and the large number of cytochromes (of the order of 15,000 per mitochondrion).

The normal biological interactions have a quantitatively programmed range because of the cascade regulatory systems. Parallel to this it is not unusual to observe the existence of situations (pathological, experimental, ageing) demonstrated by a topologically significant difference of the concentrations of reacting substances which does not allow an analysis based on the application of the law of mass action in a classical way.

The cases discussed in this study should be valid for enzyme (E) - substrate (S) interactions, also for hormones: protein hormones (PH), sex steroids (SS), target receptors (TR) interactions. Because of more complex situations, especially in the protein-protein interactions, for instance PH-TR, the general application of the law of mass action is convenient, but not always accurate enough.

On the basis of a mathematical analysis the application of the law of mass action will be introduced in terms of discussible points. The stress will be put on the possibility of a disturbance of the second requirement for the application of the law of mass action.

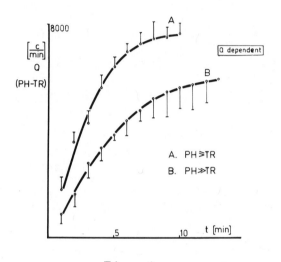

Fig. 1

This study is based on our data for PH-TR interactions (Fig.1) tha data from Everse, Barger and Kaplan /2/ (Fig.2) and the results from SS-TR (McEwen, personal communication).

The data for the complex formation between the protein hormone (PH) and its target receptor (TR) at a constant, but significantly different in the various experiments concentration of PH is shown in Fig.1. The time for the formation of a given amount of complexes is plotted along the abcissa. ICSH was used as PH and the TR was the plasma membrane fraction isolated from testicular tissue. The method is described in our previous paper /3/. Fig.2 shows the influence of the NAD^+ concentration increases the time of the reaction and influences the enzyme kinetics. Fig.3 introduces a more indirect effect: the influence of an increase of sex steroids (SS) concentration on the complex SS-TR (McEwen, 1974-personal communication).

Additions to LDH	Assay half-time (seconds)	
	H_4 LDH	M_4 LDH
Pyruvate (1mM) + NAD⁺	0,30	0,17
Pyruvate (10mM) + NAD⁺	3,00	0,27

J. Everse
R. Berger
N. Kaplan

Substrate Inhibition

Q dependent

Fig. 2

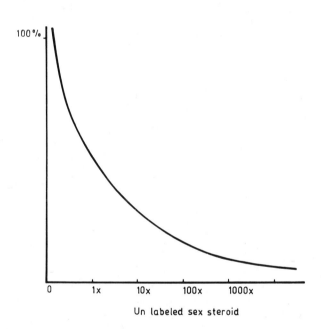

Fig. 3

The data suggest that the concentration increase of the enzyme or hormone substrate (S), to very large values does not lead, as would be expected from the first law of mass action, to a monotonous reduction of time, but to exactly the opposite. This empirical fact draws the attention to the possiblity that at large concentrations of S the validity of the first law of mass action may be disturbed, which could be due to the mutually inhibitory effect of the substrate molecules. In the present study an attempt is made to describe the cases of higher S concentration in an analytical way. The following assumptions are suggested:

1. When the concentration of S is considerable its quantity is only slighly reduced during the course of the reaction in comparison to its initial value S_o for the time T necessary for the establishment of a dynamic equilibrium, i.e. for $0 \leq t \leq T$, $S \approx S_o$.

2. The movement of the molecules of S at the first stage: approach and contact between S and E (according to Koshland's terminology) is described by the Brownian model.

The following considerations were taken into account for the application of Brownian model:

- The substrate molecules are predominantly globular (e.g. LH, SS);

- The dimensions of the substrate molecules are relatively small (by one order at least) in comparison with the total length of their transition to the formation of the enzyme complexes.

On the basis of these data it might be assumed that the conditions of the Brownian model are almost satisfied: a) the particles (in this case the molecules of S) should effect short casual shifts (wandering) for short periods of time; b) the probability of hitting a given point of space at a given time depends only on the previuos position of the particle. The initial condition: the probability $w(\vec{r}_o, t_o)$ the particle at $t_o=0$ to be at a given point of space with a radius $\vec{r}_o$ in this case is described by the δ function (i.e. the particle has a fixed position for $t_o=0$). Also, on account of assumption 1 it may be assumed that there are no limits of absorption and reflection.

3. The events: the performance of a free transition along the axes x, y and z with a length of x_{tr}, Y_{tr} and z_{tr} by a given molecule of S (on condition that in its proximity there is no other molecule of S) and the establishment of a contact with any of the surrounding molecules of E (after the performance of the transitions x_{tr}, Y_{tr} and z_{tr}) are equiprobable and independent (there are grounds for such an assumption as far as no evidence exists about an expressed selectivity of the molecules

SUBSTRATE CONCENTRATION AND THE LAW OF MASS ACTION

of S for those of E).

4. The length x_{tr}, Y_{tr} and z_{tr} of the free transition along the axes x, y and z is a function of the concentration S of the substrate and of the concentration C of the other components of the medium where the reaction takes place:

$$\left| \begin{array}{c} X_{tr} \\ Y_{tr} \\ Z_{tr} \end{array} \right| = \frac{\delta_{x,y,z}}{\sqrt[3]{S+KC}}, \quad \delta_{x,y,z} = \text{CONST}_{x,y,z},$$

i.e. the effects of (S) and (C) on x_{tr}, Y_{tr} and z_{tr} are independent and weighed with a coefficient of weight K.

5. The interactions between S and E are monovalent (one molecule of S binds to one molecule of E).

6. The transitions of the various molecules of S along the axes x, y and z are equiprobable. This presupposes a statistically mean homotropy of the space where the reaction takes place, i.e. equal Brownian models along x, y and z for the motion of the different molecules of S.

7. The events: the formation of complexes (ES) (I stage) which are equiprobable and independent at a given moment of time in relation to the different molecules of S.

8. The quantity of complexes (ES) formed for a unit of time (I + II stages) is proportional to the probability P_E any molecule of S to bind to any molecule of E, i.e.

$$(ES)_+ = \beta P_{E,S} = \beta P_I P_{II}(II/I),$$

where β is a coefficient of proportion, P_I is the probability of completion of stage I, and according to assumption 1, the conditional probability $P(II/I)$ for the completion of II stage, on condition that I stage has set in, might be assumed as independent (or slightly dependent) on the concentration S.

As already known (5) the qualitative equation describing the reversible cycle of enzyme substrate interaction is the following:

$$E + S \rightleftharpoons ES \longrightarrow E + P,$$

and the quantitative expression of this cycle is given by the equations:

/1/ $(E_o) = (E) + (ES), (S_o) = S + \int_0^t k_a (ES) dt + (ES)$

/2/ $(ES)_+ = F[(E), (S)],$

/3/ $(ES)_- = K_-(ES)_+,$

/4/ $(ES)_a = K_a (ES) = E_a + P$

where (E) is the enzyme concentration, (S) is the substrate concentration, (ES) is the concentration of the complexes enzyme-substrate. $F[(E),(S)]$ is the function of the compleces (ES) formed for a unit of time in relation to (S) and (E), K is the dissociation rate constant, K_a is the dissociation rate constant of the desintegration of complexes (E) after the completion of the process of synthesis of the product P in the enzyme-substrate complex. Equations /1/ express the mass balance of E and S, equation /2/ gives the quantity of complexes ES formed for a unit of time, equation /3/ indicates the quantity of the complexes ES decomposed for a unit of time after the completion of the synthetizing process of product P.

Function $F[(E),(S)]$ in Eq./2/ is usually expressed through the first law of mass action:

$$F[(E),(S)] = k_+(E)(S),$$

where k_+ is the association rate constant. This law presupposes that the events "formation of complexes ES (I stage)" are not only equiprobable and independent of the concentration of S (according to assumption 7) in relation to the different molecules of E, but are also independent of the concentration of S. The latter statement seems acceptible in a small concentration of S, but in large concentrations it might be disturbed and that precisely could be the reason for the deviations observed in the time for the formation of one and the same quantity of complexes ES at different concentrations of S. This explains why it is of great importance to clarify the type of function $F[(E),(S)]$ on condition that the events "formation of complexes ES" are not independent of the concentration of S, namely:

a) The probability any molecule of S to meet in volume dV any molecule of E (with assumption 5) is equal to

/5/ $\qquad f_E = dV_E / dV = \alpha . E$

where dV_E is the volume of the molecules of E in volume

SUBSTRATE CONCENTRATION AND THE LAW OF MASS ACTION

dV of the reaction space, d is the coefficient of proportionality;

b) The probability density of a given molecule of S to effect a transition along axis x for time τ, taking into consideration assumption 2, is equal to:

/6/ $\quad \Psi_x = \dfrac{1}{\sqrt{2\pi k_2 \tau}} \exp\left[-\dfrac{(x-k_1\tau)^2}{2 k_2 \tau}\right]$,

where

/6a/ $\quad K_{1x} = \lim\limits_{\substack{\Delta x_{tr} \to 0 \\ \Delta t \to 0}} \dfrac{(p_x - q_x)\Delta x_{tr}}{\Delta t}$,

$\quad K_{2x} = \lim\limits_{\substack{\Delta x_{tr} \to 0 \\ \Delta t \to 0}} \dfrac{\Delta x_{tr}^2}{\Delta t}$,

where p_x represents the probability of effecting a transition of a magnitude of Δx along the axis x for time Δt, $q_x = 1 - p_x$ is the probability of the backward transition. Eq. /6/ represents the law for normal distribution with an arithmetical mean $\mu_x = K_1 \tau$ and dispersion $\sigma_x^2 = K_2 \tau$. The spatial probability for effecting a transition along the three axes x, y and z to a distance $\geq x_{tr} \geq y_{tr} \geq z_{tr}$ for time τ on condition that for $t = t_o$ the molecule of S is localized in points x_o, y_o, z_o and the fulfillment of assumptions 2 and 6, is equal to:

/6b/ $\quad \theta(t) = \int\limits_{x_{tr}}^{\infty} \int\limits_{y_{tr}}^{\infty} \int\limits_{z_{tr}}^{\infty} \Psi_x \Psi_y \Psi_z \, dx \, dy \, dz = \left\{ \dfrac{1}{(2\pi K_2 \tau)^{1/2}} \int\limits_{x_{tr}}^{\infty} \exp\left[-\dfrac{(x-K_1\tau)^2}{2 K_2 \tau}\right] dx \right\}^3$

c) The probability the ith molecule of S to effect for time τ the transition $\bar{r}(x_{tr}, y_{tr}, z_{tr})$ to volume V and after arrival to V to meet any molecule of E of this volume (realization of I stage) according to assumption 3 equals:

/7/ $\quad P_{1i} = f_E \theta_i = d \cdot (E) \cdot \theta_i$,

d) The probability any molecule of S to complete the I stage of forming complexes ES, i.e. probability of events "either-or" is equal according to assumption 7 ($Q_i = Q =$ idem) to:

/8/ $\quad P_{SE} \sum\limits_{i=1}^{N} P_{SE_i} = d(E) N Q P(II/I) = d \mu P(II/I)(E)(S), N = \mu(S)$

where μ is a coefficient of proportionality, and N is the number of molecules of S taking part in the reaction. Therefore, according to assumption 8, the quantity of complexes (ES) formed for a unit of time is equal to:

/8a/ $(ES)_+ = \beta P_{SE} = K_+ \cdot (E)(S)\theta$, $K_+ = \alpha \beta \mu P \; (^{II}/_1)$

In this case, taking into consideration assumption 8 and Eqs./1-7/ the resultant quantity of complexes (ES) formed for a unit of time is expressed by:

/9/ $\dfrac{dz}{dt} = -k_a z + \dfrac{k_+(1-k_-)}{(\sqrt{2\pi\tau})^3}(E)(S)\dfrac{1}{K_2^{3/2}}\left\{\int_{x_{tr}}^{\infty} \exp\left[-\dfrac{(x-k_+\tau)^2}{2k_2\tau}\right]dx\right\}^3$

where: $Z = (ES)$, $\quad S = S_0 - Z - \int_0^t K_a Z \, dt$

If in Eq./9/ the value of Δx in K_1 and K_2 is replaced according to assumption 4 by $\Delta x = \dfrac{\delta x}{\sqrt[3]{VS+KC}}$ the following equation is obtained:

/9a/ $\dot{Z} = \dfrac{\alpha\beta\mu(E_0-Z)}{(2\pi)^{3/2}(\delta\sqrt{N})^3}(S)(S')\left\{\int_{x_{tr}}^{\infty}\exp\left[-\dfrac{\left(x-(p-q)\dfrac{\sqrt[3]{N\delta^*}}{\sqrt[3]{(S')}}\right)^2}{2(\delta\sqrt{N})^2}\right]\right.$ X

$\left. \times (S')^{2/3}\right] dx \bigg\}^3 - (K_a + K_-) Z$,

where $N = \tau/\Delta t$, $(S') = (S)+KC$ (N is the number of free transitions in realizing x_{tr}).

Eq. /9a/ has no solution in squares yet for the qualitative estimation of $\dot{z}$ as dependent on (S), and therefrom of $t = \int \dfrac{dz}{\dot{z}}$ the derivative $d\dot{z}/dS$ might be investigated. With significant values of S_0, according to assumption 1, $(S) \approx (S_0) \approx (S')$ and then:

$\dfrac{\partial \dot{z}}{\partial s} = \dfrac{\alpha\beta\mu}{(2\pi)^{3/2}(\delta\sqrt{N})^3} t e^{-Bt} SJ^2\left\{2J - \exp\left[-\dfrac{(x\sqrt[3]{S}-(p-q)N\delta^*)^2}{2(\delta\sqrt{N})^2}\right]\right\}$,

$J = \int_{x_{tr}}^{\infty}\exp\left[-\dfrac{(x\sqrt[3]{S}-(p-q)N\delta^*)^2}{2(\delta\sqrt{N})^2}\right]dx$, $\quad B = \dfrac{\alpha\beta\mu}{(2\pi)^{3/2}(\delta\sqrt{N})^3} S^2(J+K_-+K_a)$

Then, from the condition of extremum of $\dot{z}(S)$:

$\dfrac{\partial \dot{z}}{\partial s} = 0$

(as far as S and I are not identically equal to naught, the following equation is obtained:

/10/ $2J = \exp\left[-\dfrac{(x\sqrt[3]{S}-(p-q)N\delta^*)^2}{2(\delta\sqrt{N})^2}\right]$.

SUBSTRATE CONCENTRATION AND THE LAW OF MASS ACTION

If it is assumed that $\eta = \dfrac{(x\sqrt{S}-(p-q)N\mathscr{E})^2}{2(\mathscr{E}\sqrt{N})^2}$,

then condition /10/ obtains the form

/11/ $\quad \sqrt{\dfrac{\pi}{2}}\, G\phi(\eta)\Big|_{x_{tr}}^{S} = \exp(-\eta_{tr}^2/2)$

where $\phi(\eta) = \int \exp(-\eta^2/2)\,d\eta$. Functions $\phi(\eta)$ and $\exp(-\eta^2/2)$ have one intersection point only and therefore with a fixed x_{tr} there exists the possibility that /11/ might not be satisfied. However, for a definite magnitude of S /11/ cannot be satisfied for high values of x_{tr}. Since the value of x_{tr} is in inverse proportion to the concentration of E, then it follows that at a given critical concentration of E_{cr} of magnitude of x_{tr} becomes such that at no concentration of E the satisfaction of /11/ is possible, i.e. the rate $\dot{z}$ is a constantly decreasing function of S (that is so because the difference $I - \exp(-\eta^2/2)$ is negative for small values of $\mathscr{E}\sqrt{N}$). If however x_{tr} is small enough, then condition /11/ is realized for definite values of S and the dependence $\dot{z}(S)$ has a maximum, resp. $t(S)$ has a minimum.

Since the values x_{tr}, $\mathscr{E}$, N are difficult to be determined experimentally, the determination of S_{extr} may be achieved by varying E and S and the empirical determination of function $\dot{z}(S)$.

All considerations up to this point were made on the assumption that the concentration of S is significant. When $S \to 0$, then $S' \approx KC$ according to /9a/ and Q of /6b/ does not depend on S, and Eq. /8a/ obtains the form of the first law of mass action: $(ES)_+ = k_+(E)(S), k_+ = \text{const}$. However, if it is assumed that this ratio is valid also for high concentrations of S (when according to assumption 1 $(S) \approx (S_0) \approx (S')$), then as it follows from /9a/ that $\dot{z}$ is a linear function of S with a positive coefficient, wherefrom for the time $t = \int \dfrac{dz}{\dot{z}}$ it is concluded that it represents a monotonously decreasing function of S, which does not agree with the experimental data.

From the above exposition it can be seen that:

1. The first law of mass action cannot explain satisfactorily the experimental data available for the function $t(S)$ for the time t for the formation of one and the same quantity of complexes (ES) resp.(PH-TR), depending on the concentration S resp.PH of the substrate in

high values of the latter.

    2. A possible explanation of the course of the experimental data is obtained if the Brownian model for the motion of the molecules of S is used. Naturally, the disturbance of the first law (which is essentially expressed by the factor k_+ in Eq./8a/ it not being a constant) may be explained for instance with the inactivating action (direct or indirect) of the product P on the activity of the complex ES formation. Nevertheless, when the dissociation rate of the complex after the formation of the product is noncommensurably less than that of the formation of the complex a case met more often, i.e. when $K_+ \gg k_a$, then it seems more probable that such a comparatively slow feedback could hardly be of a decisive importance with the high concentrations of S.

REFERENCES

1. Milsum, J.H., in: Biological Control Systems Analysis, McGraw-Hill Book Co., New York, 1966.

2. Elswere, J., Borger, R., Kaplan, N., Science, <u>168</u>, (1970), No 3936, 1236.

3. Vassileva-Popova, J., Proceed. Vth Asia & Oceania Cong. Endocrin., v.I, Chandigarh, India, 1974, p.242.

4. Rodbard, D., in: Receptors for Reproductive Hormones, Eds. B. W. O'Malley & A.R.Means, Plenum Press, New York-London, 1973.

5. Kazakov, V. A., in: Introduction to the Theory of Markovian Processes, Sovietskoe Radio, Moscow, 1973 (in Russian).

INFORMATION TRANSFER IN AGEING:

AN ALLOSTERIC MODEL OF HORMONAL REGULATION

Julia G. Vassileva-Popova and T. K. Yanev

Central Laboratory of Biophysics
Bulgarian Academy of Sciences
1113 Sofia, Bulgaria

The aim of this study is to investigate the character of the information transfer in time between a central regulatory protein i.e. a protein hormone (PH) and target receptors (TR) recognizing this signal. In search for a possible solution of this question the following limitations were selected:
- The experimental analysis includes only the first steps of recognition introduced as complex formation between PH and TR (adenohypophysial PH and gonadal TR).
- Because of not yet enough evidence for the age alteration of the hormonal structure during the ontogenesis the focus of this study will be on the TR ageing transformation; especially the changes of the TR ability for recognition of the central signal (CS) incorporated in the PH.

The investigation of the ageing changes of the message transfer between the PH and the TR became of interest for us because of new data available from the advance in ageing research and the related topics.

Symbols used: Protein hormone (PH); Interstitial cell stimulating hormone (ICSH); Central signal (CS); Sex steroid (SS); Target receptor (TR) in this case for PH; Active state of TR (A); Semipassive state of TR (P); Enzyme (E); Substrate (S); Free amino acids (FAA); Central nervous system (CNS)

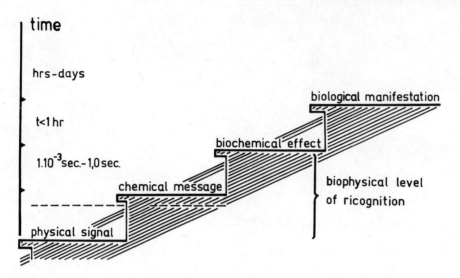

Figure 1

In a schematic way this is:
-The advanced work on biological recognition and the information transfer in living systems.(Freedman, 1974);
-The possibility of involvement of biophysics in biological recognition in ageing (Fig.1);
-The molecular level of the ageing process investigation;
-The possibility for a prolongation of human life with a preservation of the physiological and intelectual capacities (economical, psychological and theoretical considerations);
-The new data available from the modern laboratories recently oriented to the topic of ageing.

APPROACH

The present study consists of experimental results and the theoretical approach. The elements of the method used: experimental work, mathematical analysis, estimation of the constants, working hypothesis, bibliographic survey. The mathematical analysis made and the estimation of the constants allow the distinction and characterization of the curves with similar configuration.

As a result a possible model with selected limits is proposed. The concept following from this combined analysis will be discussed.

A SUMMARY INTRODUCTION TO THE AGEING COMPLEXITY

The approach for study in ageing research is experimental and theoretical at different levels: quantum, molecular, cellular, physiological and psychological.

The role of the free radical systems in ageing has become of interest (Harman, 1956, 1961, 1962, 1968; Tappel, 1968; Emanuel, 1972). Molecular gerontology is developing simultaneously with molecular biology itself. For the explanation of the findings in ageing research the genetical background and the study of nucleic acids is tempting. In this respect are the works of Vaniushin et al. (1969), Srivastava and Caaudhary (1969), Grebennikva and Gal (1974), Klimenko and Tupchienko (1974). The influence of amino acid content on the intensity of the DNA replication (possible feedback of protein synthesis and replication of DNA) in the genom, Lark, (1969) is yet an open point for the possible application of this kind of investigations in the ageing process regulation. Ageing is described as genetically programmed (Glückmann, 1951) with the possibility to operate at the level of transcription from DNA to mRNA as a gene dominated process (Bullough, 1967; Hahn, 1970; Medvedev, 1967, 1973). In this line is the work on the age changes in the DNA double helix (Hahn, 1970) and especially the changes in the nucleoprotein complex (Hermann et al. 1969; Shirey and Sobel, 1972). About the ageing of chromatine the data are not in full concordance. The observations of Andrew (1969) and Popova and Mironenko (1974) indicate an age alteration in the chromatine. Other authors consider that there is not enough evidence for accepting the possibility for primary age changes of the chromatine properties (Gutman and Hanzlikova, 1972). Klimenko (1974) found that the histones do not have an ageing alteration.

The complex ageing process is considered as an error accumulation (Curtis, 1966; Strehler, 1967; Hahn, 1970; Holliday, 1973). It still remains uncertain whether the only explanation of the complexity of the ageing process belongs to the gene level.

Another aspect of the study in this topic is the enzyme activity and the protein synthesis. There is no clear evidence for accepting either alterations of enzyme activities (Finch, 1972) or changes of the protein synthesis capacity at the senile stage. It seems that some ageing changes of the protein (Gutman and Hanzlikova, 1972) and amino acids transport rate take place (Mishchenko, 1974). For glucose-6-phosphate dehydrogenase Kohn et al. (1973) found age-dependent changes in the isoelectric behaviour of the enzyme. Our study on the developmental enzymology (Setchenska et al., 1975) shows an age-dependent inhibitory effect of dehydroepiandrosterone on glucose-6-phosphate dehydrogenase and a change of its isoenzyme characteristics. It looks that band C of the enzyme is more sensitive to sex steroid inhibition. This work is not completed yet by a study on enzyme activity in senile age.

In a more complex level the indication is that the increase of the physical tension and the neurotrophic optimization in old age produce an activation undistinguished form those in the developmental age. There is no evidence for ageing changes of the capacity for adaptation (Grebennikova and Gal, 1974; Silin et al., 1974 and Silin, 1974). From an enzymological point of view the process of "reversibility" or the absence of changes of the enzyme activity during ageing is difficult to be explained only by a gene-clock programmation. It looks that a number of enzymes do not change their characteristics during ageing (Silin, 1974; Pashkova et al., 1974; Bulankina and Machikhina, 1974; Bulankina et al., 1974; Shevtsova, 1974). For some enzymes a limitation of the activity in ageing is found (Parina et al., 1974; Driniaev et al., 1974; Anderson, 1974). A decrease or no change in the activity of CN phosphohydrolase is also observed during development (Dalal et al., 1973; Einstein, 1974). In connection with the enzyme study in ageing research attention is paid to lysosome system (Strehler, 1964).

There are suggestion that reparative enzymes and and nucleic acids could be applied for a kind of stimulation of aged tissues (Nikitin, 1970; Vilenchik, 1970). On the basis of not fully understood behaviour of enzymatic activity in senile age the study on this topic looks promising.

Another line of interest in ageing research is the study of the free amino acids (FAA) in the central nervous system (CNS). It is known that the FAA have a regulatory role in the CNS. Besides the involvement of the amino acids in the intermediary metabolism and the tricarboxylic acid cycle, in some respect they take part in the glucometabolism. More interesting for the present study is the neurotransmitter function of the FAA in the CNS. The change of the ratio of FAA in the CNS depends upon the period of development and it is characteristic for the definite part of the CNS (Hudson et al.; 1969; Gordienko, 1974).

Ageing affects the FAA distribution in the CNS. An increase of alanine and a decrease of aspartic acid concentration in the CNS is observed (Timiras et al.,1973). Alanine is proposed as an inhibitory neurotransmitter and aspartic acid as an exciting one. As a result of this a rearrangement of the neurotransmitter signals in the CNS in aged cases is observed.

Of interest for the present study are the age changes found in the phosphatidyl ethanolamine level in the plasma membrane fraction (Rubin et al., 1973). This point is of special interest for this study as far as the protein hormone receptors are also part of the plasma membrane fraction. Popova and Bondarenko (1974) have found that the change in the phospholipids is related to the membrane changes in old age. Tupchienko (1974) found an increase of liver phosphoproteins during ageing.

If we try to analyse the ageing situation on the cellular level it looks that the problem is tuned to the programmed number of cell divisions and the definite lifetime of diploid cells and a relatively fixed life-time in tissue culture (Hayflick, 1970; Hayflick et al., 1973). The change of phospholipids and ethanolamine during ageing in human diploid cell culture (Kritchevsky and Howard , 1970) is a kind of support in this respect.

There is also an indication that the mechanism controlling the rate of cell division and the cell dimensions is connected to the age regulation of the whole organism (Nikitin, 1965). In this respect it seems that the differentiation is closely related to cell ageing and that the ageing in vivo is not always analogous to the in vitro ageing (Franks, 1973).

A comprihensive study of the ageing process is performed also on a more complex level: the physiological one. The age changes in the different organs and systems is an active field of research on ageing. Despite the physiological homeostasis the age changes of the different systems and organs pose a very individual time course (Gutman and Hanzlikova, 1972).

Probably we could say that despite the relatively high degree of differentiation the CNS and the germinal gonadal tissue show a more constant behaviour in old age in comparison to the vascular system for instance. A stimulating topic of research is the ageing characteristics of the cardio-vascular system in the bio-medical asppect (Tchebotarev and Frolix, 1967). Attention is also paid to the neurotrophic function in ageing and new concepts are presented on this topic (Chernigovski, 1960; Gutman and Hanzlikova, 1972). The physiological point of ageing is demonstrated often in quite a broad aspect as a medical advice for different groups of people (Trankvillitati, 1966; Human, 1972; Alexandrova, 1974).

The more closely connected interest for the present study is the research axis between endocrinology and gerontology. A relatively direct connection with the present topic has the work done in endocrinology with respect to development and especially to ageing. Instead of very elegant and profound studies on this topic some discredited concepts are also shown.

One of the central points in developmental regulation and endocrinology belongs to steroids. This should be influenced by the recent progress made on the role of steroids on the genom (O'Malley et al., 1973) and the steroid receptor transformation and synthesis of RNA in the cell nucleus (Jensen and DeSombre, 1973). The study of the mechanism of the SS action in development (Kaye et al., 1972, 1973; Plapinger and McEwen, 1973; McEwen and Pfaff, 1973; Plapinger et al., 1973; Denef et al., 1974) is mainly in the focus in comparison to those in ageing. In the field of sex steroid research the investigation on the steroid receptor itself and the receptor transformation (Jensen et al., 1973) will provide helpful explanation of the age-dependent alteration of the hormone receptors.

Nevertheless, the application of the sex steroid and the antimetabolites in geriatric medicine becomes larger parallel with the application of steroids in developmental and contraceptive medicine. An application of hydrocortisone for studying the development and ageing processes in cell cultures is also made (Macieira-Coelho and Loria, 1967; Cristofalo, 1970). In this connection the field of research is the ageing response-capacity of the cortex and age-dependent corticosteroid pattern (Friedman et al., 1969; Stavitskaia and Shamarina, 1974; Stavitskaia, 1974).

The mechanism of protein hormone action with respect to age changes is actively investigated (Korner, 1964; Pegg and Korner, 1965; Widnell and Tata, 1966; Kessler, 1971; Parlow et al., 1973). According to the work of Galavina (1974) the possible mechanism of interaction of the GH active site with the nucleotides has no age-dependent difference, but the opportunity for a dose response is not excluded. The possible point raised is the reason for this kind of direct protein hormone-DNA interaction.

On the basis of our knowledge of the adenopituitary protein hormone concentration in old age it seems that the opinions in favour of no change or some increase of the protein hormone concentration in the hypophysis and in the circulation are predominant. In this respect is the study of Cartlige et al., (1970). The works of Dilman (1972) and Druzhinina (1974) do not indicate the decrease of the protein hormones during old age. The work on the influence of protein hormone on elastin suggests that the hypophysectomy protects the decrease of elastin in the skin (Moros, 1974). In this line the work of Everitt (1973) proposes the idea that the hypophysectomy slows down the ageing process. Of course in all this judgments the increase in pathological cases in senile age has to be taken into account. Despite the progress made, the role of protein hormones in senility is not yet fully understood.

The interest in developmental and ageing research is also centred in the field of thyroid hormone mechanism of action and some related compounds (Kovacs et al., 1969; Hudson et al.,1974; Tata, 1970, 1975). Recently investigations have been performed on the role of the thymus hormone in ageing (Bach and Dardenne, 1973). Probably the "clock" hypothesis for nonhomeostatic lower species is acceptable. In this respect the life-time should be timed by depletion of juvenile hormone (Lockshin and Williams, 1964). The explanation and discussions on the points of hormone action and ageing are to be found in the monographs and works of Parhon (1959), Korenchevsky (1961), Nagorny et al. (1963), Adelman (1973).

From the survey presented it might be concluded that not much work has been dedicated to hormone information transfer in ageing and especially to hormone receptor biorecognition. Having on one hand the inconclusive results about the age-changes at the hormonal level and on the other hand the complex nature of the target receptors, it is to be expected that the age changes in the hormonal signalization axis are centred mainly at the target receptor level. The age-dependent changes in the phospholipids in the plasma membrane fraction (Rubin et al., 1973) should be expected to reflect the changes in the receptor biopolymer complex. Instead of a phospholipid component some protein hormone receptors have also glycoproteins (Blecher, 1973 and our unpublished results). Glycopeptides are also found in the surface membrane (Glick et al., 1973). In this respect the receptor with a more complex composition should provide the conditions for a higher sensitivity in time in comparison to the hormone itself.

We are in accord with the opinion of Hall (1975) that this Colloquium focused the attention of an interdisciplinary audience on the ageing changes at the receptor level. The age characteristics of the hormonal signalization, especially its reception at the target level is tempting for future research.

MATERIALS AND METHODS

The empirical data involve the isotope ligand-receptor assay. The PH in our case is ICSH (rat and human), supplied kindly by the NIAMD, NIH, Bethesda. Rat and human ICSH were used because there is evidence of specific binding sites for gonadotrophins present in the out-species target tissue (Catt et al., 1972; Rao and Saxena, 1973). The tracer used was ^{125}I and the isotope conjugation to the ICSH was made according to the method of Hunter and Greenwood (1962). The labelled ICSH was isolated from the mixture by gel filtration on a 1cm x 40 cm column of Sephadex G-100 in a medium of 0,85g sodium chloride in 100 ml distilled water containing 0.7% of bovine albumin at pH 7.2. The specific activity and the percentage of hormone damaged were estimated. Many details of the method were carried out according to Rao and Saxena (1973), Saxena et al., (1974) and Vassileva-Popova (1974) with some necessary laboratory adaptation.

The plasma membrane fractions were obtained from testicular tissue. The source of the testes was rats of different age, controle and treated groups consisted of 10 animals. The plasma membrane fraction from adult testes was considered as a controle group (caseI). The aged group (case II) consisted of rats over 14 months. Has to be noted that the ageing of laboratory animals is expected to take place quite early and not in a typical way. In addition to this male animals were out of normal sexual contact. The third group (case III) was composed of normal adult (4-6 months) animals treated with ICSH. The treatment was performed twice in fortnight interval with 2000 mIU of ICSH and the plasma membrane fraction was isolated five days after the second injection. The IV group (case IV) represented aged (case II) animals treated with testosterone propionate, 90 mg depot per 100 gr body weight. The estimation was made two weeks after the treatment. The plasma membrane fraction was prepared according to Neville (1968) with a slight adaptation to the conditions of the experiments. The works of other authors connected to the materials and the techniques used (Saxena et al., 1974; Roth, 1973; Kahn et al., 1973 ; Gospodarowicz, 1974; Freychet et al., 1971; Cuatrecasas, 1972; Tsuruhara et al., 1972; Catt et al., 1972, 1973; Dufau et al., 1972; Blecher, 1974) were taken into consideration. The protein concentration of the fractions was estimated by the method of Lowry et al. (1951). After a density gradient centrifugation the samples were stored at 4°C. The isotope ligand-receptor assay was performed by the introduction of a constant amount (100 μg protein) of plasma membrane fractions to the increasing concentrations of ^{125}I tracered ICSH. The incubation conditions: time - 10 min; temperature - 37°C; medium - Krebs-Ringer bicarbonate buffer containing 1% of bovine albumin, pH 7.2. The separation of the complex (^{125}I.ICSH - TR) in plasma membranes was carried out by cellulose acetate filters by means vacuum pressure. The filters washed were counted and the amount of isotope labelled hormone accepted by the plasma membrane fractions (presumable target receptors) was estimated.

For assurance about the principle results, after the incubation the separation of the ^{125}I.ICSH incorporated into the plasma membrane fractions was detected through resuspending, washing and counting of the ^{125}I.ICSH residue in plasma membranes. As a test for specificity the replacement with unlabelled ICSH and the comparison of the acceptance of the labelled ICSH by nontarget (muscle) plasma membrane was made.

RESULTS AND DISCUSSION

On the ordinates of the coordinate system the relative values δ are given. Every point of δ is the ratio between the quantity of the $(PH - TR)_i$ complex to the $(PH - TR)_{sat}$, obtained in a state of saturation in controls. The ratio (I : I) of PH to TR is assumed. On the abcissa the increasing concentrations of labelled PH are presented (Figs.2-5).

$$\delta_{1i} = \frac{(PH - TR)_{1i}}{(PH - TR)-s.c.}$$

$$\delta_{2i} = \frac{(PH - TR)_{2i}}{(PH - TR)-s.c.}$$

$$\delta_{3i} = \frac{(PH - TR)_{3i}}{(PH - TR)-s.c.}$$

$$\delta_{4i} = \frac{(PH - TR)_{4i}}{(PH - TR)-s.c.}$$

s.c. - saturation in control

The values obtained for α and y in cases I - IV are given in Table 1 (eq.1).

Table 1

case I		case II		case III		case IV	
α	y	α	y	α	y	α	y
5	0,143	17,0	0,273	14,0	0,165	11	5,5
10	0,322	33,5	0,492	28,5	0,350	22	18,2
20	0,588	50,0	0,630	43,0	0,500	33	33,2
30	0,727	67,0	0,715	57,0	0,605	44	47,0
40	0,808	84,0	0,772	71,5	0,681	55	58,5
50	0,852	100,0	0,812	86,0	0,736	66	67,3
60	0,882			100,0	0,775	77	74,5
70	0,902					88	79,7
80	0,918					99	84,0
90	0,928						
100	0,938						

The sigmoidal character or the curves allows the application of the Monod, Wyman and Changeux model (1965) for allosteric interaction between enzyme (E) and substrate (S).

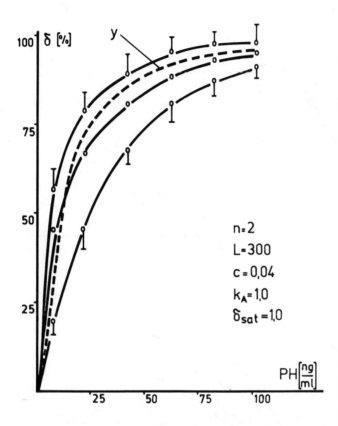

Fig.2. The experimental (case I) and model curves indicate a good agreement between the experimental data and the model proposed.

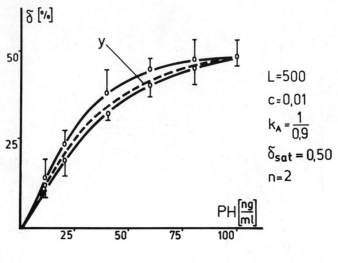

Figure 3

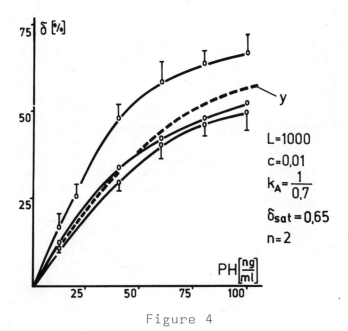

Figure 4

As a working hypothesis we introduce the following assumptions:

1. The TR is an allosteric complex biopolymer with two basic states: active (A) and semipassive (P). The ligand PH has an affinity for these two states. The affinity of the PH is different for the A and P states. As a result of this complex (PH - TR) in state A is more stable in comparison to the same complex in state P of the TR.

2. Same circumstances should allow the PH to form a complex with the SS (Vassileva-Popova and Dimitrov, 1975). The biological conditions assumed for the PH - SS interaction could exist in the sites with a relatively high concentration of PH and SS*.

3. The SS in a direct or indirect way mediates the activation of the TR for binding to the PH. One hypothetical mode of this activation should be the linkage of the SS to the presumable allosteric sites of the TR.

4. In the conditions of high PH concentration there is a possibility of mutual inhibition through the competition between the PH molecules for the interaction with their TR.

At this stage it is not possible to give an experimental confirmation of all the assumptions introduced. Nevertheless, the good agreement between the model proposed with this assumption and the experimental data is support for our hypothesis.

On the basis of assumption 1 the curve of saturation Y is described by the equation:

a) Homotropic effect without the inhibitors and activators

$$y = \frac{Lc\alpha(1+c\alpha)^{n-1} + \alpha(1+\alpha)^{n-1}}{L(1+c\alpha)^n + (1+\alpha)^n} \quad (1)$$

$$c = \frac{K_A}{K_P}, \quad K_A = \frac{k_{-A}}{k_A}, \quad K_P = \frac{k_{-P}}{k_P}, \quad c = \frac{k_{-A} \, k_P}{k_A \, k_{-P}} \quad (2)$$

where: k_{-A}, k_A, k_{-P}, k_P are the dissociation, resp. association, rate constants for A and P states.

---
*This point will be discussed in the final section of this study

n - the number of the homologous sites of the TR

$L = \dfrac{[P_o]}{[A_o]}$, allosteric equilibrium constant of the transition:

$A_o \leftrightarrow P_o$, transition state in the absence of PH (3)

$[A_o]$, $[P_o]$, the given concentration of A_o and P_o

$\alpha = \dfrac{[PH]}{K_A}$, independent variable (4)

In the heterotropic effect in the presence of inhibitors and activators the equation will be:

$$y = \dfrac{L'c\alpha(1+\alpha)^{n-1} + \alpha(1+\alpha)^{n-1}}{L'(1+c\alpha)^n + (1+\alpha)^n} \quad (5)$$

$$L' = \left(\dfrac{1+\beta}{1+\rho}\right)^n \cdot L \quad (6)$$

where:

$\beta = \dfrac{(I)}{K_I}$, $\beta = \dfrac{(SS)}{K_{SS}}$ (in our case)

$\rho = \dfrac{[A_c]}{K_{Ac}}$

The values of the equilibrium constant K_A are determined through the values of α_{sat} and $(PH)_{sat}$ for the state of saturation. Very likely in real conditions the saturation of the TR does not reach 100 per cent. The determination of K_A is necessary for a comparison of the model curves $y(\alpha)$ with the experimental δ (PH)(tabl.1).

It follows from Eq. (2) that the constant K_p is determined if the values for K_A and $\underline{c}$ are known. The determination of the L, $\underline{c}$, n was sought in the intervals:

$1 \leq n \leq 4$, $0 \leq c \leq 0,04$, $0 \leq L \leq 10\ 000$ (7)

The considerations for this were as follows:

a) With $n \geq 5$ and $L > 10\ 000$ some control calculations indicated that the model curves are considerably different and out of the limits of the experimental curves.

b) For the constant $\underline{c}$ it was calculated that for the values $\underline{c}>4$ the constant L had to have negative values so as to satisfy the experimental curves. As far as the negative values of the constant L have no physical meaning we have varied the constant $\underline{c}$ in the limits $0 \leq \underline{c} \leq 4$.

c) For n the whole range from $n=1 \div n=4$ was estimated. The results from the calculations in the selected range (7) indicated that the best agreement of the theoretical with the experimental curves was achieved with $n=2$. From the biological point of view $n=2$ and $n=4$ is more acceptable. For the TR $n=2$ is also assumed by other investigators (Rodbard, 1973).

The satisfactory values of constants: L, $\underline{c}$, K_A and $\bar{\delta}_{Sat}$ were estimated. (Table 2).

Table 2

Case	L		c	K_A	$\bar{\delta}_{sat}$
I	300		0,04	1,00	1,00
II	500		0,02	1,10	0,50
III	1000		0,01	1,40	0,65
IV	PH	L	0,00	1,1	0,85
	0 - 40	4000			
	50	3000			
	60 - 100	2000			

For the case IV, aged treated with SS (Fig. 5), it was necessary to assume the heterotropic effect and the PH was accepted as an activator of this process. On the basis of this the best agreement between the experimental and model curves was found. According to the values of the constants given in Table 2 an increase of the allosteric equilibrium constant in all cases (II, III, IV) in comparison to the control (case I) was determined. This should be due to an increase of the number of molecules in state P of the TR in comparison to these of state A (Eq./3/) in all cases besides the controls. Opposite to that the constant $\underline{c}$ shows a decline in all cases in comparison with the controls. This change in the value of the constant c should influence also the decrease of K_A.

From the constants and parameters introduced it looks that δ_{sat} is an informational criterion. The highest value of δ_{sat} belongs to case I: the controls. A decrease of δ_{sat} is to be found in cases II and III. It is interesting to analyse the increase of δ_{sat} in case IV: aged treated with SS in comparison with case II (Table 2).

<u>Treatment of the controls with PH; Possible physical meaning of the constants.</u> It should be noted that the treatment (pretreatment) of the control with PH made the curves comparable in the form and the degree of saturation $(\delta)_{sat}$ to those obtained from the aged (case II). Also as mentioned above the treatment with SS gives a close value of the degree of saturation δ_{sat} (Fig.4) to those of the control (Fig.2). This is the limit of the outward resemblance between cases II and III. At the same time quite different values of the constants L and c were found. An attempt to explain this phenomenon will be introduced.

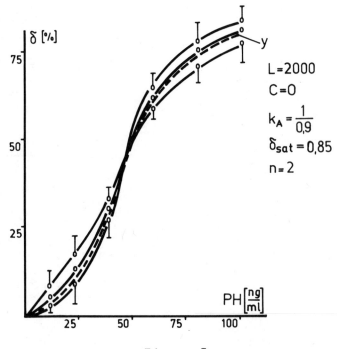

Figure 5

The analysis of the results of case III (Fig.4) is discussed because of the external simmilarity with the case II: aged. In some way the pretreatment of the controls with an additional amount of PH produces a situation similar to case II and should create a condition of an excess of PH and the subsequent reflection of this into the process of PH-TR complex formation. The considerations for this are: The decrease of the value of the rate constant c may be explained according to assumption 4. The inhibitory effect of self competition between the PH molecules in an excess state strongly influences the TR molecules, especialy in the P state of TR. As a result of this the dissociation of the amount of complex PH-TR increases and the constant c reduces its value. The decrease of δ_{sat} follows logically from the assumption 2 and 3 because of a decrease of the TR molecule in state A. In addition the increase of the value of the constant L is a result of the possibility for complex formation with the SS molecules (PH-SS) which according to assumption 2 is a possible factor influencing the reduction of the number of molecules of the TR in state A.

Treatment of aged with SS; Possible physical meaning of the constants. As a result of the SS treatment (explained in the methodological part) an increase of δ_{sat} is observed (Fig.5). This suggests an increase of the effective affinity of the aged TR for bindings to the PH. At the same time in the small PH concentrations the SS has an inhibitory effect and the increase of the value of δ_{sat} in comparison with δ_{sat} in case II and case I shows specific hysteresis. It looks that part of the negative effect of the SS treatment of the aged TR is due to a too high SS dose. The partial compensation of the negative effect of the relatively high SS dose is achieved via an increase of the PH concentration (second part of the curve). It is possible to conclude that the SS has a positive effect on the aged TR if it is introduced in a relatively small dose. In higher SS dose treatment the negative effect produced has to be neutralized with a PH treatment.

According to assumption 3 it follows that the increase of the amount of SS mediates a decrease of the TR molecule in state P possibly via a direct or indirect activation of the active site of the TR molecules in the presence of SS. This should be an explanation of the necessity for the presence of SS for the PH-TR interaction. This effect should be applied in the case of ageing with the possible role of the SS on the neutralization of the ageing screening factors which interfere with the PH-TR interaction. Another possibility of explaining the activatory role of the SS on protein hormone receptors should deal with an involvement of the TR presumable reserves in the PH-TR interaction. These cases are a possible explanation of the processes which take place during the treatment of aged TR with a proper amount of SS. Because of some evidence for hormonal oscillation the time of treatment also should play some role. In the circumstances of an excess of SS, e.g. pretreatment (possible in our case) then a more complex state should appear. During the activation of the TR from SS a dynamic ratio of activation and simultaneously an inhibition of the PH-TR complex formation is expected. According to assumption 2 an excess of SS will produce a screening effect which interferes with the contact between PH and TR. As a result of this a decrease of the efficacy of the state A of TR is expected. This change of the ratio of A and P states of the TR is reflected on the constant L as an increase. This is a kind of explanation of the increase of the value of L observed. The sigmoidal character of the saturation curves shows an active inhibitory effect on SS when the PH concentration is relatively low. The inhibitory effect of a high SS concentration is fully eliminated in the region of relatively high PH concentrations.

The increase of δ_{sat} after treatment of aged with SS should be explained with the combined action of an increase of $\underline{c}$. At the same time as a result of the increase of the $\overline{PH}$ concentration the value of the constant L decreases and there is also a reduction of the value of constant $\underline{c}$. The decrease of constant L in case IV does not reach the values of L in the controls (Table 2). At the same time $\delta_{sat.IV}$ becomes quite close to $\delta_{sat.I}$ probably due to an decrease of the constant $\underline{c}$ in case IV. This should be explained with a relative increase (towards state P of the TR) of the resistance of the complexes (PH-TR) in state A. The positive role of the SS in a high PH concentration is in the reduction of the screening effect for the PH-TR complex formation.

Comparative analysis. The comparison between case II and III (aged and PH-treated controls, described above) indicates some outstanding similarities between aged and controls pretreated with PH. These similarities are expressed as a decrease of the effective affinity of the TR for binding to the PH when to the normal working receptors we add an excess of PH. In this respect case III is a pilot imitation of the ageing state. The PH pretreatment and the TR ageing should be accompanied by a situation of "substrate inhibition", because of a relatively high PH concentration. Some theoretical aspects of this substrate inhibition are discussed in this volume. But the differences in the values of the constants between cases II and III indicate that there is a descrete dissimilarity in the mechanism of depression of the TR affinity for the PH binding in these two cases. At the same time the comparative analysis between cases III and IV (control treated with PH and aged treated with SS) indicates a possibility the differences in case II (aged) to be explained by an excess of PH in the close vicinity of the TR.

It should be pointed out that the PH and SS treatments have a simultaneous positive and negative side of its action at the receptor level. The dynamic balance of the positive and negative effect of the PH and SS treatment should strongly depend on the present status of the TR and the quantity of the hormones. The positive effect of the SS treatment of aged cases demonstrated as an increase of the TR efficacy for binding to the PH is possibly due to an increase of the number of the TR molecules in state A. The inhibitory effect of the SS treatment (dose-dependent) is reflected into the constant $\underline{c}$ as an increase of its value due to an increase of dissociation of the complexes (PH-TR) in state A (Eq.3) accompanied by an increase of complexe formation in state P. In other words the SS in relatively large amounts does not allow the full realization of state A of the TR, (produced by the SS), because of an increase of the effective dissociation. An increase of the PH amount produces a partial neutralization of the negative side of the SS effect on the TR. The increase of δ_{sat} in case IV (aged treated with SS) may be discussed as a potential of the TR reversibility. There is an opportunity the gonadal TR to be quite resistent to the ageing process as a "slow kinetic process". The decrease of $\delta_{sat.\,II}$ (aged) and the increase of $\delta_{sat.\,IV}$ obtained the SS treatment suggest that the TR is probably not genetically altered, but the mechanism of the PH-TR interaction is changed because of the appearance of a kind of screening effect. In this respect it is noticed that the P state of the TR is the weakest point of the PH-TR complex formation in comparison with state A of the TR. All kinds of influences accumulated in time should affect strongly K_p which contributes with a small weight value in the PH-TR interaction in comparison to K_A. Nevertheless, the analysis of K_p is theoreticaly meaningful for the molecular mechanism of the hormone-receptor complex formation. At the present time there are some suggestions but not full evidence indicating that the TR operates with a reversibility which the SS or other intrinsic or extrinsic factors should help to be applied in the further experimental and practical work.

GENERAL DISCUSSION

The attempt made for the further understanding of the ageing process through the experiments and proposal of a mathematical model raised some questions which we would like to discuss in a more broad way in the last part of the present study.

It looks that recognizing the hormonal signal the TR posesses the ability to control the hormonal action via a quick tuning mechanism. In this respect the TR has to be assumed as a real regulatory site of hormonal action. In other words the quantitative programme for hormonal action is build in the TR. There is not enough evidence yet for a "qualitative programme" for the realization of hormonal action in the TR. It seems that the hormonal signal is only quantitative (discussed later) and the receptor regulatory mechanism has combined quantitative and qualitative nature. Evidently the answer of the TR to CS is of type "yes" or "no" and the net action is accomplished by proportion of A to P states of the TR. This kind of explanation has a weak point coming from the possibility of control of the hormonal action via periodic or nonperiodic changes of the hormonal concentration. An easy but too mechanical explanation is that the quantity gradient in the central hormonal signal might switch off different programmes at the TR, i.e. the TR "can see" the quantity gradient in the CS and should respond by an adequate programme. A preliminary study on the possibility for an allosteric PH-TR interaction and the regulation of the hormonal action at the TR level was presented in our previous studies (Vassileva-Popova, 1972, 1973, 1974 and Vassileva-Popova et al., 1973).

Concerning the hormonal signalization some results suggest (Sonenberg, 1974) that just a few PH molecules are sufficient for the response cascade reaction in the TR. This concept looks economical from a bioenergetic point of view. In this connection there is an indication that the PH is not neutralized, on the contrary it increases its activity after its binding to the TR. As a result of that after the (PH-TR) complex dissociation the PH are capable of a 2nd step specific binding to the TR (Dufau et al., 1972). Another probable "economy" of the hormonal compound is to be found in the chemical nature of some hormones. It seems possible that in a comparative way the turnover of glycoproteins is longer (Preston, 1973). In this respect glycoprotein hormones might have longer life-time in comparison to other proteins (a possible role of sugar in the protein molecule). From the the present study the question arises: if the hormonal information transfer is fully realized via its small quantity of hormone, why relatively so large an amount of hormonal compound is present in the circulation.

In hormonal information transfer to the TR the hormone changes its quantity and the TR recognizes this periodic or non-periodic sporadic quantity gradient and responds to this signalization in an adequate way. What has happened in this regulatory axis during ageing? The data discussed in the first part suggest that the hormonal signal does not change in quantity or quality in the last period of the life span. There is not enough evidence yet for a significant change of the PH level and the chemical structure of the hormones in ageing, i.e. the hormones are the same in developing and senile subjects. In the elegant works on the chemical structure and biological activity of the PH subunits (Saxena et al., 1973; Bishop and Rayan, 1973; Basudev and Parlow, 1973; Catt et al., 1973; Reichert Jr., 1973; Dixon, 1973; Shome and Parlow, 1973) there is no evidence for an age change of the hormonal chemical structure. This means that the hormonal signal changes its quality as a regulatory programme, but not its quality as a chemical message. A different kind of changes might be suspected through a dynamic reararngement of the and subunits between related adenopituitary hormones during translation and recognition of the hormonal signal. If the hormonal signal does not change its quantity and characteristic feature in ageing it has to be accepted that the TR mechanism for the hormonal recognition and the specific information transfer is altered. At the same time the possibility for the reversibility of this age alteration of the TR is shown in this study. In this connection it looks stimulating to apply in ageing research the knowledge of protein conformation during hormonal recognition (Schwizer, 1973).

On the basis of the allosteric model of enzyme-substrate interaction, the hormone was assumed instead of the S and the receptors instead of the E. At the same time it should be noted that if we accept an allosteric interaction between the PH and TR, then we deal with two relatively big molecules and a protein-protein interaction should take place. The open question which does not yet cause complications at this stage of investigation is: if the PH takes the place of S we are dealing with a peculiar long life-time S which does not degenerate after its interaction with the TR, as happened in the E-S interaction, where the degree of S degradation indicated the regulatory protein (E) activity. Are we right to accept that the hormone is a substrate accepting on the concept of the substrate-protein interaction? Another open issue is the subunit nature of hormones and the possibility for a functionally dynamic and topologically partial substrate role. There exists also a theoretical possibility that the part of the receptor as a complex biopolymer plays the role of substrate for the hormone action. The complex nature of the TR biopolymer allows another functional characteristics to be expected. In this respect another possible role of the steroids is the transformation of state P into state A in the TR, results from a probable direct interaction of the SS with the allosteric site of the TR. This assumption will be discussed in two aspects. Being a complex biopolymer the TR has the opportunity to operate as a TR for the SS and as a TR for the PH. The so-called in this paper "allosteric site" for SS binding should play the role of a SS-receptor. The binding of the SS to the presumable al-

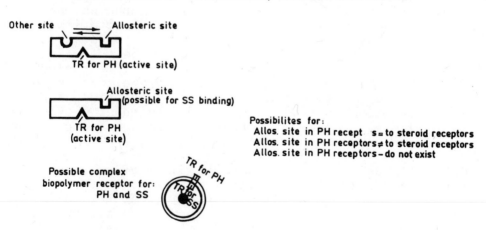

Figure 6

losteric site in the complex TR biopolymer should induce state A in the TR and increase the binding affinity of the TR to the PH. If we discuss the possibility for the functional complex-"common" receptors for the protein and steroid hormones there is another type of question. How is it to be explained that the PH receptors are on the cell membrane and the TR for SS are into the cell if the receptor biopolymer complex is a kind of "common" receptor for SS and PH. It is not so simple to say that this is "out" and "in" cell-oriented functional complex in some way similar to hydrophilic, resp. hydrophobic, membrane orientation. It has to be confirmed that this concept is just a possible way of thinking (Fig.6). The hypothesis assuming a possible allosteric site in the receptors for the PH possibly acting as some kind of acceptors for the SS is not in disagreement with the existence of a separate TR for the SS and of the finding of the individual protein, the steroid receptor (Jensen, 1974) and also the classical mechanism for PH action (Schwartz, 1973).

It is important to underline that this hypothesis is suggested from our experimental and mathematical modelling analysis, but is not the only obligatory way of

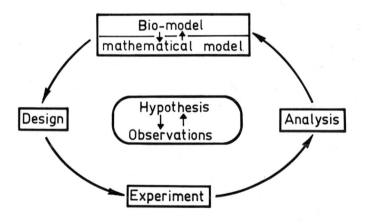

Figure 7

explaning our findings. And the observation and the hypothesis are different, but not isolated stages of the process of investigation (Fig.7).

Which are the more pragmatic questions raised by the study made? Our findings indicate a limitation of the senile TR for the recognition of the hormonal signal demonstrated as a decrease of the hormonal effect. At the same time there is no evidence for a decline of the concentration and the chemical characteristics of the hormone in ageing.

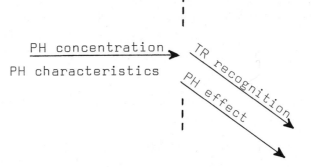

Following this logics we expect the age changes to be demonstrated mainly at the receptor level: possibly as an inhibition of the recognition or the appearance of the screening effect. Some suport for this is that if the age change was due to an insufficiency of the hormonal signal (relative limitation in the endocrine sphere) the replacement therapy will take off such problems we are still dealing with. The present study stresses the point of dose response and hormonal treatment in ageing. It seems that the senile receptors do not need a large dose of regulatory compounds. It is quite easy to have a pretreatment instead of the necessary treatment. There is a suggestions that unutilized hormone already is commulated of the target level which is a kind of screening effect for hormonal recognition. It looks also that because of the hormonal oscillation the time of treatment is quite an important parameter. In the case of sex steroids pretreatment (as we had in our case) a compensation with a simultaneous treatment with the steroidogenetic ICSH has to play a positive role.

The study on the age changes of the transmitters (Wayner and Emmers, 1958; Rockstein et al., 1971) indicate that ageing produces an accumulation of neurotransmitters in the nerve terminals. It seems that the problem of age changes of the regulatory substance utilization at the target level is a process characteristic for ageing, dealing with a kind of hormonal information non-transfer (Northcutt, 1975). It is not surprising that there is agreement that the search for age changes should be entirely oriented to the receptor level (Hall, 1975). In this light trying to understand ageing as a "slow kinetic" process the detailed investigation should be focussed on the life-time of the regulatory, recognizing receptor proteins and on the "quick kinetic" process of the hormone-receptor interaction.

The search for correlation between cell ageing and death, cell to cell communication and the connection of the processes on this level to the biosystem will stimulate a new way of thinking. (Sheldrake, 1974).

Instead of the research orientation to the TR, the dynamic of the formation of protein hormone receptors will become of interest. The work of French and Ritzen (1973a, 1973b), French et al., (1974) has indicated that FSH is responsible for the synthesis of androgen binding protein. In this connection the question for a possible inductor of the dynamic formation of protein hormone receptors will be of interest.

The possibility for the existence of a hormone-hormone interaction is applied in this study. The PH and the SS might interact in the gonads where there is an opportunity for a high concentration of PH molecules, because of an "attraction" of the PH by the TR of the PH. Since the gonads are also the place for SS formation in the hydrodynamic conditions the opportunity exists for a relatively high SS concentration there. This might provide the conditions possible for the PH-SS complex formation with relatively high PH and SS concentration. At the cellular level, as the target cell membrane consists of PH receptors and at the same time the hydrophobic part of the membrane provides conditions for the SS attachment this should result in a simultaneous accumulation of PH and SS in the same cell location. Concluding we may say that in the present study the focus is on the receptors recognizing the hormonal signal and the possible mechanism explaining the hormonal information non-transfer in ageing and its possible reversibility. It is not surprising the future aim of ageing research to be quite apart from that.

On the basis of the combined approach made (the experimental work, mathematical analysis, and mathematical model proposed) let us complete this study by stressing the point that the model constants presented are not to replace the further experiments, but they provide evidence for a theoretical prediction of the process characteristics. At the same time to have the opportunity to estimate the model constants on the basis of more comprehensive experiments is tempting.

ACKNOWLEDGEMENT

We would like to thank the NIAMD at the NIH, Bethesda for the kind supply and Doctor M. Cotes from the NIBSC, London for the consultation, the Department of Biochemistry, Imperial College, London and the Population Council at the Rockefeller University, New York for the facilities provided.

We express our deep gratitude to Doctor S. Moore from the Rockefeller University, New York, and Doctor J. R. Tata from the NIMR, London for the stimulating discussion, to all the colleagues at the Colloquium '74 for the helpfull discussions and criticism. Our thanks are extended as well to all who submitted reprints of their papers.

REFERENCES

Adelman, R.C., Molecular and Cellular Mechanisms of Ageing, INSERM, 5 - 7 December, Paris, 1973.

Alexandrova, M.D., in: Problems of Social and Psychological Gerontology, Ed., N.N.Krupinova, Leningrad University, Leningrad, 1974, (in Russian)

Anderson, P.J., Biochem.J., $\underline{140}$, 341, (1974)

Andrew, W., in: The Biological Basis of Medicine, $\underline{1}$, Eds., E.E.Bittar and N.Bittar, Academic Press, London-New York, 1969.

Bach, J. and Dardenne, M., Molecular and Cellular Mechanisms of Ageing, INSERM, 5 - 7 December, Paris, 1973.

Basudev, Sh. and Parlow, A.F., J. Clin. Endocrinol.Metab., $\underline{36}$, 618, (1973).

Bishop, W.H. and Rayan, J., Biochemistry, $\underline{12}$, 3076, (1973)

Blecher, M., 1973 (personal communication).

Blecher, M., paper presented at the First International Colloquium on Physical and Chemical Information Transfer and Regulation of Reproduction and Ageing, 7 - 14 October, Varna, 1974.

Bulankina, N.I. and Machikhina, R.G., in: Problems of Age Physiology, Biochemistry and Biophysics, Eds., V.N.Nikitin et al., Naukova Dumka, Kiev, 1974, (in Russian).

Bulankina, N.I., Mitriaev, A.B. and Ushakova, I.D., in: Problems of Age Physiology, Biochemistry and Biophysics, Eds., V.N.Nikitin et al., Naukova Dumka, Kiev, 1974, (in Russian).

Bullough, W.S., in: The Evolution of Differentiation, Academic Press, London, 1967.

Cartlige, N.E.F., Black, M. M., Hall, M.R.P. and Hall, R., Gerontol. Clin., $\underline{12}$, 65, (1970).

Catt, K.J., Dufau, M.L. and Tsuruhara, T., J.Clin. Endocrinol. Metab., $\underline{36}$, 73, (1973).

Catt, K.L., Tsuruhara, T. and Dufau, M.L., Biochim.Biophys. Acta, $\underline{279}$, 194, (1972).

Chernigovski, V.N., Interoreceptors, Medgis, Moscow, 1960, (in Russian).

Cristofalo, V., in: Ageing in Cell and Tissue Culture, Eds., E.Holeckova and V.Cristofalo, Plenum Press, New York-London, 1970.

Cuatrecasas, P., Proc.Nat.Acad.Sci. USA, $\underline{69}$, 318, (1972).

Curtis, A.J., in: Biological Mechanisms of Ageing, Ed., C.C.Thomas, Springfield, 1966.

Dalal, K.B., Petropoulos, E.A. and Timiras, P.S., Environ.Physiol.Biochem., $\underline{3}$, 117, (1973)

Denef, C., Magnus, C. and McEwen, B.S., Endocrinology, $\underline{94}$, 1265, (1974).

Dilman, V.M., in: Why Death Appears, Meditsina, Leningrad, 1972, (in Russian)

Dixon, J.S., in: Peptide Hormones, Eds., S.A.Berson and R.S.Yalow, North-Holland Publ.Co., Amsterdam-London, 1973.

Driniaev, V.A., Shabanova, N.A. and Onoprienko, V.D. in: Problems of Age Physiology, Biochemistry and Biophysics, Eds., V.N.Nikitin et al., Naukova Dumka, Kiev, 1974, (in Russian).

Druzhinina, M.P., in: Problems of Age Physiology, Biochemistry and Biophysics, Eds., V.N.Nikitin et al., Naukova Dumka,

Kiev, 1974, (in Russian).

Dufau, M.L., Catt, K.J. and Tsuruhara, T., Proc. Nat. Acad. Sci. USA, 69, 2414, (1972).

Einstein, E.R., in: Drugs and the Developing Brain, Eds., A. Vernadakis and N. Weiner, Plenum Publishing Co., New York, 1974.

Everitt, A.V., Exp. Gerontol., 8, 265, (1973).

Emanuel, N.M., Int. Congress of Gerontology, I., Naukova Dumka, Kiev, 1972, (in Russian).

Finch, C.E., Exp. Gerontol., 7, 53, (1972).

Franks, L.M., in: Molecular and Cellular Mechanisms of Ageing, INSERM, 5 - 7 December, Paris, 1973.

French, F.S. and Ritzen, E.M., Endocrinology, 93, 88, (1973a).

French, F.S. and Ritzen, E.M., J. Reprod. Fert., 32, 479, (1973b).

French, F.S., Nayfer, S.N., Ritzen, E.M. and Hansson, V., Research in Reproduction, Ed., R.G. Edwards, 6, 4, (1974).

Freychet, P., Roth, J. and Neville Jr., D.M., Proc. Nat. Acad. Sci. USA, 68, 1833, (1971).

Freedman, R., New Scientist, 21, 560, (1974).

Friedman, M., Green, M.F. and Sharland, D.E., J. Gerontol., 24, 292, (1969).

Glick, M.C., Kimhi, Y. and Lihauer, U.Z., Proc. Nat. Acad. Sci. USA, 70, 1682, (1973).

Galavina, O.I., in: Problems of Age Physiology, Biochemistry and Biophysics, Eds., V.N. Nikitin et al., Naukova Dumka, Kiev, 1974, (in Russian)

Glückmann, A., Biol. Rev., 26, 59, (1951).

Gordienko, A., in: Problems of Age Physiology, Biochemistry and Biophysics, Eds., V.N. Nikitin et al., Naukova Dumka, Kiev, 1974, (in Russian)

Gospodarowicz, D., 1974, (manuscript).

Grebennikova, N.P. and Gal, T.G., in: Problems of Age Physiology, Biochemistry and Biophysics, Eds., V.N. Nikitin et al., Naukova Dumka, Kiev, 1974, (in Russian).

Gutman, E. and Hanzlikova, V., Mech. Age. Dev., 1, 327, (1972).

Hahn, H.P., Exp. Gerontol., 5, 323, (1970).

Hall, M.R.P., in this volume, 1975.

Harman, D., J. Gerontol., 11, 298, (1956).

Harman, D.,J. Gerontol., 16, 247, (1961).

Harman, D., Radiation Res., 16, 753, (1962).

Harman, D., J. Gerontol., 23, 476, (1968).

Hayflick,L., Exp. Gerontol., 5, 291, (1970).

Hayflick,L.,Schwartz,B.D., Smith,J.R.,Stein,G.H.and Wright,W.E., Molecular and Cellular Mechanisms of Ageing, INSERM, 5 - 7 December, Paris, 1973.

Hermann,R.L., O'Meara, A.R.,Russel,A.P.and Dowling,L.E., 8th Int.Congr.Gerontol.,Washington, 1, 141, (1969).

Holliday,R., Molecular and Cellular Mechanisms of Ageing, INSERM, 5 - 7 December, Paris, 1973.

Hudson, D. B., Merrill, B.J. and Sands, L.A., in: Drugs and the Developing Brain, Eds., A. Vernadakis and N. Weiner, Plemum Publishing, New York, 1974.

Hudson, D., Vernadakis, A. and Timiras, P.S., Brain Res., 23, 213, (1969).

Human, A., in: Penguin Books Australia, Ltd., Ed., S.M. Chown, Ringwood, Victoria, Australia, 1972.

Hunter, W.M. and Greenwood F.C., Nature, London, 194, 495, (1962).

Jensen, E.V. and DeSombre,E.R., Science,182, 126, (1973).

Jensen, E.V., Mohla,S., Blecher,P.I. and DeSombre,E.R., in: Receptors for Reproductive Hormones, Eds., B.W.O'Malley and A.R.Means, Plenum Press, New York-London, 1973.

Jensen, E.V., paper presented at the First International Colloquium on Physical and Chemical Information Transfer and Regulation of Reproduction and Ageing, 7 - 14 October, Varna, 1974.

Kahn, C.R., Neville Jr. D.M. and Roth, J., J. Biol.Chemistry, 248, 244, (1973).

Kaye, A.M., Sheratzky, D. and Lindner, H.R., Biochim. Biophys. Acta, 261, 475, (1972).

Kaye, A.M., Sömjen, G., King, R.J.B. and Lindner,H.R., Annual Plenary Meeting of the Israel Endocrine Society, 26 - 27 April, 1973.

Kessler, B., Biochim.Biophys. Acta, 240, 496, (1971).

Klimenko, A.I., in: Problems of Age Physiology, Biochemistry and Biophysics, Eds., V.N.Nikitin et al., Naukova Dumka, Kiev, 1974, (in Russian).

Klimenko, A.I. and Tupchenko, G.S., in: Problems of Age Physiology, Biochemistry and Biophysics, Eds., V.N. Nikitin et al., Naukova Dumka, Kiev,1974, (in Russian)

Korenchevsky, V., Physiological and Pathological Ageing. Basel-New York, 1961.

Kohn, A., Meienhofer, M.C., Vibert, M. and Dryfus, J.C., Molecular and Cellular Mechanisms of Ageing, INSERM, 5 - 7 December, Paris, 1973.

Korner, A., Biochem. J., 92, 449, (1964).

Kovács, S., Cocks, W.A. and Balazs, R., J. Endocrin., 45, IX, (1969).

Kritchevsky, D. and Howard, B.V., in: Ageing in Cell Tissue Culture, Eds., E. Holečková and V.J.Cristofalo, Plenum Press, New York, 1970.

Lark, K.G., Ann.Rev. Biochem., 38, 569, 1969.

Lockshin, R.A. and Williams, C.M., J. Insect Physiol., 10, 643, (1964).

Lowry, O.H., Sosebrough, N.J., Farr, A.L. and Randall, R.J., J. Biol. Chem., 193, 256, (1951).

Macieira-Coelho, A. and Loria, E., in: Aspects of the Biology of Ageing, Symp.Soc. Exp. Biol., 21, Ed., H.W.Woolhouse, Academic Press, New York-London, 1967.

McEwen, B.S. and Pfaff, D.W., in: Frontiers in Neuroendocrinology, 1973, Eds., W.F. Ganong and L. Martini, Oxford University Press, Inc., 1973.

Medvedev, Zh.,A., in: Aspects of the Biology of Ageing, Symp. Soc. Exp. Biol., 21, Ed., H.W.Woolhouse, Academic Press, New York-London, 1967.

Medvedev, Zh. A., in: Molecular and Cellular Mechanisms of Ageing, INSERM, 5 - 7 December, Paris, 1973.

Mishchenko, V.P., in: Problems of Age Physiology, Biochemistry and Biophysics, Eds., V.N.Nikitin et al., Naukova Dumka, Kiev, 1974, (in Russian)

Monod, J., Wyman, G. and Changeux, J.R., Molec. Biol., 12, 88, (1965).

Moros, J.A., in: Problems of Age Physiology, Biochemistry and Biophysics, Eds., V.N. Nikitin et al., Naukova Dumka, Kiev, 1974, (in Russian)

Nagorny, A.B., Nikitin, V.N. and Bulankin, I.N., in: Problems of Ageing and Longevity, Ed., V.N.Nikitin, Medgiz, Moscow, 1963, (in Russian).

Neville, D.M., Biochim.Biophys.Acta, 154, 540, (1968).

Nikitin, V.N., in: Proc.Symp.of Basic Problems of Developmental Physiol. and Biochem., HGU, Harkov, 1965, (in Russian).

Nikitin, W., in: Ideen des Exakten Wissens, DVA - Anstalt, Stutgart and APN - Pressagentur Novosti, Moscow, 1970, (in German).

Northcutt, R.C., in this volume, 1975.

O'Malley, B.W., Schrader, W.T. and Spelsberg, T.C., in: Receptors for Reproductive Hormones, Eds., B.W.O'Malley and A.R. Means, Plenum Press, New York-London, 1973.

Parhon, K.I., Ageing Biology, Meridiani,Bucharest, 1959.

Parina, E.V., Driniaev, V.A. and Hotkevich, T.V., in: Problems of Age Physiology, Biochemistry and Biophysics, Eds., V.N.Nikitin et al., Naukova Dumka, Kiev, 1974, (in Russian).

Parlow, A.F., Coyotupa, J. and Kovacic, N., J. Reprod. Fert., 32, 163, (1973).

Pashkova, A.A., Gorbachev, V.I., Kurtach,E., in: Problems of Age Physiology, Biochemistry and Biophysics, Eds., V.N.Nikitin et al., Naukova Dumka, Kiev, 1974.

Pegg, A.E. and Korner, A., Nature, 205, 904, (1965)

Pierce, J.G., in: Peptide Hormones, Eds., S.A.Berson and R.S.Yalow, North-Holland Publ. Co., Amsterdam-London, 1973.

Plapinger, L.and McEwen, B.S. and Clemens,L.E., Endocrinology, 93, 1129, (1973)

Plapinger, L.and McEwen, B.S. Endocrinology, 93, 1119, (1973)

Popova, L.Y.and Bondarenko,V.A.in: Problems of Age Physiology, Biochemistry and Biophysics, Eds., V.N.Nikitin et al., Naukova Dumka, Kiev, 1974, (in Russian).

Popova, L.Y.and Mironenko,T.G., in: Problems of Age Physiology, Biochemistry and Biophysics, Eds.,V.N.Nikitin et al., Naukova Dumka, 1974, (in Russian).

Preston, R.D., 1973, (personal communication)

Rao, Ch.V.and Saxena, B.B., Biochim.Biophys.Acta, 313, 235, 1973.

Reichert Jr., L.E., in: Peptide Hormones, Eds., S.A.Berson and R.S.Yalow, North-Holland Publ. Co., Amsterdam-London, 1973.

Rockstein, M., Groy, F.H. and Berberian, P.A., Exp. Gerontol., 6, 211, (1971).

Rodbard, D., 1973, (personal communucation).

Roth, J., Metabolism, 22, 1059, (1973).

Rubin, M.S., Swislocki, N.I. and Sonenberg, M.,
 Soc.Exp.Biol.and Med., 142, 1008, (1973).

Saxena, B.B., Rathnam, P. and Rommler, A.,
 Endocrinology, 7, 19, (1973).

Saxena, B.B., Hasan, S.H., Haour, F. and Schmidt-Gollwitzer, M., Science, 184, 793, (1974).

Schwartz, I. and Walter, R., in: Peptide Hormones,
 Eds., S.A. Berson and R.S. Yalow,
 North-Holland Publ. Co., Amsterdam-London, 1973.

Schwizer, R., Proceedings of the Fourth International
 Congress on Pharmacology, 14-18 July, Basel, V, 196, 1973.

Setchenska, M., Russanov, E. and Vassileva-Popova,J.,
 FEBS Letters, 49, 297, (1975).

Sheldrake, A.R., Nature, 250, 381, (1974).

Shevtsova, M. Ya., in: Problems of Age Physiology, Biochemistry and Biophysics, Eds., V.N.Nikitin et al.,
 Naukova Dumka, Kiev, 1974, (in Russian).

Shirey, T.L.and Sobel, H., Exp. Gerontol., 7, 15, (1972).

Shome, B. and Parlow, A.F.,
 J. Clin. Endocrinol. Metab., 36, 618, (1973).

Silin,O.P., in: Problems of Age Physiology, Biochemistry and Biophysics, Eds., V.N.Nikitin et al.,
 Naukova Dumka, Kiev, 1974, (in Russian).

Silin,O.P., Chupygina, A.N. and Svistun, L.Ya., in:
 Problems of Age Physiology, Biochemistry and Biophysics,
 Eds., V.N.Nikitin et al., Naukova Dumka,Kiev, 1974,
 (in Russian)

Sonenberg, M., 1974, (personal communication)

Srivastava, U. and Caaudhary, K.D.,
 Canad. J. Biochem., 47, 231, (1969).

Stavitskaia, L.I. and Shamarina, V.P., in: Problems of
 Age Physiology, Biochemistry and Biophysics, Eds.,
 V.N.Nikitin et al., Naukova Dumka,Kiev, 1974, (in Russian)

Stavitskaia, L.I., in: Problems of Age Physiology, Biochemistry and Biophysics, Eds., V.N.Nikitin et al.,
 Naukova Dumka, Kiev, 1974, (in Russian)

Strehler, B.L., in: Aspects of the Biology of Ageing,
 Symp.Soc.Exp.Biol., 21, Ed., H.W. Woolhouse,
 Academic Press, New York, 1967.

Strehler, B.L., in: Advances in Gerontological Research, 1, Ed., B.L. Strehler, Academic Press, New York, 1964.

Tappel, A.L., Gereatrics, 23, 97, (1968).

Tata, J.R., Nature (London), 227, 686, (1970).

Tata, J.R., in this volume, 1975.

Tchebotarev, D.F. and Frolix, V.V., in: Cardio-Vascular System in Ageing, Meditsina, Moscow, 1967, (in Russian)

Timiras, P.S., Hudson, D.B. and Oklund, S., in: Progress in Brain Research, 40, Elsevier Scientific Publ. Co., Amsterdam, 1973.

Trankvillitati, A.N., Transactions of the Moscow Society of Naturalists, XVII, 273, 1966.

Tsuruhara, T., Van Hall, E.V., Dufau, M.L. and Catt, K.J., Endocrinology, 91, 463, (1972).

Tupchienko, G.S., in: Problems of Age Physiology, Biochemistry and Biophysics, Eds., V.N. Nikitin et al., Naukova Dumka, Kiev, 1974, (in Russian).

Vaniushin, B.I., Korotaev, G., Masin, A. and Berdishev, G., Biokhimia, 34, 191, 1969 (in Russian).

Vassileva-Popova, J.G., European Congress of Sterility, 1 - 14 October, Athens, Greece, 1972.

Vassileva-Popova, J.G., Report presented at the Annual General Meeting of the British Soc. for Research on Ageing, January, 1973.

Vassileva-Popova, J.G., Proc. Fifth Asia and Oceania Congress of Endocrinology, I, Ed., G.K. Rastogi, Endocrine Soc. of India, 1974.

Vassileva-Popova, J.G., Chuknijsky, P. and Yanev, T., 9th International Congress of Biochemistry, Stokholm, 1973.

Vassileva-Popova, J.G., and Dimitrov, D., in this volume, 1975.

Vilenchik, M., in: Molecular Mechanisms of Ageing, Nauka, Moscow, 1970, (in Russian).

Wayner Jr., M.J. and Emmers, R., Amer. J. Physiol., 194, 403, (1958).

Widnell, C.C., Tata, J.R., Biochem. J., 98, 621, (1966).

Physical Mechanisms of
Information Transfer

THE CONFORMATIONAL CHANGES OF BIOPOLYMERS RETAINING THEIR INITIAL STRUCTURE: A NECESSARY STAGE OF INTRACELLULAR INFORMATION TRANSFER

Ya.M. Varshavsky

Institute of Molecular Biology
Academy of Sciences of the USSR, Moscow, USSR

There are two widely different types of information in the living organism. On the one hand there is the genetic information, carried by the chromosomes or more precisely the DNA, and on the other, episodic information about the environment, received by the organism in the form of input signals. Both genetic and environmental information are effected through complex processes passing through various stages such as: chemical reactions catalyzed by specific ferments, electron and ion transfer, different gradients (notably the pH gradient) across the intracellular membranes etc. A large number of monomer and polymer compounds, some of them specifically produced by the organism, participate in these processes.

Thus, for instance, in genetic information transfer a key role is played by information RNA specially synthesized to this end at some sections of the DNA molecules with the assistance of a special enzyme: RNA polymerase. It is known that the transfer of information about such complex processes as cell differentiation, growth and structure of the organism an essential role is performed by the hormones which promote or inhibit enzyme activity and in this way control the functioning of the organism within given parameters. The data about the mechanism of information transfer available to date prove convincingly that all of these mechanisms cannot dispense with the help of biopolymers: nucleic acids and proteins. The work of the complex system of the living cell is ultimately aimed at producing enzymes necessary for the synthesis of compounds which the cell needs at that moment. That applies both to the synthesis of non-protein compounds: nucleic acids, hormones etc., and to the ribosome synthesis of enzymes i.e. proteins.

An important feature of genetic information transfer processes is the continuous feed-back control coupled with the wide range of speed variability. These properties are ensured mainly by the enzymes which can be reversibly inhibited or promoted as a result of their reaction with specific substrates or the change of the physico-chemical properties (pH, ionic strength etc.) of their environment. The essential point is that in both cases the change in the functional activity of the enzymes and substrates is accompanied by changes of their conformational characteristics. Thus, the conformational changes of biopolymers represent a very important and necessary stage of the process of information transfer within the cell. What is more, the same applies to nucleic acids, proteins, and various kinds of low-molecular compounds which may play the role of substrates.

In this paper I would like to dwell on the problem of the conformational mobility of biopolymers, chiefly proteins, and describe some of the results obtained at our laboratory through the methods of hydrogen isotope exchange kinetics and infra-red (IR) spectroscopy.

In a solution of protein in heavy water (D_2O) the observed attenuation rate of the intensity of the oscillation band of Amide P ($\sim 1\,550\ cm^{-1}$) in the IR spectrum, corresponding to the valency variation of the peptide (non-lateral-NH-groups, can be used as a measure of the rate at which the hydrogen atoms of these groups are being replaced by the deuterium atoms from the heavy water.*

Experiments have shown that in the case of denaturated proteins having no spatial structure and short peptides, all hydrogen atoms in the peptide NH-groups are re-

---

*In actual fact when observing the deuterium exchange rate in proteins, it is not simply the Amide-P band intensity that is registered but its relation to the Amide 1 ($\sim 1\,650\ cm^{-1}$) band intensity which corresponds to the oscillation of the peptide C=O group and remains practically unchanged after the displacement of hydrogen by deuterium in the NH-groups. In this way the Amide 1 band serves as an internal standard which helps avoid errors entailed by the eventual changes of the concentration and the thickness solution layer observed.

placed very fast by deuterium atoms, while in the native proteins in part of the peptide NH-groups replacement takes place as quickly as in the denaturated proteins (in 2 - 3 minutes), in another part the replacement is considerably slower (several hours) whereas the rest do not react with the D_2O for 10-20 hours or even more. Since it is known that the NH-groups of the short peptides exchange their hydrogen atoms very fast, then evidently the slowing down of the exchange in the case of the native proteins is due to their great compactness, which prevents the contact of the water molecules with the NH-groups, located in the "nucleus" of the protein globule. Since the exchange of hydrogen between the water and part of the "inside" NH-groups still takes place even though slowly, the contact between the D_2O molecules and the sections of the polypeptide chain located deeply in the hydrophobic regions of the protein globule should be be explained yet.

K.Linderstrom-Lang when he first discovered the slow-down of deuterium exchange in proteins suggested that it was caused by the difficult diffusion of water molecules inside the protein molecules. By means of physical methods as well as by detailed analysis of deuterium exchange data for a large number of proteins at various pH values. It was proved, however, that D_2O molecules actually do not reach the hydrophobic "nucleus" of the protein globule. Now then D_2O molecules and the inside sections of the protein polypeptide chain do get into contact without which no exchange can occur. We shall not review in detail the large number of works devoted to this interesting and important question. Suffice it to note that the polypeptide chains of compact protein molecules have been shown to display a high degree of conformational mobility which is ultimately due to the possibility of free rotation round the ordinary peptide C-C-bonds. Thanks to the synchronization of such rotations some sections located inside the hydrophobic regions are able to overcome the steric hindrance and now and then "emerge" to the surface of the protein globule or get deep into the wide "interstices". In this way the NH-groups of these sections are made accessible to the D_2O molecules. It should be noted that such changes in the protein molecules which facilitate the contacts between the inside sections of the polypeptide chain and the water molecules may occur spontaneously, i.e. due to the thermal motion (kT) or may be caused by various outer influences on the protein molecule. Protein therefore may be regarded as a statistical ensemble comprising a great number of interconvertible conformational isomers, (molecules having different conformation but the same primary structure) which are usually called conformers. The statistical ensemble has a dynamic nature: each of the conformers can be characterized by its average life and stationary concentration. The life-time of each conformer is determined by the barrier amount of energy which should be overcome in its formation.

The stationary concentration is determined by the relative stability: the higher the free energy of a conformer, (i.e. the lower its relative stability) the lower its stationary concentration.

Thus the physico-chemical characteristics of the biopolymer related to its functioning within the cell cannot be explained only by its primary, secondary, and tertiary structure. According to the dynamic conception, at physiological temperatures the biopolymer molecules are in a state of constant movement, moreover besides the transitions between the different electronic, vibrational, and rotational energy levels they are characterized by the transitions between confomational energy levels. In other words besides electronic, vibrational, and rotational spectra biopolymers should also have conformational-excited-state spectra; the so called conformational spectra. Evidently the best approach would be to measure directly the conformational spectra as is the common practice with electronic, vibrational, and rotational spectra in the UV, visible, and IR regions. In the case of conformational spectra however, direct measurement is impossible at present. On the one hand they are located in the low-frequency bands where loud noises would jam the very weak signals, and on the other, such spectra consist of a great number of close or even coinciding lines, corresponding to the separate conformational transitions, which cannot be resolved by means of the now existing instruments. That is why the study of conformational spectra should resort to roundabout methods. Biopolymer conformational spectra have yet another important feature which renders research particularly difficult. That is the extremely low stationary concentration of molecules in a state of conformational excitation. The latter is due to the fact that the dependence of the stationary concentration of the conformer on free energy which is necessary for its formation has a non-linear, expotential character. Consequently small differences in the energies of two conformers correspond to very large differences in their stationary concentrations. Actually, the majority of the biopolymer molecules are in a state corresponding to the minimum of free energy at the moment, while an extremely small share of them happen to be in a state of conformational excitation. That is why the optical and hydrodynamic methods which are widely used for the study of polymer conformations as well as the method of X-ray structure analysis cannot be applied to detect and describe the energetically unfavourable conformationally--excited states as they always yield a picture which is averaged with respect to all conformers and is essentially relevant to the most energetically favourable conformations. There are however, sound grounds to assume that the conformationally excited biopolymer molecules, despite their very low concentration, are playing a major role in the functioning of the biopolymer molecules by forming transitional complexes with their "partners" and in this way securing the necessary speed of functioning. In the case of enzymes the partners are the molecules of the substrate, the cofactor or the inhibitor; in the case of nucleic acids: the active polymerase and synthetase sections; in the case of receptor proteins: the hormone molecules, etc.

At present the only available method of determining the stationary concentration of the polymers whose inside polypeptide chain sections are open to contacts with the molecules of the solvent is the method of the kinetics of hydrogen isotop exchange between heavy water and the protein peptide NH-groups. The study of deuterium exchange rates are various pH values yielded the unambiguous conclusion that the contacts between the polypeptide chains located deep into the protein globule and the heavy water molecules should not be explained by the diffusion of the water molecules inside the protein globule but by the internal movements of the polypeptide chain which secure brief contacts between the peptide NH-groups located deeply in the hydrophobic centre of the protein globule and the water molecules. It should be noted that the above conclusion concerning the mechanism of deuterium exchange between water and proteins was reached after the study of the deuterium-exchange kinetics of a very large number of various proteins. This kind of experiments were carried out in many laboratories and the above conclusion can be considered as proved beyond doubt.

Figure 1 illustrates schematically the mechanism of hydrogen isotop exchange between water and the peptide NH-groups located inside the protein globule. It also shows the ratios between the rate constants of the sepa-

closed open closed

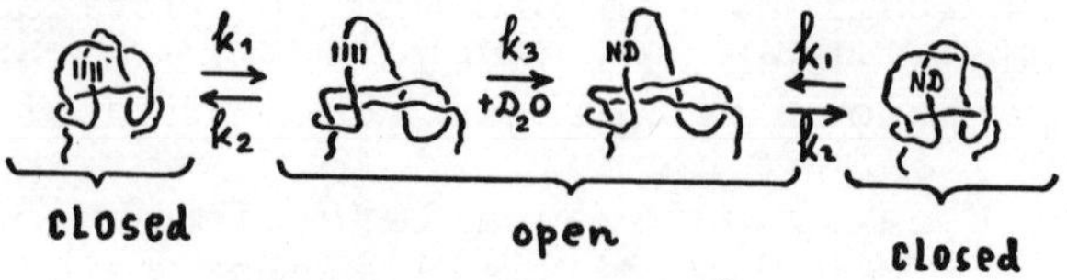

$k_1 \ll k_2$ $\mathcal{K} = \dfrac{k_1}{k_2} \ll 1$

$$\beta = \dfrac{k_1 \cdot k_2}{k_1 + k_2 + k_3}$$

$k_1 < k_3$; $\beta = k_1$ $k_1 > k_3$; $\beta = \dfrac{k_1}{k_2} \cdot k_3 \; \mathcal{K} \cdot k_3$
(EX_1-mechanism)   (EX_2-mechanism)

$$\Delta F = -RT \ln \mathcal{K}$$

Figure 1

The mechanism of deuterium-exchange between heavy water (D_2O) and peptide NH-groups located in the nucleus of the protein globule

rate stages of the exchange reaction. In the first boundary case (mechanism EX_1) when the rate of deuterium exchange between the NH-groups and the D_2O molecules in the conditions of full accessibility of the NH-groups (k_3) exceeds considerably the rate-constant of the conformational transition, leading to the "emergence" of the peptide NH-group on the surface of the globule (k_1), then the value of the exchange rate constant in the protein (β) is numerically equal to the rate-constant of the "emergence" of the peptide NH-group to the surface of the globule, i.e. $\beta = k_1$. In the second boundary case (mechanism EX_2) corresponding to a very high intramolecular mobility of the protein, when the rate-constant of the "emergence" of the peptide NH-group to the surface of the globule (k_1) exceeds considerably the rate of deuterium exchange between the D_2O and the NH-groups in the conditions of complete accessibility, (k_3), then

$$\beta = \dfrac{k_1}{k_2} k_3 = K k_3 .$$

The kinetic analysis of deuterium-exchange rate data for a large number of proteins and within a wide pH range (i.e. under conditions where the k_3 value undergoes substantial changes) has shown that the exchange in proteins follows the EX_2 mechanism. Cosequently, if we change β under given conditions (pH, t°, etc.) knowing the k_3 value under the same cnditions (from the deuterium-exchange data in short peptides and model compounds) and using the above equation we shall be able to find the value of K.

Thus, even though the deuterium-exchange method does not enable us to determine the rate of the conformational transitions in the protein*, still it makes possible to determine with a sufficient degree of precision the ratio between the rate-constant of the "opening" (k_1) and the "closing" (k_2) of the protein globule. Knowing the value of this ratio, which is numerically equal to the constant of the equilibrium between the "open" and the "closed" conformations of the protein (K), one may compute the free energy change (ΔF) corresponding to the transition of the protein globule from a more favourable "closed" state to a less favourable "open" state. This specific feature of the deuterium-exchange method allows for the construction of the protein conformational spectrum which would be characterized by the relative "ease" (in ΔF units) with which the separate sections of the polypeptide chain are rendered accessible to the water molecules. Defining the protein conformational spectrum practically boils down to determining experimentally the exchange rate at varying temperatures and pH values within the limits where the conformational characteristics and the biological activity of the protein are preserved. Ideally the protein conformational spectrum should consist of a set of ΔF values with exact indications to which of the peptide

---

* The fact that deuterium-exchange in proteins happens according to the EX_2 mechanism indicates that the average rate-constant value of the conformational transitions of the protein molecules from "closed" to "open" state (k_1) should significantly exceed the deuterium-exchange rate-constant under conditions of unimpeded contacts between the NH-groups and the D_2O molecules (k_3), which for neutral pH values ranges between 10^{-2} and 10^{-4} sec. The latter compares favourably with the available data about the rate of conformational changes obtained by NMR and flash-photolysis according to which the time for the structural rearrangement of the protein globules ranges from 10^{-6} to 10^{-7} sec.

NH-groups (i.e. to which amino acid-residue number) each value corresponds.

Owing to various reasons however, the deuterium-exchange method cannot ensure such a high resolving power and in practice the conformational spectra obtained contain only information about the share of the peptide NH-groups of the polypeptide chain, characterized by the ΔF values. Using the X-ray structure analysis data one may occasionaly interpret the conformational spectrum or some of its regions, i.e. determine the numbers of the amino acid residues to which some of the obtained ΔF values correspond. Unfortunately such an interpretation is not always unique.

Proceeding from deuterium-exchange kinetics data Dr.L.Abaturov from our laboratory succeeded in constructing the spectrum of haemoglobin. This is shown in Fig.2.

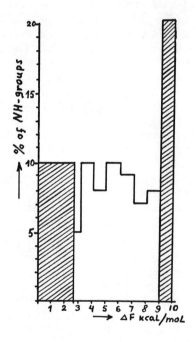

Figure 2

The conformational spectrum of haemoglobin

The hatched left-hand part of the spectrum corresponds to the sections of the haemoglobin polypeptide chain sections located in the hydrated regions where contact with the water molecules is relatively easy and does not require large free energy expenditure ($\Delta F < 2,5$ Kcal/mol). The deuterium-exchange rate in such NH-groups is so high, that with the presently used variant of infra-red spectroscoping it was only possible to measure its lower limit. As is evident from Fig.2 the share of such NH-groups is approximately 25%, which tallies with the X-ray structure data according to which the share of hydrated amino acid residues in haemoglobin is 20-25%. The right-hand side of the spectrum corresponds to the high free energy values ($\Delta F > 10$ Kcal/mol.), i.e. very low exchange rates. The deuterium-exchange rate of the corresponding NH-groups located deep in the centre of the protein globule is very low and practically cannot be measured by IR spectroscopy. As shown in Fig.2 the share of these NH-groups is approximately 20%.

The ΔF values in the 2,5-10 Kcal/mol interval correspond to the remaining 55-60% of the haemoglobin peptide NH-groups, whose deuterium-exchange rate can be measured by means of infra-red spectroscopy within the temperature and pH range where haemoglobin retains its native conformation.

Evidently the comparison of the conformational spectra of groups of cognate protein allows for determining the relationship between their functional activity and the intramolecular mobility. For that purpose Dr.L.Abaturov made a comparative analysis of the conformational spectra of four cognate proteins: haemoglobin, two of its subunits (α- and β- chains), and globin. The results are given by Fig.3.

The comparison of the spectra indicates that the splitting off the prosthetic group (haem) from the haemoglobin, entailing its inactivation is accompanied by an abrupt rise of the exchange rate which testifies to the "loosening" of the protein molecule and an increase of its intramolecular mobility. This change of structure which is characterized by the drop of the average ΔF value from 6,6 Kcal/mol in haemoglobin to 3,5 Kcal/mol in the case of globin is manifested by the practically almost complete disappearance of the sections of the polypeptide chain which are unable to come into contact with the water molecules as well as by the increase of the share of sections accessible to the water molecules. This particular example illustrates the deep conformational changes that can occur in the protein molecule as a result of the local change of structure in a relatively small section which is responsible for the globine-haem

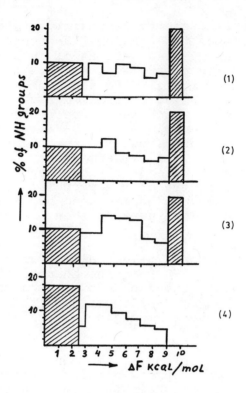

Figure 3

Comparison of the conformational spectra of a group of cognate proteins of varying functional activity.   1 - haemoglobin; 2 - β-chains; 3 - α-chains;   4 - globin

bond. This kind of behaviour of the biopolymer molecules is determined by their characteristic properties. It is of major biological significance and plays a key role in the mechanisms of the specific interactions of the receptor-proteins with hormones which secure the possibility of directed changes in their enzyme activity.

The comparison of the conformational spectra of haemoglobin and its subunits (α- and β-chains), containing prosthetic groups and having higher affinity to oxygen in the instance in point made it possible to assess the changes of the intramolecular mobility accompanying the general changes of the structure and the functional activity of the protein.

* * *

The data reported in the present paper clearly testify to the conformational mobility of the protein molecules which may be characterized quantitatively by their conformational spectra. In principal this conclusion also applies to the nucleic acids. This is confirmed by numerous data about the hydrogen isotops exchange kinetics between water and various types of nucleic acids. The difference between proteins and nucleic acids however, consists in the fact that in the case of the latter and particularly in DNA the basic (non-exited) state of the system is relatively more stable chiefly owing to the greater degree of co-operativity of their structure. It is precisely the high conformational mobility of biopolymers-proteins and nucleic acids - combined with the uniqueness and the high degree of compactness of their structures that ensure their effective participation in the intracellular information transfer processes without changing their primary structure.

FLEXOELECTRIC MODEL FOR ACTIVE TRANSPORT

Alexander G. Petrov

Institute of Solid State Physics
Bulgarian Academy of Sciences, Sofia, Bulgaria

A general model of active transport is proposed, based on the flexoelectric membrane properties, with regard to liquid crystalline membrane structure. The physical principles of the flexoelectric effect are briefly reviewed and it is admitted that the spherical deformation of planar membrane produces a flexoelectric polarization which results in depolarizing electric field. The orientational state of integral protein globuli in this field is considered and it is found to be similar to that of a trigger. It is shown that a change of the globula dipole moment by adhering to the negative end of a proton may cause its reorientation accompanied with translocation of the proton to the outer membrane side. The theoretical formulae describing such a mechanism are derived and both the value and the sign of transmembrane potential are calculated as a function of flexoelectric membrane coefficients. Two chanal feedback control leading to stable self-regulation both of the membrane curvature and transmembrane potential, is involved in this model. In view of such mechanism a natural explanation may be given for a number of active transport features: curved membrane sectors are found to be metabolitically active; transport energy is secured by membrane itself; assymmetry and vector character of the transport are clearly demonstrated. There exist at least two experiments which are consistent with the notion of flexoelectricity and permit in principle to measure the value of flexoelectric membrane coefficient.

INTRODUCTION

When in 1969 R.Meyer predicted the existence of a peculiar kind of flexoelectricity in liquid crystals he pointed out its importance for the biostructures, possessing liquid cristalline properties. Caserta and Cervigni (1972) made a hypothesis for the universal role of this kind of effects upon biological membrane functioning. As a concrete application of this hypothesis they proposed a transducer model for phosphorylation in photosynthetic membrane in terms of the classical piezoeffect, connected with the contraction of the membrane thickness upon illumination (Caserta & Cervigni, 1973). What is more, the authors cited stressed the possible application of the peculiar flexoelectric properties, caused by the liquid crystalline character of biomembranes.

As a biophysical application of our own studies of flexoelectric properties of liquid crystals we show in this paper how the idea of membrane flexoelectricity made it possible for us to work a general model of active transport, based on exact theoretical relationships and manifesting possibilities of stable self-regulation.

LIQUID CRYSTALS AND BIOMEMBRANES

Let us begin with a general characteristic of the liquid crystals. These anisotropic structures are divided in two large groups according to the anisotropic arrangement observed at solid crystal melting - the so called thermotropic liquid crystals (Gray, 1962), or to the dissolution of substances with anisotropic amphiphilie molecules in water or other polar solvents - liotropic liquid crystals (Winsor, 1968). The biological liquid crystalline (LC) structures belong to the second group. It is remarkable that the first known LC structures are from biological origine (the well known studies of Wirchow on the myelin from the middle of the last century).

Biological membranes are a typical example of the LC structure. They are built up from lipid molecules arranged in a characteristic manner for smectic liquid crystals - a fact that is already included in the classical bilayer membrane model of Danielli and Davson (1935). Liquid crystalline properties of model lipid-water systems are quite well examined: Chapman's studies on phase transitions in such systems (1971). In the model systems however an important feature of the intact membrane is absent - the intercalation of globular protein compounds in the lipid bilayer.

The fluid mosaic membrane model of Singer (1972) is presented in Fig.1. It is the basis of our consideration. The bimolecular lipid layer with its long range order is shown. This layer is the liquid crystalline scaffold of

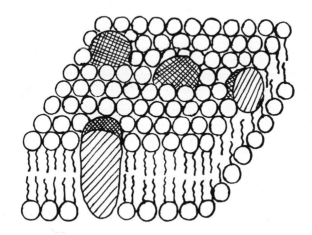

Figure 1
Fluid lipid-globular protein membrane model of Singer (1972). Transport functions are ascribed to integral proteins, protruding from the two membrane interfaces.

the membrane in which integral protein globuli are embedded. The polar and ionic hydrophilic portions of integral proteins are protruded from the one or two interfaces of the membrane. The lipid molecules possess a big constant dipole moment and bilateral asymmetric form, which is of great significance for the flexoelectric membrane properties, as shown below.

ELASTICITY AND FLEXOELECTRICITY OF BIOMEMBRANES

Let us consider the elastic and piezoelectric membrane properties in connection with their liquid crystalline structure. From the point of view of micropolar medium which manifests orientational elasticity. For this kind of elasticity "deformations" represent the local changes in orientation of the rodlike molecules and gives rise to torques which tend to restore the parallel orientation corresponding to minimum elastic energy. This problem is discussed in details by Frank (1958) and recently by Nehring and Saupe (1971). Analogical remarks are valid for the piezoelectric effects, which also have orientational character (Meyer, 1969). Therefore talking about elastic and piezoelectric properties we have in mind those determined by changes in relative orientation of building membrane elements and not by changes in distance as in the case of classical translational elasticity and piezoeffect.

Let us consider the molecular structure of lipids composing the membrane. We concentrate our attention mainly upon lecithin - a phospholipid found in great ammounts: 40 - 60 % (Borovyagin, 1971) in various types of membranes (Fig.2a). In this conformation the molecular form is highly asymmetric.

Due to the fast rotation of lipids around the long axis (Schindler & Seelig, 1973) the effective molecular form may be represented by a cone (Fig.2,b). Moreover, lecithin molecules have a big constant dipole moment ∼24D (Pennock, Golman, Chacko & Chock, 1973) due mainly to the positive charge of the holin radical and the negative of

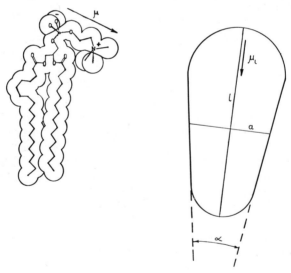

Figure 2

The lecithin molecule and its conical representation. For the given interrelation of conical asymmetry and axial dipole moment, piezocoefficient of the membrane must be negative.

FLEXOELECTRIC MODEL FOR ACTIVE TRANSPORT

oxygen, so in this conformation its axial component points towards the top of the cone (Fig. 2). Such relation between cone asymmetry and dipole moment results in a negative flexoelectric coefficient (Meyer, 1969).

As seen the planar membrane is a symmetric construction built up from asymmetric elements. That is why the screened asymmetry shows even at slightly asymmetric actions, e.g. deformation of the membrane causes its flexoelectric polarization. A membrane cross-section illustrating the mechanism of this effect is shown in Fig. 3.

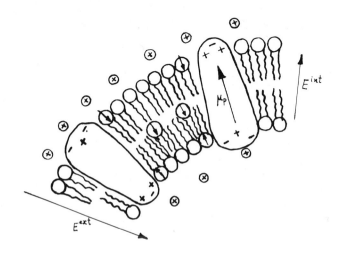

Figure 3

Cross-sectional view of curved membrane region. Due to splay deformation a definite number of lipid molecules reorient from inner to outer single layer which causes membrane polarization and produces internal electric field. On the other hand pH-gradient causes external electric field.

When the membrane is deformed (in terms of liquid crystalline elasticity such a fan-like deformation in the lipid molecules orientation is called splay) the one-molecule orientational potential changes. An orientation with the thick end of asymmetric lipid molecules towards the center of the curvature becomes energetically unfavourable. Then a certain number of molecules from the inner layer reorient. This number is statistically controlled by Boltzmann distribution (see section 4). The appearance of uncompensated dipole moments polarizes the membrane. The polarization is proportional to the deformation expressed in this case by inverse value of the curvature radius R (Meyer, 1969):

$$P = e \frac{1}{R} \qquad (1)$$

e is the flexoelectric coefficient of the membrane. When it is negative the polarization vector points towards the curvature center. For the normal component of the electric displacement in and out of the membrane when external electric field is lacking from the continuity condition we get

$$D^i = E^i + 4\pi P = D^e = 0 \qquad (2)$$

(assuming $R \gg d$, d - membrane thickness)

Thus, depolarizing electric field in opposite direction to the polarization appears in the membrane

$$\vec{E^i} = -4\pi\vec{P} \qquad (3)$$

After the pioneering work of Meyer the flexoeffect in thermotropic LC is the object of theoretical (Helfrich, 1971; Derzhanski & Petrov 1972; Petrov 1974) as well as experimental investigations (Schmidt, Schadt & Helfrich, 1972; Derzhanski, Petrov, Khinov & Markovski, 1973). As a result of these investigations theoretical relations of piezocoefficients with molecular asymmetry and dipole moment are obtained and the real existence of the flexoelectric effect is demonstrated. The typical order of magnitude of piezocoefficients is determined $e \sim 10^{-5}$ dyne$^{1/2}$. Obviously it is actual at present to investigate flexoelectric properties of liotropic LC. As the dipole moment in biomembranes is large we may expect a piezocoefficient of a greater order of magnitude: $e \sim 10^{-4}$ dyne$^{1/2}$.

MECHANISM OF ACTIVE TRANSPORT

Let us consider now the orientational state of integral proteins which have a fundamental role in our model. The protein globuli also may have asymmetric form and constant dipole moment, even greater than 100D (Edsall, 1953). In accordance with the bimodal concept of Green (1972) the electrically charged groups of the proteins are placed in hydrophylic portions which are protruding from the membrane. Then the electric field caused by membrane deformation tends to orient the globuli in such a manner, that their dipole moment will coincide with the direction of the depolarizing electric field. It may turn out that in this orientation which corresponds to minimum electric energy, the elastic energy connected with the asymmetric form of the globula, may be maximal (Fig. 3). In this case the orientational state of the globula is similar to a system with two stable states, e.g. a trigger. Decrease of the dipole moment can make such orientation energetically unfavourable and lead to a reorientation of the globuli.

If the dipole moment decreases through the adhering of a positive ion to the negative globula end, after the reorientation it will come out on the other side of the membrane. Release of the ion restores the initial dipole moment value and the globula reorients back to its initial position. This is the gist of the proposed mechanism of active transport. Naturally at such reorientations the globula may be considered as a rigid dipole only approximately. The most general description of this phenomenon involves conformational change of the globula as a result of the membrane electric field.

Another possibility leading to reorientation is the increase of globula asymmetry as a result of conformational change which occurs through the adhering of transportable particles.

A characteristic feature of this two possibilities is the asymmetric action - the reverse transport on the basis of the same mechanism is excluded. The reorientation of the globula back to its initial position is possible only upon releasing of transportable particle. Thus the globula is capable of performing a new elementary act of transport.

THEORETICAL DESCRIPTION

In the case of an arbitrary deformed membrane surface the flexoelectric polarization in any point is given by:

$$P = e \left(\frac{1}{R_1} + \frac{1}{R_2} \right) \quad (4)$$

where R_1 and R_2 are the two principle radii of the curvature at this point. Let us consider a spherically deformed membrane sector. Then

$$P = e \frac{2}{R} \qquad (5)$$

In our previous investigations (Derzhanski & Petrov, 1972) a formula for flexoelectric coefficient was derived, as follows

$$e = K \frac{\mu_1 S_1}{kT} \qquad (6)$$

kT being the energy of the thermal motion
μ_1 axial component of lipid dipole moment
S_1 asymmetry factor of lipid molecule. Approximately

$$S_1 = \frac{\alpha}{b} \qquad (7)$$

where α is the cone angle (Fig.2) and b is the mean distance between two adjacent molecules. The lipid molecule asymmetry shows a rather strong dependance on its real conformation. Assuming $\alpha = 0,07$ rad $\approx 4°$ and taking b the square root of one molecule area ($45 Å^2$ in accordance with the data of Vanderkooi, 1972), i.e. $b = 7 Å$, we obtain $S_1 = 10^{-6}$ cm^{-1}.

In (6) K is the elastic constant for splay deformation. An estimation according to one dimensional formula of Lubenski (1970)

$$K = \frac{1}{b^2} \frac{kT}{(1 - a/b)^2} \qquad (8)$$

(l - the lenght of a rigid part of the molecule, a - molecular width) shows that the order of magnitude is the same as in thermotropic LC

$$K \sim 10^{-6} \text{ dyne } (kT \sim 10^{-14} \text{ erg, } l = 20 Å)$$

The total dipole moment of lecithin is $\mu \approx 24$D. Taking for the axial component in the conformation displayed on Fig.2 $\mu_1 = 8$D and $kT = 4 \cdot 10^{-14}$ erg, which corresponds to room temperature, we come to

$$l = 2 \cdot 10^{-4} \text{ dyne}^{1/2} \qquad (9)$$

As can be seen, the large dipole moment of lecithin predetermines the strong flexoelectric properties. These

FLEXOELECTRIC MODEL FOR ACTIVE TRANSPORT

considerations apply also for other types of lipids with a constant dipole moment. So it may be concluded that the piezoeffect of membranes is of an order of magnitude greater than that of thermotropic LC.

Now it is possible to calculate the inner depolarizing field. From

$$E^i = -4\pi P = -8\pi e/R \qquad (10)$$

we have at $R = 1000\text{Å}$, $E = 500$ cgs units $= 150$ kV/cm - a significant electric field caused by the deformation 1/R characteristics of living cells. This electric field is quite capable of acting as a motive power of the active transport as we suppose in section 3.

When the ion concentration at the two membrane interfaces is different, external electric field E^e also appears. Then specifying our early formula (Derzhanski & Petrov, 1972) for the difference between the two opposite orientations of protein globuli and taking into account both the electric and the elastic parts we have

$$\Delta U = \Delta U_E - \Delta U_R = 2\mu p (E^e + E^i) - \frac{2KSp}{N} \frac{2}{R} \qquad (11)$$

μp is the dipole moment of the globula
Sp is asymmetric factor for the globula
N is the volume density of lipid molecules.

Langevin's formula for bistable system applies to the relative part of the globuli with a dipole moment oriented along the electric field.

$$W = \frac{n^+ - n^-}{n^+ + n^-} = \frac{1}{2} \frac{\Delta U}{kT} \qquad (12)$$

Since the breath of the globuli is 10 times greater (Vander - Kooi, 1972) we expect that their asymmetric factor is of the order of magnitude smaller than that of the lipids

$$Sp \sim 10^5 \text{ cm}^{-1} \qquad (13)$$

So at a volume density of lipid 10^{20} cm^{-3} we compute

$$\Delta U_R = 1 \% \ kT \qquad (14)$$

i.e. the difference between the elastic energies of the two opposite orientations is negligible.

On the other hand assuming μ_p = 120 D, when E^e = 0, we have
$$\Delta U_E = 3kT \tag{15}$$

This difference is cosiderably greater and thus the electric energy entirely controls the orientation of the globuli.

Assumed value of μ_p combined with globula lenght m=100Å corresponds to two charges $\pm q/4$ (q - electron charge) located at the two opposite ends of the globula. Then the adhering of a proton to the negative end produces there a summary positive charge of 3/4 q. As a result μ_p calculated relatively to the center of the globula reverses its direction retaining its magnitude:

$$\mu_p = \frac{1}{4} q \frac{m}{2} - \frac{3}{4} q \frac{m}{2} = -120D \tag{16}$$

According to (11) and (12) this reversion leads most probably to reorientation of the globula, accompanied with translocation of the proton from the inner to the outer membrane side. Its releasing rest ores the initial direction of μ_p and after back reorientation the globula is capable of performing a new elementary act of transport.

This mechanism provides a difference in proton concentration which increases the external electric field directed opositely to the internal one. The mechanism is blocked when the probability of reorientation is annulled (12) i.e. when ΔU becomes zero. Then from (11) we have (if R $\geq$ 1000Å)

$$E^e = -E^i + \frac{K\,S_p}{N\,\mu_p} \cdot \frac{2}{R} \approx -E^i \tag{17}$$

Thus a considerable pH-gradient is the final result of the action of the proposed mechanism. It corresponds to transmembrane potential of 200 mV when the thickness of the membrane is 75A. In our case of negative flexocoefficient the outer surface is charged positively with respect to the inner. Our result $E^e \approx E^i$ indicates that transmembrane potential is proportional to membrane curvature - it decreases upon increasing of R.

DISCUSSION

As shown the proposed mechanism acts as a proton pump. The translocation of protons against the concentration gradient is a type of active transport. Here the protein globuli play the role of Maxwell demons, able to create concentration difference. This process acts against diffusion, aiming to divert the system from thermodynamic equilibrium i.e. this is a typical nonequilibrium process.

Due to encreasing entropy of the system such a process requires energy. As shown this energy may be provided from the electric energy of flexoelectric polarization, which was not taken into account in our former papers.

Here we want to stress an important feature of our model - its possibility of self-regulation. There are at least two canals for such self-regulation:

1) Diffusion tends to equilize the concentration difference. The uncompensated internal electric field arises however and the proton pump restores the difference corresponding to the given curvature.

2) The external electric field created tends to deform the membrane because of converse flexoeffect. This field has such a direction that supports the initial curvature. Let us calculate the curvature which this field is able to produce. From the constitutive equation of the elastic stresses (Meyer, 1968), assuming that the membrane is free to deform to a state with vanishing stresses, we have

$$K \frac{2}{\widetilde{R}} - eE^e = 0 \tag{18}$$

so

$$\frac{2}{\widetilde{R}} = \frac{e}{K} E^e = \frac{4\pi e^2}{K} \cdot \frac{2}{R} \approx \frac{2}{R} \tag{19}$$

when $e = 3 \cdot 10^{-4}$ dyne $1/2$. At such a flexocoefficient value the resulting curvature is equal to the initial. So the external field supports entirely the initial curvature - once deformed the membrane does not straighten out by accidental circumstances. At a greater value of the flexocoefficient there is an interesting possibility for self-increase of the curvature to extremely large values, when the elastic term in (17) becomes important and limits the decrease of R.

Thus, from a cybernetical point of view the model represents a system with two inputs - electrical and deformational and one output at which we obtain the pH gradient. The rate of the active transport may be controlled independently by two inputs. There exists a feed-back control also - pH gradient creates external electric field which is applied at the electrical input and via converse flexoeffect at the deformational input as well (Fig.4).

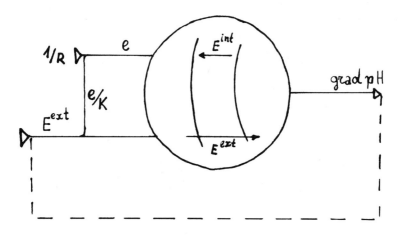

Figure 4

The model as a cybernetical system with two inputs (deformational and electric) and one output. The two canal feed-back control which secures stable self-regulation, is represented schematically.

These two possibilities of self-regulation affect the final result of the transport. Other ways for local change of the transport rate also exist as well as such for the reversion of its direction. These are based on local changes of lipid composition or lipid conformation in a definite membrane sector and are connected with change of the value and eventually the sign of the flexoelectric coefficient.

Finally let us say something about the relation of this model with the experimental data. Up to now at least two experiments exist which are consistent with the notion of lipid membrane flexoelectricity (Pasechik, V.I. & Sokolov, V.S. 1973; Ochs, A.L. & Burton R.M., 1974).

In these experiments both responses to vibration of modified (Pasechnik et al.) and unmodified (Ochs et al.) black lipid membranes are studied in the presence of transmembrane potential difference. In both cases generation of a.c. potential difference is registered with the same frequency of membrane vibration in the case of zero external potential. In Ochs and Burton's experiments this effect is called "subharmonic noise", while in the first case it is called flexoeffect. Similar experimental arrangement involving spherical deformation of planar membrane permits in principle to measure the value of flexocoefficient. The data on the piezo electric properties of proteins presented in this volume (Vassileva, Vassilev, Iliev, 1974) are of interest in this field. The possible curvatures however which may be obtained in such an experiment are rather small. In this sence it is interesting to study the spherical deformation of planar membrane caused by the action of external potential difference. The liposomes are other artificial structures with probably strongly manifested flexoelectric phenomena. There is a number of structures containing very strongly deformed membranes in living cells. Mitochondria are a typical example.

CONCLUSIONS

1) The notion of membrane flexoelectricity due to its liquid crystalline properties permits to work out a general model of active transport providing possibilities of stable self-regulation. The action of this model is described on the basis of exact theoretical relationships.

2) Bearing in mind such a mechanism a natural explanation may be given for a number of active transport features. For example the energy for the transport is secured by the membrane itself. The elastic membrane properties determine both the value and the sign of the transmembrane potential. It is also shown that curved membrane sectors are metabolically active. The asymmetry and vector character of the transport are clearly demonstrated.

3) This notion suggests a experimental programme with a goal to provide a confirmation of the hypothesis for the unifying role of the flexoelectric effect in the physics of living matter.

ACKNOWLEGEMENT

The author is indebted to Miss E. Loshchilova for suggesting the idea of this work and to Dr. A. Derzhanski, Dr. S. Stoilov, Dr. B. Atanasov and Mr. B. Markovski for the very helpful discussions.

REFERENCES

Borovyagin, V.L. (1971) Biophysica (in Russian) 16, 746

Caserta, G. & Cervigni, T. (1972) Proprint CNEN RT/BIO (72) 76

Caserta, G. & Cervigni, T. (1973) J. Thoor. Biol. 41, 127

Chapman, D. (1971) Simposium Faraday Soc. 5 (1971) 12

Danielli, J.F. & Davson, H.A. (1935) J. Cell. Comp. Physiol. 5 495

Derzhanski, A. & Petrov A.G. (1972) Compt. rend. Acad. Bulg. Sci. 25, 167

Derzhanski, A., Petrov, A.G., Khinov, Kh. & Markovski B.L. (1973) Bulg. J. Phys. (to be published)

Edsall, J.T. (1953) In the Proteins: Chemistry, Biological Activity and Methods (H. Neurath & K. Balley, ed.) vol. I, Part B, pp 688-694 New York: Academic Press Inc.

Frank, F.C. (1958) Discussions Faraday Soc. 25, 19.

Gray, G.W. (1962) Molecular Structure and the Properties of Liquid Crystals, London & New York: Academic Press

Green, D.E. (1972) Ann. N.Y. Acad. Sci. 195, 150

Helfrich, W. (1971) Z. Naturforsch. 26a, 833

Lee, J.D. & Eringen, A.C. (1971) J. Chem. Phys. 54, 5027

Lubenski, T.C. (1970) Phys. Letters A 33A, 202

Meyer, R. (1969) Phys. Rew. Letters 22, 918

Nehring, J. & Saupe, A. (1971) J. Chem. Phys. 54, 337.

Ochs, A.L. & Burton, R.M. (1974) Biophys. J. 14, 473.

Passechnik, V.I. & Sokolov, V.S. (1973) Biophysica (in Russian) 18, 655.

Pennock, B.F., Goldman, D.E., Chacko, G.K. & Chock, S. (1973) J. Phys. Chem. 77, 2383

Petrov, A.G. (1974) Thesis, Sofia.

Schindler, H. & Seelig, J. (1973) J. Chem. Phys. 59, 1841.

Schmidt, D., Schadt, M. & Helfrich, W. (1972) Z. Naturforsch. 27a, 277.

Singer, S.J. (1972) Ann.N.Y.Acad.Sci. 195, 16

Vanderkooi, G. (1972) Ann.N.Y.Accad.Sci. 195, 6.

Vassileva-Popova, J.G., Vassilev, N., Iliev, R., Proceedings of the Ist Inter. Colloq. on Phys. and Chem. Inform. Transfer, (in press)

Winsor, P.A. (1968) Chem.Rev's 68, 1.

ON THE ROLE OF THE CONFORMATION CHANGES AT THE ACTIVE SITE FOR ALLOSTERIC INTERACTIONS

V. Kavardjikov and I. Zlatanov

Central Laboratory of Biophysics, Bulgarian Academy of Sciences, 1113 Sofia, Bulgaria

In the existing theoretical works on the allosteric interactions in proteins only the subunit interactions are considered without taking into account the conformation changes in the active sites themselves. The present paper is an effort to take into consideration the role of the fine conformation changes in the allosteric interactions at the active sites. The respective correction factors to the rate constants of ligand binding in the equation of saturation (Eq. of Adair) have been found. The mathematical formalism for the fluctuation equilibrium between a number of conformation states of the active site with different ligand affinities has been obtained.

This report should be considered as a preliminary communication about the theoretical work being done in this field which in a more detailed and explicit form will be presented in a future publication. There are many works on the allosteric interactions in proteins (1 - 5) and one of the latest publications on this problem is that of Ricard et al. (6). Their theoretical considerations have permitted the formulation of a saturation equation in a structural and kinetic approximation by the definition of thermodynamic parameters bearing directly on the subunit interactions. This has been obtained on the following basis: a chemical process is considered where the active substance is a polymeric enzyme composed of four identic subunits. The chemical process may include various actions: ligand binding on the protein, ligand releasing from the enzyme-ligand complex, product formation in the active sites or enzyme isomerization. The free energy of activation $\Delta G^{\neq}$ is split up into four components:

1) The intrinsic component, $\Delta G_L^{\neq *}$, corresponding either to the activation free energy of the substrate binding on a protomer, to the substrate release from that protomer, or the intramolecular rearrangements of bonds, apart from the effects of conformation changes and subunit interactions.

2) The intrinsic transformation component, $\Delta G_T^{\neq *}$, that is the free energy of activation for a conformation change of a protomer, without taking into account the energetic contributions of binding, release, rearrangements of bonds, or subunit interactions. Therefore, the sum of these two components of free energy $\Delta G_L^{\neq *} + \Delta G_T^{\neq *}$ is the ideal free activation energy for an isolated protomer.

3) The contribution of the various subunit interactions to the free energy of activation for either substrate binding, release or catalysis, apart from the conformation changes of a given subunit as a whole, is registered by another component:

$$\sum \Delta G^{\neq}(\text{int.L})$$

4) The contribution of the subunit interactions to the free energy of activation for the intersubunit processes of conformation changes is registered by the term:

$$\sum \Delta G^{\neq}(\text{int.T})$$

The sum of all components gives the free energy of activation:

$$\Delta G^{\neq} = \Delta G_L^{\neq *} + \Delta G_T^{\neq *} + \sum \Delta G^{\neq}(\text{int.T}) + \sum \Delta G^{\neq}(\text{int.L}) \qquad (1)$$

The first and the fourth addends correspond to the free activation energy for an ideal polymeric enzyme undergoing no total conformation changes. In the work reviewed (6) it is assumed that the direct neighbourhood of the active site is not modified by subunit interactions, i.e.

$$\sum \Delta G^{\neq}(\text{int.L}) = 0 \qquad (2)$$

That means the subunit interactions should not affect

CONFORMATION CHANGES AT THE ACTIVE SITE

the binding, the release, or the catalytic process at the active site. According to us this is an approximation which does not allow to take into account the role of the steric position of the groups in the active site itself, and this is of considerable importance for the proceeding of the processes investigated. Our work is based on the statement that

$$\sum \Delta G^{\neq}(\text{int L}) \neq 0 \qquad (3)$$

The process of the substrate binding on the tetramer enzyme with identical subunits is of interest for us, and may be given the following diagrammatic form:

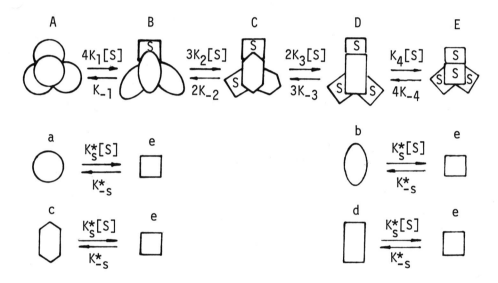

Figure 1

We shall concentrate our attention on one subunit of the tetramer. At the various stages of the course of the process the subunits considered will be different, but since we have made a priori the assumption that all the subunits are equal, then this will not bring any ambiguity in the discussion.

Let the tetramer be in state A. All subunits in it are in the elementary state "a". If we consider any subunit taken separately (in dissociated state), having no bonds with the others, the change of its effective free energy $\Delta G^{\neq\ast}$ in the process:

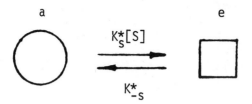

Figure 2

$$\Delta G^{\neq *} = \Delta G_L^{\neq *} + \Delta G_T^{\neq *} \tag{4}$$

The rate constant of this process is:

$$K^* = \frac{K_B T}{h} e^{-\frac{\Delta G_L^{\neq *} - \Delta G_T^{\neq *}}{RT}} \tag{5}$$

where K^* means either K_S^* or K_{-S}^* according to the given process. However, this is an idealized situation. In fact the subunit is in a bound state in the tetramer. The interactions with the other subunits affect the state of the active site. These interactions change the conformation state of the active site of the substrate-free subunit. The following diagram of the first binding stage of the substrate on some of the identical subunits in the enzyme represents a situation closer to the actual one.

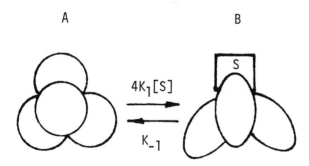

Figure 3

Let us concentrate on a subunit that has completed the transition a ⟶ e (Fig.2), this time not independently,

CONFORMATION CHANGES AT THE ACTIVE SITE   131

but as a part of the whole protomer. The free energy
change in such a process (Fig.3) will have the form:

$$\Delta G^{\neq *} = \Delta G_L^{\neq *} + \Delta G_T^{\neq *} + \sum \Delta G^{\neq}(int.L) + \sum \Delta G^{\neq}(int.T) \quad (6)$$

The last two addends in (6) register the fact that the
subunit interacts with the other subunits. The rate constant of such a process cannot still be K^*(Eq.5) and will
need some corrections.

Fig.2 shows that in a tetramer the interactions between the subunits are of the type "aa" for state A and
"bb" and "be" for state B. (In an analogous way in Fig.1
to states C, D, and E correspond the interactions "cc"
and "ce", "de", "ee"). Let us designate with 1, m, n, p,
q, r, s the number of interactions between the subunits
of a given type at any fixed state. These numbers (quantities) will be positive integers located in the interval (0,6). Thus, there will be also interactions of the
type bb ⟶ m."bb", etc. (n"be", p"cc"; q"ce"; r"de";
s"ee"; l"aa"). The first correction registering the interaction between the subunits will be:

$$e^{-\frac{\Delta G_{aa}^{\neq}(int.T)}{RT}} \quad (7)$$

This term should be considered as a correction factor
in K^* (5). It registers the influence of the subunit interactions on the transglobular processes, but not the
direct effect of the interaction on the conformation of
the active site. We introduce still another correction
factor registering that (7) does not do. It is related
to the corresponding conformation state of the separate
groups in the active site, and for a subunit in state "a"
in tetramer   A by analogy with (7) it has the form:

$$e^{-\frac{\Delta G_a(int.L)}{RT}} \quad (8)$$

When the protomer has passed into state B (Fig.1) the
same subunit is already not in the elementary state "a",
but in "e". The correction factor of K_{-s}^* i.e. the disso-

ciation constant, will have the form:

$$e^{-\frac{\Delta G_e^{\neq}(int.L)}{RT}} \tag{9}$$

Taking into consideration (5) and (8) we can express the corrected $K_1 \neq K^*$ for state A (Fig.1)

$$K_1 = \frac{h}{k_B T} e^{-\frac{(\Delta G_L^{\neq *} + \Delta G_T^{\neq *})}{RT}} \cdot e^{-\frac{\Delta G_{aa}^{\neq}(int.T)}{RT}} \cdot e^{-\frac{\left[\sum_{i}^{N}\Delta G_i^{\neq}(int.L)\right]_a}{RT}} \tag{10}$$

Where the summation is affected with $i = 1, 2 \ldots N \rightarrow$ number of chemical groups in the active site. Let us assume

$$e^{-\frac{\Delta G_{aa}^{\neq}(int.T)}{RT}} = \alpha_{aa}^6$$

$$e^{-\frac{\left[\sum_{i=1}^{N}\Delta G(int.L)\right]_a}{RT}} = \beta_a \tag{11}$$

With the new designation for K_1 we obtain

$$K_1 = \alpha_{aa}^6 \beta_a K_s^* \tag{12}$$

Analogously for the B state we obtain for K_{-1}

$$K_{-1} = \alpha_{bb}^3 \alpha_{be}^3 \beta_e K_{-s}^* \tag{13}$$

where

$$\alpha_{bb}^3 = e^{-\frac{\Delta G_{BB}^{\neq}(int.T)}{RT}}$$

$$\alpha_{be}^3 = e^{-\frac{\Delta G_{Be}^{\neq}(int.T)}{RT}}$$

$$\beta_e = e^{-\frac{\left[\sum_{i=1}^{N}\Delta G_i^{\neq}(int.L)\right]_e}{RT}} \tag{14}$$

CONFORMATION CHANGES AT THE ACTIVE SITE

From (12) and (13) we obtain the equilibrium constant $\bar{K}_1$ of the first step taking into account the stoichiometric coefficients indicated in Fig.1

$$\bar{K}_1 = \frac{4K_1}{K_{-1}} = 4\frac{\alpha_{aa}^6}{\alpha_{bb}^3 \alpha_{be}^3} \cdot \frac{\beta_a}{\beta_e} \cdot \frac{K_s^*}{K_{-s}^*} = 4\frac{\alpha_{aa}^6}{\alpha_{bb}^3 \alpha_{be}^3} \cdot \frac{\beta_a}{\beta_e} \bar{K}^* \quad (15)$$

By analogy with (12) and (13)

$$K_2 = \alpha_{bb}^3 \alpha_{be}^3 \beta_B K_s^* \quad (16)$$

and

$$K_{-2} = \alpha_{cc} \alpha_{ce}^4 \alpha_{ee} \beta_e K_s^* \quad (17)$$

From (16) and (17) we have

$$\bar{K}_2 = \frac{3}{2} \cdot \frac{\alpha_{BB}^3 \alpha_{Be}^3}{\alpha_{cc} \alpha_{ec} {}^4\alpha_{ee}} \cdot \frac{\beta_B}{\beta_e} \cdot \bar{K}^* \quad (18)$$

Analogously:

$$\bar{K}_3 = \frac{2}{3} \cdot \frac{\alpha_{cc} \alpha_{ec}^4 \alpha_{ee}}{\alpha_{ee}^3 \alpha_{de}^3} \cdot \frac{\beta_c}{\beta_e} \cdot \bar{K}^* \quad (19)$$

$$\bar{K}_4 = \frac{1}{4} \cdot \frac{\alpha_{de}^3 \alpha_{ee}^3}{\alpha_{ee}^6} \cdot \frac{\beta_d}{\beta_e} \bar{K}^* \quad (20)$$

The equilibrium constants expressed in this way are replaced in Adair's saturation equation

$$Y = \frac{\bar{K}_1[S] + 2\bar{K}_1 \cdot \bar{K}_2[S]^2 + 3\bar{K}_1 \cdot \bar{K}_2 \cdot \bar{K}_3[S]^3 + 4\bar{K}_1 \bar{K}_2 \bar{K}_3 \bar{K}_4[S]^4}{4\{1 + \bar{K}_1[S] + \bar{K}_1 \bar{K}_2[S]^2 + \bar{K}_1 \bar{K}_2 \bar{K}_3[S]^3 + \bar{K}_1 \bar{K}_2 \bar{K}_3 \bar{K}_4[S]^4\}} \quad (21)$$

With the modern experimental technique such parameter could be found that would allow the corrections β to be given real content. This problem will be dealt upon in greater detail in our future publication.

Fig.1 expresses the idea of a "fixed" situation at every active site. The same situation might be defined by another statistical approach which is more adequate.

Let us denote with a', b', c', d', e' the elementary states of the subunits in the tetramer corresponding to such a dynamics in their active site which at a given moment results from the probability of realization of some of the elementary states a, b, c, d, e (Fig.1). The state of the active site at a given moment will depend on the magnitude of the statistical weights with which each of the states a, b, c, d, e participates. Then, the correction factor $\beta_{a'}$ might be written as follows:

$$\beta_{a'} = g_{aa} \exp\left\{\left[-\sum_{i=1}^{N} \Delta G_i^{\neq(int.L)}\right]_a / RT\right\} + g_{ab} \exp\left\{\left[-\sum_{i=1}^{N} \Delta G_i^{\neq(int.L)}\right]_b / RT\right\} +$$
$$+ g_{ac} \exp\left\{\left[-\sum_{i=1}^{N} \Delta G_i^{\neq(int.L)}\right]_c / RT\right\} + g_{ad} \exp\left\{\left[-\sum_{i=1}^{N} \Delta G_i^{\neq(int.L)}\right]_d / RT\right\} +$$
$$+ g_{ae} \exp\left\{\left[-\sum_{i=1}^{N} \Delta G_i^{\neq(int.L)}\right]_e / RT\right\}$$

(22)

where g represents the statistical weights of a, b, c,... which take part in the realization of the more complex state a'. Analogously, for $\beta'_{b'}$, $\beta'_{c'}$, $\beta'_{d'}$, $\beta'_{e'}$.

We can arrange the factors g in the following table:

$$q = \begin{bmatrix} g_{aa} & g_{ab} & \cdots & g_{ae} \\ g_{ba} & g_{bb} & \cdots & g_{be} \\ \cdots & \cdots & \cdots & \cdots \\ \cdots & \cdots & \cdots & \cdots \\ g_{ea} & g_{eb} & \cdots & g_{ee} \end{bmatrix}$$

(23)

That is nothing else but a matrix and in order to have commensurate equations for $\beta'_{a'}$, $\beta'_{b'}$, ... $\beta'_{e'}$ it must not be degenerate, i.e.

$$\|g\| \neq 0 \tag{24}$$

Particularly, when $sp/g/ = 1$, then the case considered in fig. 1 for states a, b, c, d, e is realized.

REFERENCES

1. Monod J., Wyman J., Changeux J.P., J. Mol. Biol. 12, (1965) 88
2. Koshland D.E., Nematy G., Filmer D., Biochemistry 5, (1966) 365
3. Adair G.S., J. Biol. Chem. 63, (1925) 529
4. Pauling D., Proc. Nat. Acad. Sci. USA 21, (1935) 186
5. Frieden C., J. Biol. Chem. 242, (1967) 4045.
6. Ricard J., Mouttet C., Nari J., Europ. J. Biochem. 41, (1974) 479-497.

A POSSIBLE TRANSFER OF PHYSICAL INFORMATION TROUGH THE CONVERSION OF MECHANICAL ENERGY INTO ELECTRICAL: A PIEZOELECTRIC EFFECT IN PROTEINS

Julia G. Vassileva-Popova, N.V. Vassilev, and R. Iliev*

Central Laboratory of Biophysics and *Institute of Electronics, Bulgarian Academy of Sciences, 1113 Sofia, Bulgaria

In the search for a more direct convertion of one kind of energy into another in biological compounds experiments of the piezoelectric effect (PEE) in proteins were carried out.

The PEE properties were found for several compounds: phenanthrene (1), arabinose (2), acridine (2), L-leucine (3), methilamine (3), urotropin (2), Organic polymers (4), nucleic acids (5) and other substances (6). In some works the ferro electric effect is connected to the PEE (7).

On the basis of a relatively good background of the study and application of the PEE in quartz and polymers, the PEE properties of proteins are not yet established.

For the further unerstanding of the energy conversion on the level of molecular arrangement of matter the PEE becomes of interest. The relation between the anisotropic PEE and the appearance of the electric charge as a result of mechanical deformation requires a low crystal symmetry (3,8). A complete theory explaining the PEE still does not exist. Nevertheless, it is possible to indicate that in the crystals without central symmetry the PEE appears to correspond to the conditions of induced or constant dipoles (9).

Usually the PEE is introduced as a PEE anisotropic coefficient

$$d \equiv \frac{\text{surface charge K/cm}^2}{\text{mechanical force dyn/cm}^2}$$

(10). A higher PEE is found when the axis of mechanical forces is parallel to the crystallographic one (1).

MATERIALS AND METHODS

For the estimation of the PEE the crystals were obtained from albumin and the "hybrid" crystals from albumin and prolactin (P_r). The methodological suggestions from the protein crystalization were taken into consideration (11, 12).

The electronic method of mechanical resonance adapted for this purpose was applied. This was achieved with a high frequency generator with continuous control in the range between 100 kHz - 30 MHz.

The electric circuit passing through the crystal investigated is closed by a condenser. The value of the voltage (U) between the samples is measured using a voltmeter.

The mechanical resonance method is based on the fact that the PEE is phenomenologically related to the piezoresistance (1). In our method the PEE in the sample investigated is demonstrated through the change of U. The mechanical resonance corresponds to the definite value of the frequency (F). The appearance of mechanical resonance indicates the PEE properties of the crystal included in the electric circuit. In this case for given values of F an F-dependent change of U is observed. When the results are positive (PEE observed) there is an increase and a maximum of the function U(F) (Fig.4). The maximum of U corresponds to the definite value of F - F_o. An electronic model of the PEE behaviour of the crystal is an oscillating circuit.

The delicate points of the method are:
a/ the design of the crystal holders;
b/ the linking of the electrodes, and
c/ the preparation of a crystal with proper dimensions.

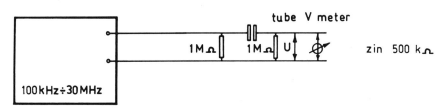

Q meter method

mechanical resonance approach

Fig. 1

$x_L = x_C$ — Resonance condition

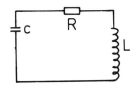

$x_C = \dfrac{1}{2\pi FC}$ $x_L = 2\pi FL$

$x_C = x_L$ $X = x_C = x_L$

$Q = \dfrac{X}{R}$

Fig. 2

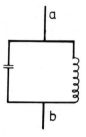

$U_{ab} = f(F)$

Fig. 3

Negative results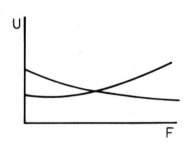

Positive results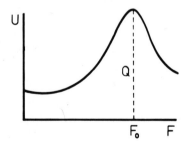

Fig. 4

In addition to this the electroconductivity produces remarkable difficulty for the PEE estimation and measurement (13,14,8). Another approach to the study of the PEE is the holographic estimation of the dimension changes occurring after the application of electric current to the crystal. This method allows a more accurate quantitative estimation of the PEE.

RESULTS AND DISCUSSION

The experimental results suggest that proteins have PEE properties. At this stage of the experiment it is not possible to give an accurate value of the PEE coefficient. The further holographic investigations will introduce more definite values. The present results indicate that the value of protein PEE is higher than 7÷9mHz. It is interesting to observe that the "hybrid" crystals (prolactin and albumin) show a higher PEE in comparison to the pure albumin crystals where the PEE is discussible. This difference is probably due to the process happening during the "hybrid" crystal arrangement. This point is still an open question. More likely the PEE nature of the "hybrid" crystal belongs to the presence of a regulatory protein in the crystal. In some way similar results were obtained when the albumin constituent was used and phosphorylase kinase was added. It seems that the amount of the regulatory protein hormone or enzyme influence the appearance of the PEE properties of the proteins investigated. The PEE nature of protein reflects the complexity of the physical changes. It may have an influence on the mobility of the free radicals producing the local space field and local space charges (15) usually with a relatively long life-time. In this connection it should be considered that the PEE play a role in the plastic transfer of mobile charges.

One of the critical points in this study is the fact that PEE is observed in proteins in dry state corresponding to the way of measuring the classical PEE. Having into consideration the postulate that the bio-action of proteins is demonstrated in the liquid state and the solid state biology as a future interest is not ready yet to be taken as a support, the speculation on the possible bio-significance of the findings has to be limited.

It is tempting just to bring as support the point of view of protein crystallography. In this respect the new concept for the possible similarity of the conformation of some biopolymers in crystals and in solutions should be of interest. The question arises whether proteins keep their PEE properties in the various biological solutions. The similarity of this problems is that the PEE measurements as in the crystallographic method are fully dependent on the crystal state of the sample investigated. In this case all methods which deal with determinations made in a non-crystal state are not able to measure the classical PEE. In this connection liquid crystals are a good example for the theoretical work of flexoelectricity in the membrane (16). A useful discussion on the nature of the liquid crystals (17) and the flexoelectric effect in liquid crystals and the membrane characteristics (18) is to be found in these works. The interest in the problems of cell-to-cell communication (18) and of semiconductor junction as possible specialized electrical properties of the biopolymers has marked a new way of thinking (19).

Coming back to the main topic in this study: the finding of the PEE of proteins, it might be concluded that the present study suggests a new physical property of proteins.

In a broader way the present work deals with the possible direct conversion of 2 types of energy: electrical into mechanical and vice versa. This economic conversion of energy possibly has an opportunity to contribute to the allosteric properties of proteins. The PEE shoud be considered as the physical level of bio-information transfer.

ACKNOWLEDGMENT

For the very helpful discussion and stimulation of this work we would like to thank Doctor David C. Phillips and to the audiance of the Winter Meeting 1974 of the British Biophysical Society. For the usefull information on crystal preparation we thank Doctor T.L. Blundell.

REFERENCES

1. Gay, G., Kara, R., Mathieu, J.P., Bull.Soc.Franc. Minaral.Crist., $\underline{84}$, 187 (1961).

2. Shumakov,A.A., Kopchik,V.A., Kristallografia, $\underline{4}$, 212 (1960), (in Russian).

3. Mason, W.P., in: Piezoelectric Crystals and Their Application to Ultrasonics, Princeton, New York (1950).

4. Pohl, H.A., Rembaum, A., Henry, A., Journ. Am. Chem. Soc., 84, 2699 (1962).

5. Polonssky, J., Dousou, P., Sardon, C., Compt. Rend., 250, 3414 (1960).

6. Rez, I.S., Sonin, A.S., Tsepelovich, E.E. and Filimonov, A.A., Soviet Phys. Crist., 4, 59 (1960).

7. Duchesne, J., Depireux, J., Bertinchamps A., Cornet, N., van der Kaa, J.M., Nature, 188, 405 (1960).

8. Arlt, G., Journ. Appl. Phys., 36, 2317 (1965).

9. Rez, I.S., Cesk. Časopis Fys., 13, 31 (1963).

10. Gutmann, F., Lions, L.E., in: Organic semiconductors, John Wiley and Sons, Inc., New York-London-Sydney (1967).

11. Blundell, T.L., Crystallization (in press)

12. Zuppezauer, M., in: Methods in Enzymology, Vol.XXII, Enzyme Purification and Related Techniques, Ed. W.B. Jakoby, Academic Press, New York and London (1971).

13. Merten, L., Zs.Naturforsch., 19a, 788, 1162 (1964).

14. Arlt, G., Journ. Acoust. Soc. Am., 37, 151, (1965).

15. Merten, L., Phys. d. Kondens. Materie, 2, 53 (1964).

16. Petroff, A., in this volume.

17. Brown, G.H., J. Soc. Amer., 63, No 12, (1973).

18. Andrew, C., Basset, L., in: The Biocemistry and Physiology of Bone, 2nd ed., Vol.III, Academic Press, Inc., New York and London (1971).

19. Cope, F.W., and Damadion, R., Nature, 228, 76 (1970).

Protein Hormones and Their Receptor Interactions

EFFECT OF PGF-2alfa ON THE SYNTHESIS OF PLACENTAL PROTEINS AND HPL

Olga Genbačev, Miodrag Krainčanic,*Vojin Sulović,
Gordana Kostić and Milić Aleksić

Institute for the Application of Nuclear Energy
in Agriculture, Veterinary Medicine and Forestry,
Zemun, Yugoslavia

*Clinic of Gynecology and Obsterics, Faculty of
Medicine, Beograd, Yugoslavia

INTRODUCTION

The synthesis and secretion of human placental lactogen (HPL) is thought to be autonomous (Friesen, 1968). An increased number of evidence has been accumulated indicating that prostaglandins are involved in the regulation of HPL biosynthesis and secretion. According to Ylikorkala and Pennanen (1973) PGF-2alfa administered by intrauterine routes appeared to have a direct effect on the placental synthesis and/or release of HPL in early pregnancies. Karim and Hillier (1970) found high concentrations of PGF-alfa in human amniotic fluid in spontaneous abortions. On the basis of this evidence it is not surprising that the plasma level of HPL is lower than normal in threatened abortion (Niven et.al.1972) even if the outcome of pregnancy is successful (Ylikorkala et.al. 1973).

The purpose of the present study was to study the role of PGF-2alfa in the regulation of placental proteins and HPL biosynthesis and secretion. This approach may lead to the better understanding of prostaglandin mechanism of action during artificial abortions or induction of labor and its role in the regulation of placental hormone synthesis and release. Measuring of HPL level during PGF infusion during abortion may provide useful tool in predicting the outcome of induction.

MATERIAL AND METHODS

Human placentas were obtained from patients (abortions induced by PGF-alfa, sections parvae and normal deliveries) having legal abortions induced mechanically or by prostaglandins injections and from women who delivered at the Clinic of Gynecology and Obstetrics of the Faculty of Medicine, Beograd.

Immediately after delivery of abortion the placenta was transported to the laboratory under ice in a thermal flask. It was dissected free of foetal membrane and major blood vessels. After repeated washing in Tris-HCl buffer, pH 7,4 and slightly removing the water excess with a filter paper, 0.5 g of tissue was weighed accurately and placed into a 10 ml glass flask in 3 ml leucine-free Eagle medium. The flask was shaken at 80 cycle/min from 15 min to 3 hours in Dubnoff metabolic shaker at 37° C in an atmosphere of 95% and 5% CO_2.

After the preincubation of 15 min, the medium was discarded and the incubation started by adding 2.5μCi of uniformly labeled ^{14}C-1-leucine (346 mCi/rmole, Radiochemical Center, Amersham, England) and 0.3 mg of prostaglandin F-2alfa in 3 ml of fresh Eagle medium. Flasks containing placental tissue fragments were set up in duplicate for each incubation time and treatment, and analysed separately. The values showed fairly good reproducibility and only the mean of these values is shown in each experiment. Incubation medium was collected and stored at -20° C.

Tissue fragments were washed and homogenized in Tris-HCl buffer, pH7.4, centrifuged for 10 min at 8000xg, the residue was discarded and the supernatant was centrifuged for 10 min at 15000 x g (Sorval, RC-2B) and at 105000 x g (Spinco, Model 1). The supernatant was stored at -20° C.

Incorporation of ^{14}C-leucine into HPL proteins was counted after precipitation by 5% trichloracetic acid (TCA), filtration and washing in 5-6 ml 5% TCA on Millipore filters (25 mm, 0.8 μ).

To measure the incorporation of ^{14}C-leucine into HPL, the precipitate obtained after the addition of an optimal amount of specific anti-HPL antiserum and sheep anti-rabbit gamma globulin (ARGO), was washed on Millipore filters soaked with 20% albumine in phosphate buffer (BBS) and counted with Nuclear Chicago Scintillation Counter. The amount of total HPL was determined by an Amersham HPL immunoassay kit (Radiochemical Centre, England).

RESULTS AND DISCUSSIONS

Effect of PGF-2alfa on the protein synthesis in the placenta during first trimester of pregnancy. - Synthesis of total proteins in the 3-month old placenta and secretion of newly synthesized proteins from control tissue and tissue incubated in the presence of PGF-2alfa are shown on Fig.1. The incorporation of ^{14}C-leucine increased progressively during the 180 min incubation. Prostaglandin inhibited incorporation of labelled amino acids with the maximal effect between 60 and 180 min. The release of newly synthesized proteins in the medium of the control tissue is relatively slow and is inhibited by prostaglandin. In placenta obtained after PGF-2alfa induced abortion (Fig.1b) the protein synthesis is about 100 time slower compared with the control placenta of the same gestational age, and about 50 times in comparison to the placenta incubated with the PGF-2alfa <u>in vitro</u>. The concentration of total proteins released in the total medium by placenta after <u>in vivo</u> PGF treatment and from controls are

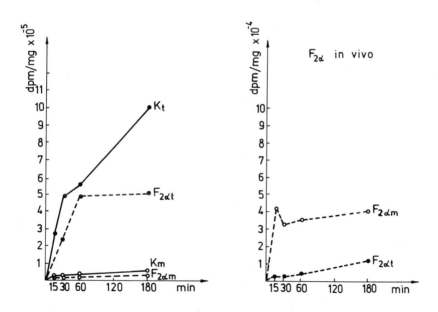

Figure 1

about the same although this concentration is reached in control after 3 hours of incubation and in PGF-treated tissue after 15 minutes. By increasing the time of incubation the concentration of released proteins is not changed.

The effect of PGF on the protein synthesis and secretion of a 4-month old placenta trated in vivo and in vitro is shown in Fig.2. Protein synthesis in control placenta of this age is more intensive than in placenta of 2,5 months. The effect of PGF in vitro is clearly expressed, while the in vivo produced effect is less pronounced than in placenta of 2,5 months. However, the synthetic activity of the PGF treated tissue is still significantly reduced compared with the corresponding control or to a tissue obtained after in vitro treatment with PGF.

At present our results do not permit to establish whether these differences are due to the difference in gestational age of placenta or individual variation sensitivity to PCF.

However, these results indicate that PGF-2alfa inhibit the protein synthesis and release in the placenta un-

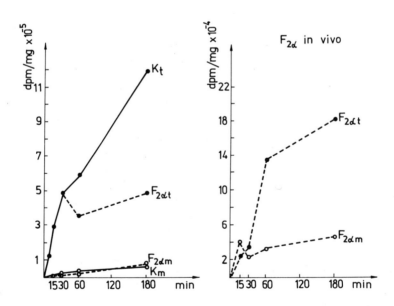

Figure 2

Table 1

No	Gestational age (months)	Treatment	Substrate	Amount of HPL in tissue and medium (μ/g tissue)			
				15min	30min	60min	180min
1	2,5	Control	Tissue	214	220	212	260
			Medium	117	122	152	236
2	2,5	PGF-2alfa in vitro	Tissue	-	-	250	318
			Medium	-	-	144	196
3	3	PGF-2alfa in vivo	Tissue	203	248	228	230
			Medium	120	124	114	120
4	4,5	Control	Tissue	210	210	223	195
			Medium	157	180	238	308
5	4,5	PGF-2alfa in vitro	Tissue	256	232	234	224
			Medium	156	154	214	292
6	4	PGF-2alfa in vivo	Tissue	240	292	245	237
			Medium	141	140	112	168

der in vivo and in vitro conditions. This fact is confirmed by decreased concentration of HPL in tissue and medium after different time intervals of incubation in the presence of PCF presented in Table 1.

The concentration of HPL is increased in medium during incubation of intact placentas of different gestational age and remain constant in the tissue.

This fact shows that incubated tissue do synthesize hormone under in vitro conditions and that this hormone is released in medium. In the PGF treated placenta the total amount of the HPL remained constant in the tissue and medium indicating that the synthesis and release are affected by the PGF. This effect is less pronounced in in vitro experiments although the total HPL released in medium is lower than in controls.

This effect of PGF on the synthesis and release of HPL is reflected in the level of HPL in the serum of women 1 and 6 hours after prostaglandin infusion during abortion induction.

In Table 2 are shown the absolute values of HPL after the injection of abortifacient.

TABLE 2

No	Gestational age	Concentration HPL µg/ml serum (time of PGF infusion)	
		15 min	6 hrs
1	3	0,76	0,43
2	4	0,48	0,38
3	3,5	1,15	1,10
4	4,5	1,80	0,90
5	3	0,42	0,17
6	3,5	0,91	0,84
7	3	0,58	0,74
8	3,5	0,50	0,40
9	4	1,10	1,40
1o	4	1,70	1,90

The HPL decrease between 15 min and 6 hrs interval after PGF infusion in 8 out of 10 examined women. In some cases this level remained unchanged or slightly increased, and these women did not abort within 12 hours after infusion or did not abort at all. The same observation was published by Ylikorkala and Pennanen (1973).

The effect of prostaglandin on the protein synthesis and HPL in the full-term numan placenta. - The biosynthesis of human placental lactogen (HPL) was studied in full--term human placenta in vitro after the addition of prostaglandin F-2alfa. The synthesis and release of total ^{14}C-protein are suppressed by PGF (Fig.3). The total

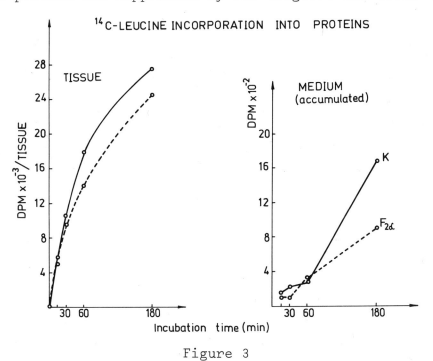

Figure 3

amount of HPL determined by radioimmunoassay, after a sharp rise during the first 15 minutes, decreases progressively in the incubation medium reaching a constant level after 60 minutes of incubation. The total synthesis of HPL in the tissue and the release of newly synthesized hormone in the medium are suppressed by PGF (Fig.4). The effects discribed are similar to results obtained with placenta from the first and second trimester of pregnancy.

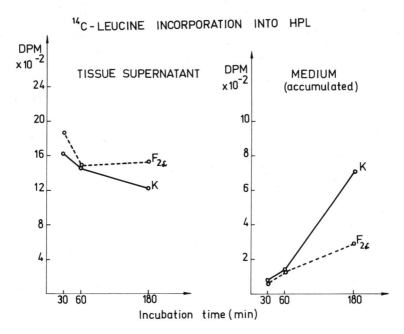

Figure 4

CONCLUSIONS

On the basis of the results obtained we can draw the following conclusions:

a) the syntitiotrofoblast of the placenta retains the ability to synthesize proteins and HPL in in vitro system during the period examined ;

b) prostaglandin PGF-2alfa inhibits the synthesis and release of total proteins and HPL after in vivo and in vitro treatment with the effect being much more pronounced after the in vivo application; the effects described are evident in all examined placentas of different gestational age;

c) the induced changes in the biosynthetic ability of placenta after PGF-2alfa injections are reflected in the decrease of HPL concentration in serum;

d) according to the correlation of the results obtained after in vivo injection of prostaglandin and in vitro produced effect are inclined to consider that the effect of PCF-2alfa is connected to the inhibition of total protein synthesis and accordingly the HPL level in serum may reflect the functional sate of placenta in these conditions.

REFERENCES

1. Friesen, H.C., Endocrinology, 83 (1968) 744.

2. Niven, P.A.R.; London, J., Chard, T. (1972) British Medical Journal, 3, 799.

3. Karim, S.M.M. (1968) British Medical Journal, 618.

4. Karim, S.M.M. and Hillier, K., J. Obstet, Gynecol. Brit. Commonw. 77.

5. Ylikokala, O. and Pennanen, S., J. Obstet. Gynecol. Brit. Commonw. 80 (1973) 927.

BIOENERGETIC EFFECT OF THE GONADOTROPIC HORMONES IN VIVO

S.Milanov, D.Panayotov, P.Kolarov,

M.Protich and Alevtina Maleeva

Institute of Endocrinology, Gerontology and Geriatrics, Medical Academy, blvd.Chr.Mikhailov 6, Sofia, Bulgaria

There are many publications on the gonade morphologic and functional stimulation with gonadotropic hormones (GH). However there are no descriptions of the bioenergetic effects of the GH on the gonade cells. It may be suggested that this has a significant effect on their stimulating action as the increased mitosis in gonade cells demands a solid supply of energy.

The authors wanted to determine if there was a GH bioenergetic effect as a possible mechanism of GH stimulation action as well as whether this effect was specific.

Bioenergetic activity in male gonades under increased GH stimulation. In order to raise the GH stimulation in male rats they were injected with gonadotropic hormones. During the different periods of stimulation, depending on the time of the stimulating action appearence, the tissue breathing and oxidative phosphorylation were followed. The results are shown in Table 1. There was a significant increase of the oxygen consumption and esterification of non-organic phosphate during the oxidative phosphorylation in male gonades. This was noted in all stages of the experiment on the 8th, 9th, 10th, 15th and 22nd days.

Changes in tissue bioenergetic action in testicles under inhibited function condition. The inhibition of GH secretion was done by a combination of estrogens and androgens. The results of the experiment about the tissue breathing and oxidative phosphorylation in male rats tes-

Table 1

Bioenergetic metabolism in rat testicles on different stages of GH increased stimulation

Group	Number	P_n^+ in µA $\bar{X}$	$S_{\bar{X}}$	O_2^{++} in µA $\bar{X}$	$S_{\bar{X}}$	P/O µA/µA/mg $\bar{X}$	$S_{\bar{X}}$	P_t
8th day of increased stimulation								
Stimulated rats +++	8	43,6	2,1	19,5	0,6	2,75	0,04	<0,001
Controls	8	31,8	1,5	15,4	0,7	2,06	0,03	-
9th day of increased stimulation								
Stimulated rats	8	57,6	2,1	20,2	1,0	2,85	0,06	<0,001
Controls	8	35,1	0,9	16,1	0,8	2,11	0,08	-
10th day of increased stimulation								
Stimulated rats	8	58,2	2,3	21,1	0,7	2,76	0,06	<0,001
Controls	8	34,0	1,1	17,1	0,9	2,05	0,02	-
15th day of increased stimulation								
Stimulated rats	8	57,1	2,2	21,0	0,5	2,27	0,05	<0,001
Controls	8	30,3	1,9	15,3	1,0	1,98	0,02	-
22nd day of increased stimulation								
Stimulated rats	8	55,2	2,3	20,0	0,5	2,76	0,06	<0,001
Controls	8	33,7	1,8	16,2	0,6	2,08	0,04	-

P_n^+ - amount of esterified non-organic phosphate

O_2^{++} - amount of oxigen consumption in oxidative phosphorylation.

Stimulated rats +++ Gonabion (ICSH)-2 U/200 g and Gestil (FSH)-2 U/200 g, given daily for 7 days and 15 days every second day.

Table 2

Bioenergetic effect in rat testicles on different stages of GH inhibition

Group	Number	P_n in μA $\bar{X}$	$S_{\bar{X}}$	O_2 in μA $\bar{X}$	$S_{\bar{X}}$	P/O μA/μA/mg $\bar{X}$	$S_{\bar{X}}$	P_t
8th day of inhibition								
Inhibited rats+	8	17,1	1,0	10,5	0,6	1,63	0,04	<0,001
Controls	8	31,8	1,3	15,4	0,7	2,06	0,03	-
9th day of inhibition								
Inhibited rats	8	16,2	0,9	11,8	0,8	1,37	0,04	<0,001
Controls	8	34,0	1,1	16,1	0,8	2,11	0,01	-
10th day of inhibition								
Inhibited rats	8	16,6	1,1	11,7	1,0	1,42	0,04	<0,001
Controls	8	35,1	0,9	17,1	0,9	2,05	0,02	-
15th day of inhibition								
Inhibited rats	8	14,1	0,7	11,1	0,7	1,27	0,04	<0,001
Controls	8	30,3	0,9	15,3	1,0	1,98	0,02	-
22nd day of inhibition								
Inhibited rats	8	16,9	0,9	12,1	0,8	1,40	0,04	<0,001
Controls	8	33,7	1,8	16,2	0,6	2,08	0,04	-

Inhibited rats - depovirin 5mg/200 g and depofemin 0,4mg/200g.

ticles showed a decrease in oxygen consumption and the esterification of non-organic phosphate in the process of oxidative phosphorylation at all the experimental stages. The results are shown in Table 2.

Testicle adenosintriphosphate level under increased of inhibited stimulation. In order to receive full information about the testicle bioenergetic metabolism under increased or inhibited stimulation, the ATP in testicle homogenate of experimental animals and controls was examined at one stage of the experiment.

The results are shown on Table 3. In animals with increased GH stimulation the testicle ATP contents raises, while in animals with inhibited condition the ATP decreases.

Table 3

Changes in ATP concentrations in testicles in M/g under increased or inhibited GH stimulation in rats

	Groups		
	Controls	Stimulated	Inhibited
Number	10	10	10
$\bar{X}$	2,89	3,27	2,11
σ	0,20	0,13	0,14
$S_{\bar{X}}$	0,06	0,04	0,04
P_t	-	0,001	0,001

Tissue bioenergetic metabolism of liver and sceleleton muscles under stimulated and inhibited testicle function. In order to find out the specific bioenergetic action of GH, the tissue breathing and oxidative phosphorylation of liver and sceleton muscles were followed, simultaneously with the testicle functional tests. The results are shown on Table 4. Increased or decreased GH stimulation do not affect the bioenergetic metabolism of liver and sceleton muscles.

Other metabolic changes, connected with testicle bioenergetic activity under increased or inhibited testicle function. The experiments were to add to the knowledge about male gonade bioenergetic metabolism under different GH activity and the reflection of these changes on the blood serum, with the aim of clinical significance.

Table 4

Bioenergetic metabolism in rat liver homogenate under increased GH stimulation or inhibition.

Group	Number	P_n in µA		O_2 in µA		P/O µA/µA/mg		
		$\bar{X}$	$S_{\bar{X}}$	$\bar{X}$	$S_{\bar{X}}$	$\bar{X}$	$S_{\bar{X}}$	P_t
8th day of treatment								
Controls	8	39,0	1,0	17,4	0,8	2,20	0,01	-
Stimulated	8	37,2	0,8	16,7	0,5	2,23	0,02	<0,5
Inhibited	8	32,7	1,5	15,0	0,5	2,18	0,03	<0,5
15th day of treatment								
Controls	8	37,6	1,1	17,1	0,7	2,20	0,01	-
Stimulated	8	36,8	0,8	16,8	0,6	2,19	0,03	<0,5
Inhibited	8	32,3	1,3	15,0	0,5	2,15	0,05	<0,5
22nd day of treatment								
Controls	8	37,6	0,9	17,5	0,6	2,15	0,06	-
Stimulated	8	33,8	1,1	16,1	0,2	2,10	0,03	<0,5
Inhibited	8	36,5	0,9	16,6	0,7	2,20	0,04	<0,5

Two transaminases (SGOT and SGPT) and aldolase were followed simultaneously in blood serum and in testicle homogenates. The results showed that the activities of the enzimes change under the stimulating action of GH: decrease under inhibited GH secretion and raise under increased GH stimulation. However, the enzyme activity in blood serum does not change significantly enough to have a diagnostic importance in determination of testicle functional condition.

CONCLUSIONS

1. GH has a bioenergetic effect on the testicles and appears as a stimulating mechanism.

2. The increased GH stimulation activates the testicle tissue breathing and oxidative phosphorylation, leading to higher levels of all factors (amount of oxigen consumption - Q_{O_2}, Q_{CO_2}, P/O_2, ATP), while the respiratory quotient does not change.

3. The inhibited GH stimulation reduces the testicle tissue breathing and oxidative phosphorylation.

4. The changes in the testicle bioenergetic process are connected to the changes in protein and carbonhydrate metabolism during GH stimulation increases or decreases.

5. The changes of testicle bioenergetic activity when influenced by GH is reflected morphologically and functionally into male gonades activity.

6. The GH bioenergetic effect is specific and is not noted in other organs (liver, sceleton muscles).

REFERENCES

1. Milanov St. et al. - Proc. Postgr. Med. Inst. (ISUL), 1969, v.XVI, 1, 57-62.

2. Milanov St. et.al. - Experim. Med. and Morphol., 1970, 2, 84-88.

3. Umbreit W., P.Burns and J.Stauffer- Manometric techniques, Burgess Publ., Mineapolis 1959, p.556.

BIOENERGETIC EFFECT OF THYREOSTIMULATING HORMONE IN VITRO

S. Milanov, Alevtina Maleeva and Nellie Visheva

Institute of Endocrinology, Gerontology and Geriatrics, Medical Academy, blvd. Christo Mikhailov 6, Sofia, Bulgaria

The stimulating effect of the thyreostimulating hormone (TSH) on the thyroid gland has been reported in literature. According to some authors, the mechanism of this effect is realized through increased oxidizing process in thyroid cells, which suggests its participations in the energetic mechanism of these cells. Concerning other organs its possible effect on lipid metabolism could be noted by the blood cholesterol level. There are no publications about TSH effect on the bioenergetic processes in organism. However, the increased level of TSH in blood of hypothyroid patients and during treatment with thyrestatic drugs is well known.

The authors wanted to evaluate the TSH extrathyroid bioenergetic effect, without connection with the thyroid hormones. They followed the effects of different concentrations of TSH on bioenergetic metabolism in the liver, the myocard, the sceleton muscles.

<u>Determination of the TSH physiologic concentration.</u>
In order to find out the TSH effect on the bioenergetic process it is necessary to determine the physiologic TSH concentrations. For this reason the authors investigated 35 healthy persons and found out the mean TSH level in serum to be 0,26 ± 0,08 mU/ml. Later on these findings were confirmed by many control investigations. The results are close to the findings of the other authors (see Table 1). Some authors consider the lower limit of TSH to be 0,00 mU/ml, which is practically impossible, because

Table 1

Normal values of TSH in blood of healthy persons, as cited by different authors (in mU/ml).

Author	Year	Number	TSH (mU/ml)
S.A. d'Angelo	1951	9	0,0-1,0
A. Querido and al.	1956	-	0,0-2,0
J.C. Gilliland and J.I. Strudwick	1956	8	0,165
A.M. di George and al.	1958	3	0,0-0,5
F.S. Guespan and W. Lew	1959	-	0,50
P.M. Bottary	1958	120	0,22
J.M. Mc Kenzie	1958	-	0,165
S.A. d'Angelo	1963	-	0,40
M. Linquette	1967	30	0,13
I.A. Eskin	1967	11	0,0-0,126
St. Milanov	1968	35	0,26±0,08

it excludes the physiologic variations. Our method determines the lower limit of TSH activity as well.

The TSH effect on tissue bioenergetic process in vitro.

The research was carried out in vitro, because there was no information of the TSH bioenergetic influence on different organs and the authors wanted to find out the TSH effect without the effect of interfering factors. On the basis of the previously found physiological level of TSH, the research was carried out with the following concentrations in incubation homogenates: 0,00 mU/ml - controls, 0,10 mU/ml - subphysiological concentration, 1,25 mU/ml - physiological, 0,50 mU/ml - above the physiological concentration, 1,0 mU/ml - much higher than physiological. The research was carried out simultaneously on the liver, the myocard and the sceleton muscles. The results are shown on Tables 2,3 and 4. They show that the TSH subphysiological concentration do not affect significantly the tissue breathing and oxidative phosphorylation activity.

Table 2
Effect of different concentration of TSH (in mU/ml) on bioenergetic metabolism in liver homogenate of healthy rats in vitro

		\multicolumn{5}{c}{TSH concentrations}				
		0,0	0,10	0,25	0,50	1,00
Non organic esteryfied phosphate µA/mg	n	10	10	10	10	10
	$\bar{X}$	38,3	38,5	46,5	31,9	19,2
	$S_{\bar{X}}$	1,2	2,0	2,1	1,2	0,8
	P_t	-	<0,5	<0,01	<0,001	<0,001
O_2 µA/mg	n	10	10	10	10	10
	$\bar{X}$	17,8	17,4	19,3	16,8	13,7
	$S_{\bar{X}}$	1,0	2,2	2,4	0,8	0,6
	P_t	-	<0,5	<0,1	<0,1	<0,001
P/O in µA/µA/mg	n	10	10	10	10	10
	$\bar{X}$	2,20	2,21	2,41	1,90	1,40
	$S_{\bar{X}}$	0,01	0,05	0,06	0,06	0,06
	P_t	-	<0,5	<0,01	<0,001	<0,001

The physiological concentration stimulates these two basic processes, the effect being higher in the process of esterification of non-organic phosphate and lower in the stage of oxygen consumption. The high and very high TSH concentrations cause inhibition of bioenergetic metabolism in organ homogenates of healthy animals.

The results of TSH extrathyroid bioenergetic effect suggests further investigations of the mechanism of TSH and its clinical significance.

CONCLUSIONS

1. TSH extrathyroid bioenergetic effect in vitro has been found out.

2. The TSH extrathyroid bioenergetic effect depends on the acting concentration:

Table 3

Effect of different concentration of TSH (in mU/ml) on bioenergetic metabolism in myocard homogenate of healthy rats in vitro

		TSH concentrations				
		0,0	0,10	0,25	0,50	1,00
non-organic esterificated phosphate µA/mg	n	10	10	10	10	10
	$\bar{X}$	26,8	27,9	38,0	18,0	11,8
	$S_{\bar{X}}$	2,1	2,2	3,0	2,1	1,0
	P_t	-	<0,5	<0,01	<0,01	<0,001
O_2 in µA/mg	n	10	10	10	10	10
	$\bar{X}$	12,5	12,6	15,7	9,6	8,4
	$S_{\bar{X}}$	0,9	1,1	1,0	0,7	0,7
	P_t	-	<0,5	<0,05	<0,05	<0,01
P/O in µA/µA/mg	n	10	10	10	10	10
	$\bar{X}$	2,14	2,21	2,42	1,88	1,40
	$S_{\bar{X}}$	0,04	0,05	0,04	0,05	0,05
	P_t	-	<0,1	<0,01	<0,05	<0,001

a) the physiologic TSH concentrations stimulate moderately the oxidative process in the respiratory chain and formation of macroergic links in oxidizing phosphorylation.

b) The subphysiologic TSH concentrations do not affect the tissue breathing activity and oxidative phosphorylation.

c) The TSH high concentrations inhibit moderately both the process of electronic transportation in the respiratory chain and the processes of oxidative phosphorylation.

Table 4

Effect of different concentrations of TSH on bioenergetic metabolism in sceleton muscles homogenate of healthy rats in vitro

		TSH concentrations (mU/ml)				
		0,00	0,10	0,25	0,50	1,00
non organic esteryficated phosphate $\mu A/mg$	n	10	10	10	10	10
	$\bar{X}$	29,4	32,8	36,2	22,5	16,9
	$S_{\bar{X}}$	2,0	2,0	1,1	0,9	1,0
	P_t	-	<0,1	<0,05	<0,01	<0,001
O_2 in $\mu A/mg$	n	10	10	10	10	10
	$\bar{X}$	14,0	15,9	16,8	12,5	10,6
	$S_{\bar{X}}$	1,2	0,9	1,0	1,5	0,8
	P_t	-	<0,5	<0,1	<0,5	<0,05
P/O in $\mu A/\mu A/mg$	n	10	10	10	10	10
	$\bar{X}$	2,10	2,06	2,15	1,80	1,60
	$S_{\bar{X}}$	0,01	0,06	0,06	0,05	0,06
	P_t	-	<0,1	<0,1	<0,1	<0,001

d) The very high TSH concentrations inhibit greately the tissue breathing activity and the oxidative phosphorylation.

3. The above described action of TSH was noted in several organs: the liver, the myocard and the sceleton muscles.

REFERENCES

1. Milanov S. - Annales d'Endocrinologie, Paris, 1970, 31, 6, 1101-1110.

2. Hunter F.E. - Methods of Enzymology, 1962, New York-London, v.177, 2, p.192.

3. Umbreit W., R.Burris, J.Stauffer - Monometric Techniques, 1959, Burges Publ., p.556, Mineapolis.

SEX-DEPENDENT PROLACTIN PATTERN DURING DEVELOPMENT

Julia Vassileva-Popova and J.R.Tata*

Central Laboratory of Biophisics, Bulgarian Academy of Science, 1113 Sofia, Bulgaria

*National Institute for Medical Research, Mill Hill, London NW7 1AA, Great Britain

INTRODUCTION

Despite its name which suggests a very specific role in the control of lactation, the hormone prolactin is now known to regulate numerous physiological functions in a wide variety of vertebrate species. Although suggestions have already been made that prolactin exerts different effects on the male reproductive tract, but it is not yet clear as to what physiological role this pituitary hormone may play in the male mammal. It seemed to us that before a precise hormonal function could be ascribed it is essential to establish quantitatively the prolactin /Pr/ level in the pituitaty and blood of male animals. Furthermore, from a consideration of the possibility of an important role for Pr during sexual maturation we decided that it would be more useful to determine Pr levels in the developing male and compare the data with the Pr level in developing females. In this paper we describe our initial findings on the comparison of circulating and hypophyseal levels of prolactin in developing male and female rats. It will be shown that Pr is present in high amounts in the neonatal male pituitary and that substantial levels of circulating hormone are maintained during the first weeks of development of the male. The pattern of changes in the hypophyseal and circulating prolactin in the male did not correspond to that in the female.

MATERIALS AND METHODS

The animals used were CD rats, male and female, estimated at three chosen stages of development: 14 days, 28 days and 42 days. The different age groups consisted of 10 animals for each experiment. Pituitary glands and blood were collected from the same animals in one assay. Five different assays were performed for each age group simultaneously for both sexes. The same numbers of the experiments in all three age-groups indicate that the results belong to one assay and have to be compared according to the given number. The blood was collected under ether anaesthesia. In some of the experiments the blood was collected immediately after the animals were killed. The pituitary glands were obtained from the same animals. After weight estimation the tissue was homogenized and the extract was prepared. The Pr concentration in the blood sera and in the hypophysis was determined by means of a radioimmuno assay. The standard used was that suplied by Endocrinology Study Section, NIH, Bethesda. The tracer used was ^{125}I and the labelling was performed according to Hunter and Greenwood,(1). The labelled Pr was purified by gel filtration on a 1cm x 20cm column of Sephadex G-100. The further steps of the radio-immuno assay method were carried out according to the classical fashion of this competitive binding test. Our previous work on the radio-immuno assaay was taken in to consideration (2). For the end separation the double antibody method and the ethanol precipitation were applied.

RESULTS AND DISCUSSION

From the five separate assays performed for this purpose (Table 1) the average value of two of them /No 1 and 2 of each group/ is plotted.(Fig.1 and Fig.2). The reason for the choice of these two results was the relatively close values and the small diviation between then. The values of the remaining three assays differ in a relatively large range of the values obtained. Nevertheless, the characteristics of the values obtained from all assays shows the same or very similar principle dissimilaryty between the groups investigated. In other words despite the variation in the Pr values the principal results from the five separate assays indicate similar or the same Pr type-face of the blood serum and the pituitary gland between the male and female in the different age groups (Tabl.1). One of the reason for this variation in the results obtained from the different as-

TABLE 1

Prolactin Content in Pituitary Tissue and in Blood Serum in Male and Female Rats During Development /Data from Five Separate Assays/

No. of experiment	Age	male		female	
		ng/ml blood serum	ng/mg pituitary	ng/ml blood serum	ng/mg pituitaty
1.	14 days	61	542	35	185
2.		55	410	25	130
3.		177	1825	66	350
4.		450	3900	243	1100
5.		13	111	7	39
1.	24 days	72	971	102	2750
2.		55	681	71	1920
3.		247	2650	365	15720
4.		500	6800	718	19250
5.		15	197	21	550
1.	42 days	78	2550	135	6520
2.		60	2725	95	7988
3.		195	3850	112	-
4.		550	26250	950	7900
5.		16	771	27	2276

All experiments with the same number was caried out simultaneously.
Each group consists of 10 animals.
Despite the value differences the principle tendency is quite similar.

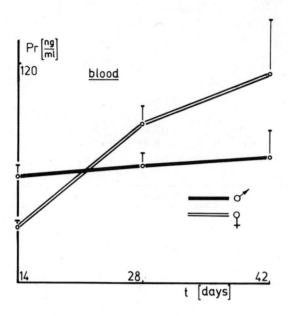

Fig. 1

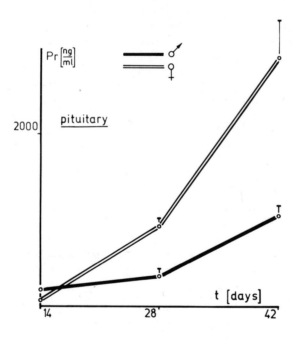

Fig. 2

says is possibly dealing with the standards used. Another reason should be the method of separation applied: ethanol and double antibody techniques. Another acceptable explanation for the difference in the Pr values is the way of blood collection (3). Differences in the Pr level will also be found in the works of other researchers on this topic (4-7).

It has to be noted that it is less likely that all the Pr amount can be extracted from the pituitary tissue and resp. measured. It is very possible that not so negligible an amount of the hormone to remain in the tissue sediment. In this respect it has to be expected that the real values of Pr in the pituitary gland to be in fact higher. At the same time there is the possibility some amount of Pr in the blood circulation to be fixed on the blood cells /our data presented in this volume/.

Pr concentration in blood serum. During the development (14-42 days) in female sex an definite increase of the Pr values is observed (Fig.1 and Table 1). Contrary to this the male sex during development shows no significant change of the Pr concentration in blood serum. An interesting point of this finding is the significant difference in the Pr values between the male and female sex in 14 days of age. The dissimilarity in the Pr concentration in blood serum is positive (double) for the male sex. It looks that the Pr concentration in male sex has a tendency of preservation at a constant level, while the female sex shows a definite increase.

Pr content in hypophysis. A well demonstrated increase of Pr, corresponding to that found in the blood serum, is observed in the pituitary gland in female sex. As a contrast to the relatively constant Pr blood serum concentration an increase of the Pr content in pituitary tissue was found in male sex (Fig.2). It is quite difficult to find an easy explanation of the contrast between the Pr concentration in the blood serum and the Pr content in the pituitary gland observed in male sex. Possibly this point should be discussed on the background of our view of the hormone utilization.

The findings in the present study indicate a well demonstrated inequality of Pr content as a function of the time of development in both sexes. The first point to be noted is that 14 days after birth the hormone content in the blood and serum and the pituitary gland in male sex is considerably higher than that in the female ($p < 0.001$). With the development the Pr content in pituitary tissue and especially in blood serum a reciprocal relationship is displayed in the two sexes. The circulating level of the hormone in the female animals paralles the sharp rise in the pituitary content observed between 14-42 days of the birth. In contrast, the blood level of Pr in the male does not parallel the elevation of its level in the hypophysis but is maintained at a fairly constant level. The absence of accordance in male sex (as was found in female) between the Pr blood serum concentration and the Pr content in the hypophysis possibly belongs to a diverse mechanism of hormone utilization in the male sex. The data describing the role of Pr in male rats indicate the synergism between testosterone and Pr for growth activation of prostate. The increase of the zinc content in accesory glands is observed also in the absence of androgens. An involvement of Pr in spermatogenesis is also observed (8). It looks that the influence of Pr on male reproduction is of interest (8-13). At the same time chlorpromazine produces a dramatic elevation in the Pr level in both sexes (2). Nevertheless, the findings suggest a probable Pr utilizational dissimilarity between male and female, which focused in the question of how the hormonal utilization reflects its function. Possibly the data for a change in the ratio: pituitary-Pr content versus Pr concentration in circulation during lactation will be of interest in this respect. The disagreement between pituitary and blood serum content of Pr is due to a high rate of hormonal utilization during development and sexual maturation of males.

The interesting point for a further study is the ratio between the hormone concentration in the circulation and its content in the pituitary tissue. For instance, in the early development the Pr pituitary content in the males is about 10-15 times higher in comparison with that in blood serum. The 42-day old rats have about 40-45 times higher content in the hypophysis in comparison with the Pr blood serum concentration. This ratio of blood-pituitary tissue content has a remarkable difference characteristics in female sex. During the early development (14 weeks after birth) the hypophysis in female sex contains only about 3-4 times more Pr than in blood circulation. This ratio increased about 25 times in the 28-day old animals and reached a value of about 50 times higher hypophyseal Pr content in comparison with the blood content in 42-day old females. The higher variation in the Pr values of 42-day old female is expected to be due to the influence of forthcoming oestrus.

Despite the dynamic dissimilarity in the Pr content in female and male sexes during development, it looks that the ratio: pituitary/blood Pr content leads to a more stable state in late development. Nevertheless, the question of the role of Pr in male growth and sexual maturation became more surprising.

REFERENCES

1. Hunter, W.M., and Greenwood, F.C., Nature (London), 194, 495 (1962).

2. Edwardson, J.A., and Vassileva-Popova, J., Proc. Second Int. Symp. Varna, Sept.13-16, 1972, Bulg. Acad. Sci. Press, Sofia, 1973.

3. Wakabayashy, I., Arimwa, A., Fed.Proc., 30, 254, (1971).

4. Amenomori, Y., Chen, C.L., Meites, J., Endocrinology, 86, 506, (1970).

5. Kwa, H.G., Verhoptad, F., J. Endocrinol. 39, 4551, (1967).

6. Mallampati, R.S., Srivastava, L.S., Fed. Proc., 30, 474, (1971).

7. Neill, J.D., Reichart, L.E., Endocrinology, 88, 548, (1971).

8. Moger, H., and Geschwind, I.I., P.S.E.B.M., 141, 1017, (1972).

9. Bartke, A., J.Endocrinol., 49, 311, (1971).

10. Hafiez, A.A., Philpott, J.E., and Bartke, A., J. Endocrinol., 50, 619, (1971).

11. Hafiez, A.A., Lloyd, C.W., and Bartke, A., J. Endocrinol., 52, 327 (1971).

12. Chase, M.D., Geschwind, I.I., and Bern, H.A., Proc. Soc. Exp. Biol. Med., 84, 680, (1957).

13. Von Berswordt-Warlrabe, R., Steinbeck, H., Hahn, J.D., and Elger, W., Experientia, 25, 533, (1969).

THE CORRELATION BETWEEN THE BINDING OF THE HUMAN CHORIONIC GONADOTROPIN TO LUTEAL CELLS AND PLASMA MEMBRANE OF LUTEAL CELLS

S. Papaionannou and D. Gospodarowicz
The Salk Institute for Biological Studies
San Diego, California

INTRODUCTION

Granulosa cells as well as luteal cells maintained in tissue culture require a low but constant level of LH or hCG to stay differentiated (1,2). It can therefore be assumed that all factors regulating the appearance or disappearance (modulation) of the receptor sites for gonadotropins are of importance for the luteotrophic function and the differentiation of luteal cells. To study the modulation of the receptor sites in luteal cells maintained in tissue culture, one needed a binding technique specific enough to detect small variations in receptor concentration under different culture conditions. Also, since luteal cells are obtained by enzymatic digestion, and since it had been shown that proteolytic enzymes can destroy receptors (3-7), one must be assured that such a treatment does not significantly alter the characteristics of the binding of gonadotropins to their receptors. In this communication we compare the binding characteristics of hCG to bovine luteal membranes and bovine luteal cells obtained by enzymatic treatment (7). We have been able to show that the binding was identical and that the enzyme treatment needed to dissociate the luteal cells before putting them into culture did not affect the receptor site for gonadotropins.

MATERIALS AND METHODS

A. Hormone Preparations and Chemicals

Highly purified hCG was obtained from Searle Laboratories. It had an activity of 10,000 IU/mg. α and β chains of hCG had respectively 0.5 % and less than 0.2 % of the activity of hCG. They were prepared by Mrs. Bidlingmeyer and Mr. Fago (Searle Laboratories). Ovine FSH and highly purified preparations of LH were obtained in this laboratory by the method of Papkoff, et al.(8, 9) from ovine pituitary glands. The LH preparations were further purified by chromatography on diethylaminoethyl cellulose to remove contaminating TSH as described by Pierce and Carsten (10). The biological activity measured by the ovarian ascorbic acid depletion assay was 2.75 U/mg (95 % confidence limits 2.1 - 3.7) referred to the standard NIH-LH-S17.

Bovine serum albumin, Fraction V was from Reheis Chemical Co., Chicago, Ill.. Concanavalin A Sepharose was from Pharmacia, New York, NY.. $Na^{125}I$ for radioiodination was from New England Nuclear, Boston, Mass.. EAWP and EGWP filters were from Millipore Co., Bedford, Mass.. Talc Tablets were from Gold Leaf Pharmaceutical Co, Inc. Englewood, NY.

B. Iodination Procedures

The iodination of hCG was performed using lactoperoxidase as the catalytic agent. All reactions were carried out in small glass tubes at room temperature. The reactants were mixed continuously with manual shaking. In order to minimize the introduction of more than 1 atom of iodine per hCG molecule, iodination was performed with an equimolar ratio of ^{125}I to hCG. To minimize the deleterious effect of the hydrogen peroxide, it was added in approximately equimolar amounts with respect to the concentration of iodine and hCG. To obtain $^{125}IhCG$ of high specific activity the reagents were added rapidly in the following order and amounts: a) 10 mCi of $Na^{125}I$ (588ng) in 25 µl of 0.1 NNaOH: b) 200 µl of 0.1 M potassium phosphate buffer, pH 7.2; c) 174 µg of hCG in 100 µl of 0.1 M potassium phosphate buffer, pH 7.2; and d) 3.75 µg of lactoperoxidase in 5 µl of the same buffer. The reaction was initiated by adding 160 ng of hydrogen peroxide in 5 µl water. To sustain the reaction, 160 ng of hydrogen peroxide was also added at 1, 2, 4, 6, and 8 min; additional lactoperoxidase (3.75 µg in 5 µl of buffer) was added at 5 min.

After the reaction was completed, final purification was achieved by gel filtration of the iodination solution on a column of Sephadex G-200 equilibrated with 0.9% NaCl and 0.01 M potassium phosphate, pH 7.3. The ability of the preparation of ^{125}IhCG to bind to talc was 90% or higher.

C. Preparation of Plasma Membrane Fraction from Bovine Corpus Luteum

Plasma membrane fraction was prepared as described by Gospodarowicz (11). Fractions F1 and F2 were used as a source of plasma membrane. Proteins were determined according to the method of Lowry et al.(12), using bovine serum albumin as a standard.

D. Preparation of Intact Luteal Cells

In the present study, corpora lutea from dairy cows of mixed breeds in the first half of pregnancy were used. The duration of pregnancy was estimated from the crown--rump lenght of the fetus, and corpora lutea corresponding to a fetal size of 6-12 cm were taken.

Slices of fresh corpus luteum were taken from the center of the gland, to avoid any contamination by thecal tissue. The slices were then minced in phosphate-buffered saline, pH 7.4 (PBS).

The tissue fragments were incubated at a final concentration of 1g in 5 ml of PBS containing 4% bovine serum albumin (BSA) and 0.2% glucose. The incubation was carried out at 37 C in a polyethylene container in an atmosphere of 95% O_2 - 5% CO_2. The incubation medium was stirred with a siliconized glass paddle driven at 150 rpm. After 10 min the fragments were washed twice in PBS containing 4% BSA, 0.2% glucose and resuspended in PBS containing 4% BSA, 0.2 glucose, 0.28% collagenase, (Worthington CLS) and 0.2% hyaluronidase (Sigma type II). The mixture was further incubated at 37 C in the presence of

95% O_2 - 5% CO_2 under constant stirring as described above.

After 45 min the dissociation medium was decanted. The fragments were washed three times in PBS to remove cellular debris and BSA as completely as possible, and incubated at 37°C in PBS containing 0.25% trypsin (Grand Island Biological Co., Type A) and 0.005% crystalline DNAse (Worthington 2.200 U/mg). Every 10 min the incubation stopped and the dissociation medium decanted and replaced by fresh medium.

After four medium renewals the incubation was discontinued. The fragments were washed three times in PBS containing 4% BSA, 0.1% lima bean trypsin inhibitor (Worthington) and 0.2% glucose.

After washing with the appropriate medium, the fragments were disaggregated by gentle agitation and the supernatant was collected. The remaining fragments were resuspended and dissociation was completed by sucking the tissue fragments up and down a 10 ml plastic pipette.

The cell suspension was then filtered through a Japanese silk screen (mesh 12). This procedure dispersed small pieces of undigested tissue and filtered out most of the stromal vascular tissue.

The filtrate was collected in a polyethylene tube and spun in a clinical centrifuge at 150 g for 4 min at 0°C. The supernatant was removed with a plastic pipette and the pellets were resuspended in the same volume of medium. This procedure was repeated three times, keeping the volume constant. The final pellet was resuspended in incubation medium to give a final concentration of 1 x 10 cells/ml. DNAse 0.005% was added to avoid clumping of the cells.

E. Binding Assay

The assay for specific binding of ^{125}IhCG to membranes was slight modification of that used to measure specific binding of insulin to intact fat cells (13) and fat cell membranes (14). Cellulose acetate EGWP filters were used. They were presoaked in a 1% albumin solution in PBS.

Briefly, membranes were incubated at 24 C to equilibrium in 0.2 ml of buffer containing 1% albumin and

^{125}IhCG (10^{-11} to 10^{-8} M). Two ml of ice-cold PBS buffer containing 1% albumin were added, and the contents were passed through the EGWP filters with vacuum. The filters were washed twice under vacuum with 2 ml of ice-cold PBS containing 1% albumin. Every determination of binding was performed in triplicate, and for every such determination parallel, triplicate samples were assayed in the presence of native hCG (12.5 μg per ml) to determine the correction for nonspecific binding of hCG. Nonspecific binding to EGWP filters in the absence of plasma membranes of cells was 0.1% of the total ^{125}I. Nonspecific binding to plasma membrane or to cells in the presence of an excess of hCG was no more than 2 to 5% that of total binding (fig 4). Plasma membranes which have been heated to 90 C for 5 min do not bind more than 0.5% of the input. Radioactivity of the filters was determined in 10 ml of toluenebased scintillation solution, made 5% with Bio-Solv Solubilizer (Beckmann). Counting efficiency was 60%.

For the dissociation studies the cells or membranes were incubated in the presence of ^{125}IhCG with or without hCG, at 25°C for 30 min. They were then filtered on millipore filter and the filters were continuously washed (under reduced pressure, with a constant flow rate and in the absence of unlabelled hCG) with PBS. At regular intervals the filters were removed and counted.

The assay for specific binding of ^{125}IhCG to luteal cells was similar to that of luteal plasma membranes except that EAWP millipore filters were used.

RESULTS

A. Physicochemical and Biological Properties
   of ^{125}IhCG Prepared by Enzymatic Radioiodination

The integrity of the ^{125}IhCG was analyzed by polyacrylamide gel electrophoresis (fig 1A). One peak of radioactivity was observed, coinciding with the hormone.

The integrity of the antigenic structure of ^{125}IhCG was confirmed by the quantitative precipitin test. The equivalence zone for hCG was the same as that of ^{125}IhCG, and over 95% of the ^{125}IhCG was precipated. When ^{125}IhCG was applied on an anti-hCG IgG Sepharose column, 93% was retained on the column, thus proving that the antigenic

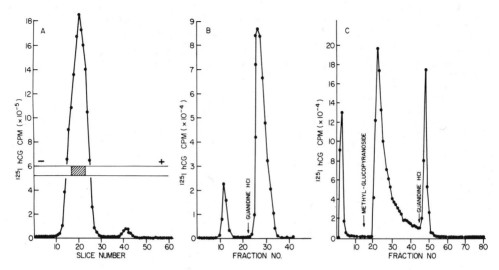

Fig. 1 A. Fractionation on 7.5% acrilamide gel, pH 9.5 (23). ^{125}IhCG (1.5×10^6 cpm) mixed with 50 µg of hCG was subjected to electrophoresis for 3 hours at 4 ma per tube. At the end of the electrophoresis, the gels were either stained with 1% Amido Schwarz in 7% acetic acid solution or frozen at -20 C on dry ice, sliced into 1mm thick discs, dissolved as described by Ward et al.(24), and counted in a liquid scintillation counter. A tracing of the gel is shown.

B. Binding and elution of ^{125}IhCG (6.8×10^5 cpm) dissolved in 0.9 M NaCl, 0.01 M potassium phosphate, pH 7.3, and applied on a column of anti-hCGIgG Sepharose (1 x 0.4 cm) equilibrated with the same buffer (25). Bound ^{125}IhCG was eluted with 6 M guanidine- HCl, pH 1.5.

C. Elution profile of ^{125}IhCG on Sepharose Concanavalin A column. 0.2 ml of ^{125}IhCG (3.1×10^8 cpm) was applied on a Sepharose Concanavalin A column (2 ml total volume) equilibrated with 1M NaCl, 1 mM MgCl$_2$, 1mM MnCl$_2$, 1 mM CaCl$_2$, 0.1 M Na acetate buffer, pH 6.0. The hCG adsorbed on the column was eluted by 1M methylglucopyranoside in the same buffer followed by 6 M guanidine- HCl, pH 1.5.

sites of hCG were intact (fig 1B). Ninety percent of the adsorbed ^{125}IhCG was eluted with 6M guanidine HCL, pH1.5. The unadsorbed fraction may have represented either iodinated peroxidase or denatured hCG (fig 1B).

The integrity of the ^{125}IhCG was further investigated by Sepharose Concanavalin A chromatography as described by Dufau et.al., (15). When ^{125}IhCG was applied on Sepharose Concanavalin A column, 10% of the radioactivity was unadsorbed (fig 1C) and had no binding activity. The labelled hormone was then eluted from the column with 1M methylglucopyranoside; 70% of the radioactivity was eluted and has binding activity. Further elution of the column by 6 M guanidine HCl, pH 1.5 resulted in the recovery of 15% of the original radioactivity. This radioactivity represented most probably the ^{125}I asialo hCG which had been shown by others (15) to stick to Concanavalin A Sepharose column very firmly.

When the biological activities of hCG and ^{125}IhCG were compared by the ovarian ascorbic acid depletion assay, their activities were indistinguishable and their ability to stimulate the release of progestins in vitro from bovine luteal cell suspensions (fig 2) was identical over a 1000-fold range of concentration.

These different criteria indicated that iodination of hCG using lactoperoxidase as a catalytic agent produced no noticeable denaturation of the hormone. This conclusion had been reached by others (16).

The final activity of hCG was 2,125 Ci per m mole corresponding to an average incorporation of 1 atom of ^{125}I per molecule of hCG. Due to virtually 100% incorporation of ^{125}I into hCG, no further purification of hCG by ion exchange chromatography was required to separate hCG from ^{125}IhCG. The only problem we encountered in preparing ^{125}IhCG was the source of ^{125}I being used. The Na ^{125}I of New England Nuclear was most satisfactory. Na ^{125}I from other sources sometimes contained minor impurities which may not have been of importance when the chloramine T method is used but which could greatly have affected the yield of ^{125}IhCG as well as the biological activity of the product when enzymatic radioiodination is being used.

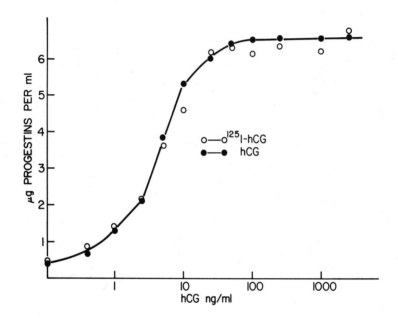

Fig. 2 Stimulation of progestin secretion (micrograms per ml) in suspensions of bovine luteal cells by hCG or ^{125}IhCG.
The cell concentration is 1×10^6 cells per ml. The incubation lasted 90 minutes. The determination of progestin concentration was performed as described in (7).

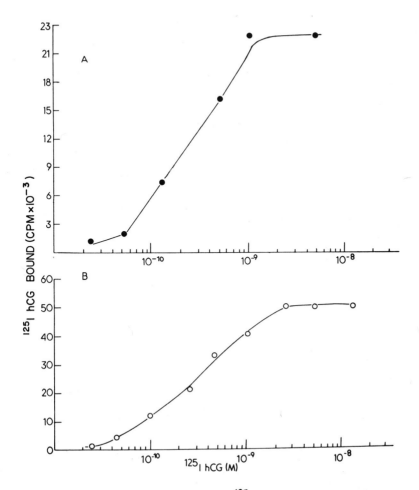

Fig.3A. Specific binding of ^{125}IhCG to luteal cells as a function of the concentration of ^{125}IhCG.
Luteal cells (1.5 x 10^5 cells) are incubated at 24 C for 20 minutes in 0.2 ml of PBS, 1% albumin, and various concentrations of ^{125}IhCG. Specific binding and correction for nonspecific adsorption of hCG is determined as described in the text.

B. Specific binding of ^{125}IhCG to bovine corpus luteum plasma membrane (200μg).
The experiment was conducted as described in A.

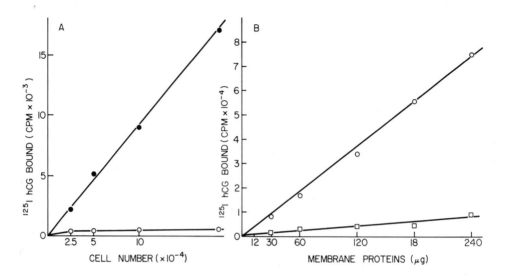

Fig. 4A. Specific binding of ^{125}IhCG to luteal cells as a function of the concentration of luteal cells in the medium.
The incubation medium contains various concentrations of luteal cells in 0.2 ml of PBS, 1% albumin, and 1.5 x 10^{-9} M ^{125}IhCG. After incubation for 20 minutes at 24C, the specific binding (● — ●) and nonspecific binding (0 - 0) of hCG is determined as described in methods.

B. Specific binding of ^{125}IhCG to plasma membrane of bovine corpus luteum as a function of the concentration of plasma membranes in the medium.
Specific binding (0 - 0), nonspecific binding (□ - □). The concentration of ^{125}IhCG was 1.5 x 10^{-9} M. The experiment was conducted as described in A.

B. Cellular Specificity of the binding of ^{125}IhCG

The specificity of the binding of ^{125}IhCG to luteal cells or to plasma membranes of luteal cells was explored by analyzing its specific binding to other cell types or tissues. ^{125}IhCG did not bind specifically to the 31A ovarian cell line nor did it bind to primary cultures of fibroblasts, endometrial cells or to the Y1 Adrenal cell line. It did not bind to a significant extent to either muscle, brain, or adrenal cortex membrane preparations (Table1).

C. Saturability and Specificity of the Binding of ^{125}IhCG to Receptor Sites Present in Luteal Cells and Plasma Membranes of Luteal Cells.

1. <u>The binding sites of hCG were saturable</u>. The specific binding of ^{125}IhCG to intact luteal cells or to purified plasma membranes obtained from bovine corpus luteum was a saturable process with respect to ^{125}IhCG concentration (fig 3A, B). Specific binding could be detected at concentrations as low as 1.8 ng of hCG per mL (5×10^{-11} M) while saturation was achieved at 92 ng/ml (2.5×10^{-9} M). In all cases the binding of ^{125}IhCG to intact cells or plasma membranes was identical: minimal effective concentration values and saturation values were the same.

The binding of ^{125}IhCG to luteal cells or to plasma membranes was directly proportional to the luteal cell or the plasma membrane concentration over the range of concentrations that could be used in these procedures (fig 4A, B). This linear relationship in the absence of saturating concentration reflects the high affinity of the interacting species and could be compared to the early portions of the curves of fig. 3 (10^{-11} to 10^{-10} M). It was impossible to achieve high enough concentrations of plasma membranes to see a plateau of binding since at high membrane of cell concentrations the filters clog.

2. <u>The binding of hCG to its receptor site was specific</u>. ^{125}IhCG specifically bound to luteal cells was displaced by increasing concentrations of native hCG in a fashion predicted by the near-identity of these 2 molecules (fig 5A). A hormone such as FSH which does not have any LH intrinsic activity appears to be 10,000-fold less potent than hCG. Its activity could be explained by its LH contamination, which was 0.005% as seen by radioimmu-

Table I

Comparison of the Binding of ^{125}IhCG to different Cell Types and Membrane Preparations

Cell Type	Specific Binding	Membrane Preparations	Specific Binding
31 A Ovarian Cells	1	Muscle	1 %
Fibroblast (human foreskin)	1	Brain	5 %
Fibroblast (mouse embrio)	1	Adrenal Cortex	2 %
Endometrial Cells	1		
Y-1 Adrenal Cells	1		

Specific binding of ^{125}IhCG to different cell types. The cell concentrations were in all cases 1×10^5 cells per 0.2 ml of incubation media. The results are expressed in % binding of ^{125}IhCG compared to the same concentration of luteal cells taken as 100%.

Specific binding of ^{125}IhCG to different membrane preparations (600 to 12,000 g precipitate). 250 µg of proteins were used in all cases. The results are expressed in % binding of ^{125}IhCG compared to the same concentration of corpus luteum membrane taken as 100 %.

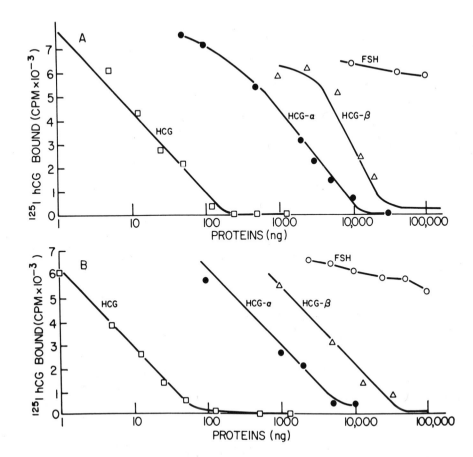

Fig. 5A. Effect of native hCG, hCG$_\alpha$, hCG$_\beta$, and FSH on the binding of ^{125}IhCG to luteal cells. Suspensions containing 1.5×10^5 cells in 0.2 ml of PBS, 1% albumin are incubated with 10^{-9} M ^{125}IhCG and the indicated amount of hCG, hCG$_\alpha$, hCG$_\beta$, or FSH for 20 minutes at 24C. Specific hCG binding is determined as described in the text and corrected for nonspecific binding.

B. Effect of native hCG, hCG$_\alpha$, hCG$_\beta$, and FSH on the binding of ^{125}IhCG plasma membranes to bovine corpus luteum. The amount of plasma membranes was 100μg and that of ^{125}IhCG 1×10^{-9} M. The experiment was conducted as in A.

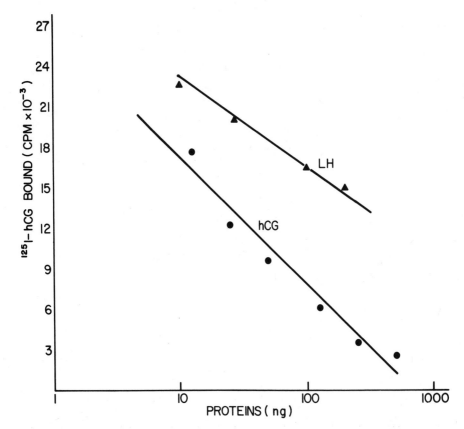

Fig. 6. Effect of varying concentrations of hCG and LH on the binding of ^{125}IhCG to plasma membranes of bovine corpus luteum. Suspensions containing 120 μg of plasma membrane in 0.2 ml of PBS, 1% albumin are incubated with 1.8×10^{-9} M ^{125}IhCG and the indicated amount of hCG or LH for 20 minutes at 24C. Specific hCG binding is determined as described in the text and corrected for non-specific binding.

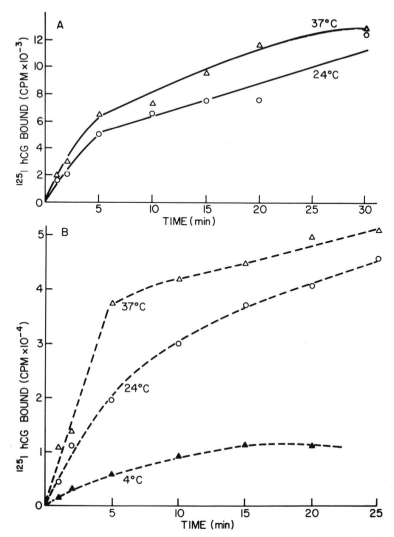

Fig. 7A. Time dependence of the association of ^{125}IhCG to luteal cells at different temperatures. Luteal cells (1.25 x 10^5 cells) are incubated at various temperatures in 0.2ml of PBS, 1% albumin, and 1.8 x 10^{-9} M ^{125}IhCG. At regular intervals the specific binding of hCG to luteal cells is determined as described in the text.

B. Time dependence of the association of ^{125}IhCG to plasma membranes of bovine corpus luteum (200 μg).
The experiment is conducted as described in A.

noassay. hCG and hCG, the two subunits of hCG, had respectively, 200- and 800-fold less activity on a weight basis than hCG (fig 5A) thus confirming the fact that to get full steroidogenic effect, the two subunits of hCG must be combined (17). When the displacement by hCG, hCG or FSH of hCG bound to luteal plasma membrane was analyzed, identical results were obtained (fig 5B). Ovine LH was able to complete for the receptor site of hCG (fig 6). However, the slope of the curve for inhibition of ^{125}IhCG binding by hCG differs from that for ovine LH by a factor of 4 at low concentrations (5 - 10 ng) and by a factor of 10 at high concentrations (50 - 100 ng), suggesting that ovine LH has less affinity for the hCG receptor site than hCG.

3. <u>The binding of hCG to its receptor site was rapid and reversible.</u> The specific binding of ^{125}IhCG to the plasma membrane was rapid. By 5 min at 37 C, most of the sites were occupied (80%). Binding then progressed at a slower rate during the next 20 min which resulted eventually in the occupancy of all the sites. With the luteal cells, a similar rate of binding was observed. However, one should note that the initial rate was not as fast as for the membranes, 50% of the sites were occupied after 5 min and during the next 20 min, the remaining sites were occupied at a slower rate (fig 7). After 30 min, no significant increase in binding was observed.

The slower rate of association of hCG to its receptor sites in whole cells as compared to fragments of plasma membranes probably reflected a greater accessibility of hCG to the sites on plasma membrane fragments rather than any alteration in the receptor site. Rapid stirring could perhaps have increased the apparent rate of association observed with the cells since the rate appeared slower than diffusion limited. However, the small volumes (0.2ml) in which the cells were suspended made it technically impractical. Another possibility which could explain the slower rate of association of hCG to its receptor sites in whole cells as compared to membrane is the faster rate of degradation of hCG when it interacts with its receptor site in whole cells as compared to membranes (18). A similar observation has also been made for other polypeptide hormones (19 - 20).

The binding of ^{125}IhCG to plasma membranes was also temperature dependent. At 4 C it progressed at a slower rate than that observed at 37 C. After 20 min, the extent of binding was only 20% of that obtained at 37C. At 24C,

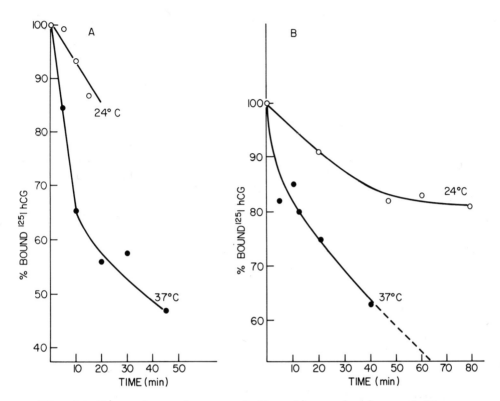

Fig. 8A. Time dependence of the dissociation of the ^{125}IhCG-receptor complexes at 24C and 37C in luteal cell suspensions.
The incubation media contained 1×10^5 cells in 0.2 ml of PBS, 1% albumin and 1.8×10^{-9} M ^{125}IhCG. After incubation for 20 minutes at 24C, the cell suspensions are filtered as described in the text and washed at 24C or 37C for various periods of time with PBS, 1% albumin. At the indicated time the washing is stopped and the quantity of ^{125}IhCG bound to the plasma membranes is determined as described in the text.

B. Similar experiment but with plasma membrane of bovine corpus luteum (150μg).

binding progressed at a rate which was intermediate between the rates observed at 4C and 37C and somewhat constant over the entire 25 min incubation. The binding ability of hCG to plasma membrane receptors at 37C or at 24C after 25 min was comparable. Similar results were obtained with the luteal cells. During the first 5 min at 37C, the initial rate was fast as observed already with the membrane. Binding slowed down though, with the completition of binding taking place at 30 minutes. Binding at 24C followed a similar but slightly slower rate than that observed at 37C.

The dissociation of ^{125}IhCG from cells or membrane fragments was a function of time and temperature (fig 8). The half life of the complex when plasma membrane was used is one hour at 37C. At 24C the dissociation rate was much slower, and the apparent half life was more than four hours. At 4C no dissociation could be seen during the time of incubation (over 6 hours). Similar results were obtained with luteal cells. The difference in slopes of the curves observed in the association and dissociation processes as a function of time probably reflected a freeze in the membrane lipids which prevented the changes in receptor conformation associated with binding. Similar observations have been made by others (21).

D. Dissociation Constant of the ^{125}IhCG Receptor Site Complexes

The dissociation constant had been computed from equilibrium data. Data from fig.3A and B had been replotted as double reciprocal curves (fig 9A and B). From the slopes, the intercepts with the vertical axis, the dissociation canstants and the number of binding sites could be calculated. For the luteal cells, KhCG equaled 5.3×10^{-10} M, and the number of binding sites per cell was 5×10^4. For the luteal membranes, KhCG was 3.8×10^{-10} M, and the number of binding sites was 140 fmoles per mg membrane protein. Binding data could also be plotted in the form of a Scatchard plot. This required the assumption that the steady state in binding was achieved after 30 min at 24C. From the slope of such plots, apparent dissociation constants were calculated for the luteal cells or plasma membranes which agreed well with the dissociation constants computed by double inverse plots or computed directly from the rate of association and dissociation of the ^{125}IhCG receptor complex.

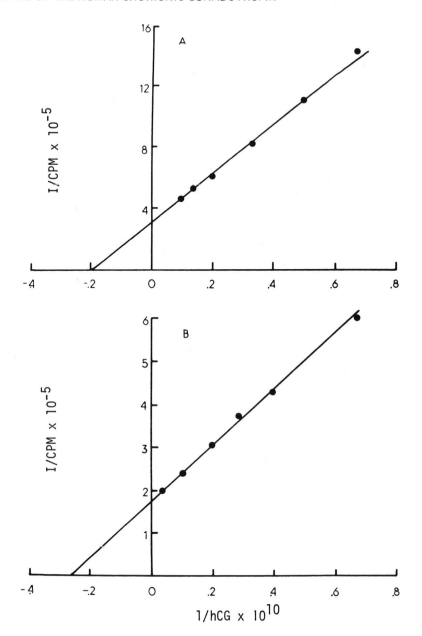

Fig. 9. Double reciprocal plot of data obtained in Fig. 3.
A) Luteal cells, B) plasma membranes. K_DhCG from the slope of the lines is 5.3×10^{-10} M for the luteal cells and 3.8×10^{-10} M for the plasma membranes.

DISCUSSION

The iodination of hCG using lactoperoxidase resulted in a product which was as active as hCG and showed similar physicochemical characteristics. The advantages of the iodination of gonadotropins by the enzymatic radioiodination as compared with the chloamine T method had already been shown by others (16) as well as by us in a previous communication (3).

The ^{125}IhCG thus obtained had a high specific activity (2,125 Ci/mM), and had a low nonspecific binding to EGWP or EAWP filters (0.1% of the input). Also the specific binding of ^{125}IhCG to the tissues was consistently 30 to 50 times than the non-specific binding levels to cells or membranes observed in the presence of an excess of unlabelled hCG (fig 4).

When the binding of ^{125}IhCG to luteal plasma membranes was compared to its binding to luteal cells obtained by enzymatic treatments, similar binding characteristics were observed. The binding process was saturable and the saturation of the receptor sites was obtained at the same concentration of ^{125}IhCG for the membranes and the luteal cells. The dissociation constant obtained for hCG with the luteal cells and luteal membranes was similar: 5.3 and 3.8×10^{-10} M, respectively. Receptor affinity for hCG was about four times higher than that observed for ovine LH (3). This is confirmed by the completition studies which show that for luteal cells, as well as luteal plasma membranes, about 4 to 10 times as much LH as hCG was required to dissociate a given concentration of bound hCG. The observation that gonadotropins could exhibit different affinities for the same receptor had been observed by several people: Rao had shown that bovine LH did not compete as well as hCG for the binding site of gonadotropins present in bovine plasma membranes (4); Lee and Ryan had observed that hCG had the higest affinity for the receptor site of LH followed by human LH and ovine LH (6, 18) and similar observations have been made by Leidenberger and Reichert (22). This can either be due to small structural differences which may be intrinsic to the molecules or caused by proteolytic degradation occuring when the hormone binds to its receptor site (18, 22).

The kinetics of the binding of ^{125}IhCG to the plasma membranes of luteal cells or to the luteal cells had been determined. Association of hCG to the receptor site at

37C took place rapidly. However, the dissociation of the complexes was slower. The half-life was approximately one hour at 37C, a value similar to that obtained by Rao (4) or Haour and Saxena (5). One could argue that the slow process of dissociation of the complexes was due to the entrance of hormone into the cells, and its subsequent entrapment. However, since the dissociation process with the membranes was kinetically similar to that observed with the cells, a more likely interpretation was that the slow process of dissociation reflects the high affinity of hCG for its receptor site. Even slower rate of dissociation for hCG and LH have been observed by Lee and Ryan (18) and Dufau et al. (15).

The bovine luteal cells and luteal membranes, when compared to other tissues, showed a high and specific uptake of hCG, whereas several other cells not expected to bear hCG receptors did not. The number of sites per luteal cell was 4×10^5, approximately 5-fold higher than that observed with the granulosa cells (17). The ability of the plasma membrane to bind hCG (140 fmoles/mg protein) was similar to that reported by others (4,5, 6). It is difficult, however, to compare the number of binding sites per mg of protein with the number of sites per cell since we do not know how much the plasma membrane acounts for in the total protein content of a cell.

In conclusion, our results showed that luteal cells isolated by a collagenase and trypsin treatment have their receptor sites, which are not strikingly different from those of plasma membrane, intact, since they bind hCG in a way similar to that of luteal plasma membranes which have not been trated by proteolytic enzymes. The high specificity of the binding of hCG to luteal cells offers a way to study the modulation of this receptor site when the luteal cells are maintained in tissue culture for prolonged periods of time.

REFERENCES

1. Channing, C.P., Endocrinology 94: 1215, 1974.

2. Gospodarowicz, D. and F.Gospodarowicz, Endocrinology submitted for publication.

3. Gospodarowicz, D., J Biol Chem 248: 5057, 1973

4. Rao, C.H.V., J Biol Chem 249: 2864, 1974.

5. Haour, F. and B.B. Saxena, J Biol Chem 249: 2195, 1974.

6. Lee, C.Y., and R. Ryan, Pro Natl Acad Sci USA 69: 3520, 1972.

7. Gospodarowicz, D., and F. Gospodarowicz, Endocrinology 90: 1427, 1972.

8. Papkoff, H., D. Gospodarowicz, and C.H. Li, Arch Biochem Biophys III: 431, 1965.

9. Papkoff, H., D. Gospodarowicz, A. Candiotti, and C.H. Li, Arch Biochem Biophys III: 431, 1965.

10. Pierce, J.G., and M.E. Carsten, IN Werner, S.G. and C.C. Thomas (eds.) Thyrotropin. Charles C. Thomas, Springfield, Illinois, 1963, p.216.

11. Gospodarowicz, D., J Biol Chem 248: 5050, 1973.

12. Lowry, O.H., N.J. Rosenbrough, A.L. Farr, and R.J. Randall, J Biol Chem 193: 265, 1951.

13. Cuatrecasas, P., Proc Natl Adad Sci USA 68: 1264, 1971.

14. Cuatrecasas, P., J Biol Chem 246: 7265, 1972.

15. Dufau, M.L., T. Tsuruhara, and K.J. Catt, Biochem Biophys Acta 278: 281 - 292, 1972.

16. Miyachi, Y., J.L. Vaitukaitis, E. Nieschlag, and M.B. Lipsett, J Clin Endocrin Metab 34: 23, 1972

17. Kammerman, S., R.E. Canfield, J. Kolena, and C.P. Channing, Endocrinology, 91: 65, 1972.

18. Lee, C.Y., R.J. Ryan, Biochemistry 12: 4609, 1973.

19. Freychet, P., R. Kahn, J. Roth, D.M. Neville, Jr., J Biol Chem 247: 3953, 1972.

20. Pohl, S.L., H.M.J. Krans, L. Birnbaumer, M. Rodbell, J Biol Chem 247: 2295, 1972.

21. Petit, V.A., M. Edidin, Science 184: 1183, 1974.

22. Leidenberger, F., L.E. Reichert, Jr., Endocrinology 92: 646, 1973.

23. Clarke, J.T., Ann NY Adad Sci 121: 248, 1965.

24. Ward, S., D.L. Wilson, J.J. Gilliam, Anal Biochem 38: 90, 1970.

25. Gospodarowicz, D., J Biol Chem 247: 6491, 1974.

ACKNOWLEDGMENT

The authors acknowledge the excellent technical assistance of Mr.J. Krausz. This work was supported by grants from the National Institute of Health (HD08-180) and the Rockefeller Foundation.

Particulate and Solubilized Hepatic Receptors for Glucagon,

Insulin, Secretin and Vasoactive Intestinal Polypeptide *

Melvin Blecher

Department of Biochemistry, Georgetown University

Medical Center, Washington, D. C. 20007

The chemistry and physiological target cells of glucagon, secretin and vasoactive intestinal polypeptide (VIP) were discussed. The characteristics of the interaction of glucagon with its receptor - adenylate cyclase complex in particulate liver plasma membrane (PM) were reviewed; a hypothetical model of this interaction was presented, based upon the Fluid Mosaic Model of Singer and Nicholson. This review established the need to solubilize the glucagon receptor - adenylate cyclase complex in order to elucidate the molecular mechanisms of such interactions.

Details were provided for the solubilization (with Lubrol-PX), identification (gel filtration following complex formation with ^{125}I-glucagon), and purification (3000-fold increase in specific binding activity) of a glucagon binding protein (molecular weight, 190,000; Stoke's radius, 42 Å), many of whose properties (kinetics, saturability, affinity constant of 10^{10} M^{-1}, reversibility, tissue specificity, and hormone specificity) were similar to those exhibited by the parent PM sites.

Partially purified glucagon "receptor" protein preparations also bound insulin, but not ACTH or epinephrine. Binding of glucagon and insulin were to separate, non-interreacting sites, both in the soluble preparation and in particulate PM. Furthermore, insulin neither influenced the adenylate cyclase activity of

---

* The work described in this abstract was supported by grant AM-05475-END from the National Institutes of Health (U.S.A.). The insulin-free glucagon was a gift from the Lilly Research Laboratories, Indianapolis (U.S.A.), while highly purified secretin and VIP were generously provided by Drs. Viktor Mutt and Sami Said of the Karolinska Institut (Stockholm).

particular PM nor inhibited glucagon's ability to activate the enzyme.

VIP and secretin also bound to particulate and soluble sites, but apparently to macromolecules different than those binding glucagon. This was demonstrated in two types of experiments: by direct binding experiments with ^{125}I-labeled hormones using both the particulate PM and partially purified, soluble glucagon "receptor" preparations; and, by competition experiments involving activation of particulate PM adenylate cyclase. The K_m for activation of particulate PM adenylate cyclase was, for secretin and VIP, ten times higher than for native glucagon. These results indicate that secretin and VIP, because of their structural similarities to glucagon, bind to the glucagon receptor; however, their structural differences result in a diminished affinity for the receptor - adenylate cyclase complex and a lesser ability to activate adenylate cyclase.

PROTEIN HORMONES BINDING TO BLOOD AND SPERM CELLS

Julia Vassileva-Popova and Sidney Shulman*

Central Laboratory of Biophysics
Bulgarian Academy of Sciences, 1113 Sofia, Bulgaria
*New York Medical College, New York, N.Y., 100029, USA

The distinction between specific and non-specific binding remains still a stimulating point for discussions in the field of hormonal inforamtion transfer. The character of the cascade answer following in the binding between a regulatory substance with the receptor-biopolimer or artifact is usually accepted as a destinguishing criterion. When we are dealing with a binding known as a specific information transfer, e.g. hormones-target receptors the search for more discrete stages of this binding is quite possible.

The object of this study is to understand some features of the contact between a protein hormone (PH) and free mobile cells (blood and sperm) not recognized as a typical target. The biological meaning of our in vitro study is the fact that in a normal bio-situation there are real conditions for a contact between the PH and the cells investigated. In this respect the present study should be accepted as an attempt for a better understanding of the possible hormone "trapping" sites.

MATERIALS AND METHODS

As protein hormones (PH) the fulicle stimulating hormone (FSH), and prolactin (Pr) i.e. luteotropic hormone (LTH) were used. The hormones were supplied by NIAMD-NIH, Bethesda and NIBSC, London. Unlabeled Pr, FSH and LH were used for the specificity test in a limited number of samples. An isotope non-competitive binding assay was applied. The tracer used for PH was ^{125}I. The labelling procedure was carried out according to Hunter and Greenwood (1962). The methodological advice and the modifications made were taken into account (Yalow and Berson, 1969, Dufau et al., 1972, Edwardson and Vassileva-Popova, 1973, Saxena et al., 1973, 1974).

Separation of the labelled PH from the free iodine and the other compounds was achieved by gel filtration on a Sephadex G-100 column (1 cm x 40 cm) conditioned with 0,14 M sodium chloride and 0,7 per cent bovine serum albumine at pH=7,3.

Two types of free moving cells were used: human blood cells and human spermatozoa. The fraction of blood cells contained all types of normal cells. A laboratory control of the ejaculate and the blood samples were made. The semen samples were estimated for: percentage of cell motility and incubation tests were performed. The concentration of the two types of cells was determined. The cells were washed and centrifugated with the buffer containing 0.1 per cent Knox Gelatine at pH=7,2.

No haemolysis was observed during the experiment. Following the washing of the cells, to the constant amount (10^5) of blood cells and spermatozoa in a volume of 0.5ml increasing concentrations of labelled FSH, respectively Pr, were added. Each value is the average of five samples. After mixing, the samples were incubated for 10 min at 37°C. Immediately after the incubation 1 ml of the same buffer at the same temperature (37°C) was added to each sample. The separation of the cells from the labelled hormone unfixed to the cells was obtained by centrifugation, and the cells were washed twice using the same buffer. The supernatant was aspired and the cell sediment was counted with a gamma scintilation counter.

As a control of the binding test part of the same cell samples were used for a 30 min preincubation at 55°C. It was desided that the comparative analysis of hormonal binding to the constant number of sperm and blood cells should be an informational criterion. Instead if the comparative approach at this stage of the study the replacement competitive test was not carried out with all samples.

RESULTS AND DISCUSSION

The data obtained indicate a significant difference between the amount of labelled FSH and Pr accepted from the constant cell concentration. At the same time there is a significant difference between the cells just washed and the so-called control with a preliminary temperature treatment.

The results are also rather different for FSH and Pr. The amount of Pr fixed on blood cells is greater in comparison with that fixed on the sperm cells. (Fig.1).

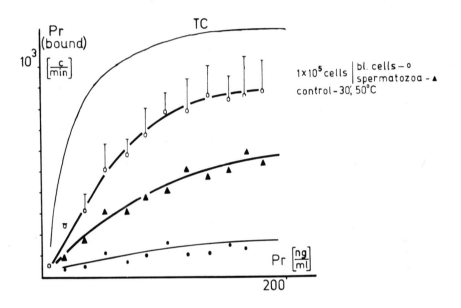

Figure 1

The preliminary temperature treatment significantly reduces the binding of labelled Pr to the cells. It looks that the increase of hormonal concentration corresponds to an increase of the amount of labelled hormone fixed on the cells. At this stage of the study it is not so clear how to explain the binding dissimilarity between the blood and sperm cells. Possibly the unequal physico-chemical and morphological nature of the cell membrane are also the reason for the different binding behaviour of these two types of cells.

Despite the not so negligible deviation in some points on the curves for the different hormonal concentrations, the assay performed (non-competitive binding test) indicated a relatively high Pr acceptance from the blood cells. It should be expected that blood cells have a trapping role for the protein hormones, and this effect should have a reflection upon the dynamic balance of the hormonal concentration in blood circulation.

In this respect, because of the cross reaction and a kind of similarity between Pr and GH the study of the character of the contact between the GH and the blood cells should be of interest. Unfotunately, we have no data from research in pathology for the further explanation of the hormonal binding at he blood cells.

The picture is quite different for the binding of FSH (in comparison to Pr) for the types of cells investigated (Fig.2). The sperm cells indicate a higher binding property than the blood cells. The temperature treated cells (so-called control) also show a lower binding when compared to the intact washed cells.

The results obtained indicate that the blood cells are not completely indifferent to the acceptance of FSH which is

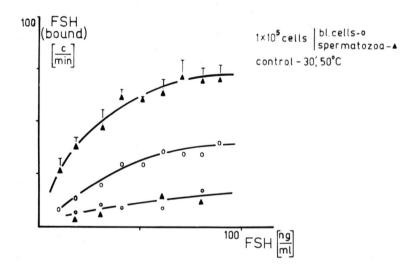

Figure 2

proportional to the amount of labelled hormone. Despite the deviation in curve points the dissimilarity between the two types of cells for FSH binding are present there. The sperm cells show a higher ability for acceptance of FSH in comparison to the blood cells. In both cases the amounts of hormone accepted from the two types of free cells correspond in some way to the hormonal concentration with the tendency for an equilibrium in the system. The possible way of explaining these results should be the mechanism of hormonal action on one hand and the difference in the membrane structure (Friend and Fawcett, 1974, 1975) on the other.

It is known that FSH has an important role on the testis germinal epithelium (McLeod, 1973) when the LH is a typical steroidogenetic hormone. Since sex steroid mediation is required for the FSH action, no clear distinction between the steroidogenetic function of FSH and LH exists, possibly because of the usual contamination of one of the hormones by the traces of the other. The explanation of the results obtained could be found mainly in the role of FSH in the spermatogenetic cycle.

As far as the FSH action is directed to the germinal epithelium it might be assumed that there is a possibility for some of the FSH receptors to remain in older spermatogenetic forms and possibly also in mature sperm cells. Another question should be: if the FSH receptors remain in mature spermatozoa after about a 75-day cycle then does the FSH binding indicate just a trace of FSH receptors or this hormone has some functional role in mature sperm cells. If we are dealing with mature spermatozoa with traces of FSH receptors from immature forms the binding ability will correspond to the age stage of the cell spermatogenetic milieu. In this connection it might be expected that the FSH binding test is an indication for the presence of the youngest sperm cells to be distinguished from the sperm cells with a long time epididymal storage. The recent data about the role of FSH in the formation of androgen binding proteins (French et al., 1974) will possibly suggest some other explanations. In this respect the work on the induction of infertility in men (Briggs et al., 1974, 1975) has to be taken into consideration.

Analyzing the possible reason for the sperm and blood cells binding to FSH and Pr it seams that this process depends upon: the physico-chemical nature of the cell membrane; the cell dimension; the biological significance of the hormonal binding to the cells investigated. It is not possible to neglect the sticking feature of the proteins and the so-called non-specific binding itself.

As a conclusion two points should be noted:
- the possible role of blood cells in the hormonal trapping mechanism and in the dynamics of the bound and free hormonal state
- the chance for a correspondence between the age stage or other characteristics of the cells investigated and the hormonal binding.

ACKNOWLEDGEMENT

We would like to thank for the help and facilities provided by the Population Council at the Rockefeller University, and to Dr. B. McEwen and Dr. G. Boll at the Rockefeller University.

REFERENCES

Briggs, M., and Briggs, M., Nature, Lond., 252, 585 (1974)

Briggs, M., and Briggs, M., Research in Production (IPPF - London) 7, 1, (1975).

Dufau, M.L., Catt, K.J., Tsurahara, T., Proc. Nat. Acad. Sci. USA, 69, 2414, (1972).

Edwardson, J.A., and Vassileva-Popova, J.G., in: Immunology of Reproduction, Proc. of the Second Int. Symp. held in Varna, Sept. 13-16, 1971, Bulg. Acad. Sci., Sofia, 1973.

French, F.S., Nayfer, S.N., Ritzen, E.M., Hanson, V., Research in Reproduction (IPPF - London), 6, 4, (1974)

Friend, D.S., Fawsett, D.W., J. Cell. Biol., 63, 6H1, (1974)

Friend, D.S., Fawsett, D.W., Research in Reproduction (IPPF - London), 7, 1, (1975).

Hunter, W.M., Greenwood, F.C., Nature, 194, 495, (1962).

McLeod, J., in: Physiology and Biophysics, T.C. Ruch and H.D. Patton Eds., W.B. Saunders Company, Philadelphia-London-Toronto, 1973.

Saxena, B.B., Rathnam, P., Römmler, A., Endocr. Exptl., 7, 19, (1973)

Saxena, B.B., Hasan, S.H., Haour, F., Schmidt-Gollwitzer, M., Science, 184, 793, (1974).

Yalow, R.S., and Berson, S.A., in: Protein and Polypeptide Hormones Proc. of the 18th Int. Symp. of Liege, Ed. M., Margoulies, Exerpta Medica Foundation, Amsterdam, 1969.

Small Molecules, Steroids, and Receptor Interactions

RECEPTOR TRANSFORMATION -- A KEY STEP IN ESTROGEN ACTION

E. V. Jensen, S. Mohla, T. A. Gorell and E. R. DeSombre

Ben May Laboratory and Department of Biophysics

University of Chicago, Chicago, Illinois 60637

INTRODUCTION

Experiments from many laboratories (1,2) have established that, for all classes of steroid hormones, the hormone-receptor complex found in the nuclei of target cells is derived from an initially formed extranuclear complex, as first proposed for estrogens in uterine tissue (3,4). Evidence is accumulating that the steroid-induced translocation of the hormone-receptor complex to the nucleus is accompanied by a hormone-mediated alteration of the receptor protein, a process that has been called receptor transformation (5-7) or activation (8,9). This paper summarizes some observations concerning transformation of estrophilin, the receptor protein for estrogens, and describes recent achievements in the purification of the transformed estradiol-receptor complex of calf uterus.

CRITERIA FOR RECEPTOR TRANSFORMATION

Sedimentation Properties

Experiments leading to the recognition of the transformation phenomenon followed important observations by Erdos (10) and by Korenman and Rao (11) that, in the presence of 0.3 or 0.4 M KCl, the 8S estradiol-receptor complex of rat uterine cytosol, originally described by Toft and Gorski (12), is dissociated into a steroid-binding subunit that sediments at approximately the same rate as the complex previously observed in extracts of rat uterine nuclei (13,14). By careful ultracentrifugation in salt-containing sucrose gradients, we could show that these two complexes are not identical (15); the nuclear complex sediments at about 5.2S, slightly ahead of a 4.6S bovine plasma albumin marker, whereas the cytosol complex sediments at about 3.8S, slightly behind albumin. For convenience these two entities are designated 5S and 4S, respectively. The foregoing observations necessitated modification (15) of the original proposal (4) for the interaction of estradiol in uterine cells; instead of being formed by degradation of the 8S complex, it now appeared that the 5S nuclear complex is produced by some more subtle alteration of the 4S unit of the cytosol receptor protein (Fig. 1).

One of the main pieces of evidence for the two-step interaction mechanism outlined in Fig. 1 is the fact that a 5S complex, indistinguishable from that formed _in vivo_, can be extracted from isolated uterine nuclei after they have been incubated with estradiol in the presence but not in the absence of uterine cytosol (4). Accordingly, the conversion of the 4S binding unit of the cytosol receptor to the 5S nuclear form was first assumed to take place in the nucleus, and it was proposed that the difference between the

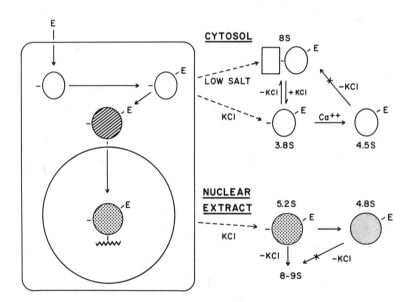

Fig. 1. Schematic representation of interaction pathway of estradiol (E) in uterine cell. Diagram at left indicates extranuclear estradiol-receptor complex undergoing transformation and entering nucleus to bind to chromatin. Diagrams at right indicate sedimentation properties of complexes extracted from the cell, before and after losing the ability to aggregate in media of low ionic strength. Reproduced from (28).

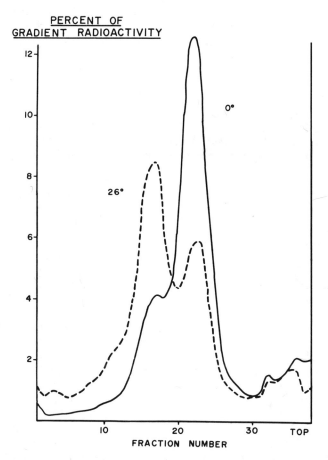

Fig. 2. Temperature-dependent transformation of the estradiol-receptor complex of rat uterine cytosol. Uterine horns from 22-day old rats were homogenized in nine volumes of 10 mM tris buffer, pH 7.5. The cytosol fraction was made 2 nM in estradiol-^{3}H and portions incubated for 30 min at either 26° or 0°C, after which 200-µl aliquots were layered on cold 5 to 20 percent sucrose gradients containing 400 mM KCl, 10 mM tris buffer and 1.5 mM EDTA and centrifuged at 2°C for 15 hr at 308,000 g. Successive 100 µl fractions were collected from the bottom for determination of radioactivity. Reproduced from (5).

nuclear and the cytosol complexes might be the incorporation of some nuclear component (15). This hypothesis was abandoned when it was found that transformation of the cytoplasmic receptor protein to the nuclear form does not require the presence of nuclei (5,16). If uterine cytosol containing tritiated estradiol is warmed at temperatures of 20-37°C and then examined for sedimentation behavior on salt-containing sucrose gradients, there is a gradual appearance of a 5S sedimentation peak accompanied by disappearance of the 4S complex until finally all the 4S receptor has been converted to the 5S form (Fig. 2). This transformation of the estrophilin requires that it be bound to estrogen; it does not take place on warming the cytosol in the absence of hormone. At low concentrations where estradiol is effective, estrone does not induce transformation, although it has been found to do so when present in higher concentrations.

Hormone-induced conversion of the receptor protein from the native (4S) to the transformed or nuclear (5S) form is strongly temperature-dependent, especially under conditions of low ionic strength where it proceeds readily at temperatures of 25° to 37°C but to no appreciable extent in the cold (Fig. 2). It is retarded by the presence of Ca^{++}, Mg^{++} and Mn^{++} ions as well as by EDTA, and it is significantly enhanced by the presence of salt. It is accelerated with increasing pH over the range 6.5 to 8.5. The molecular basis of the transformation process is not clearly understood, but under certain conditions it appears to follow second-order kinetics involving association of the binding unit of the receptor protein with some entity having a molecular weight of about 45,000 to 50,000 daltons (17,18). Second order kinetics, interpreted as dimerization, were similarly observed for the estradiol-induced, temperature-dependent conversion from a 3.5S to a 4.5S form of the estrogen-receptor complex extracted from the microsomal fraction of pig uteri (19).

Binding to Nuclei and Chromatin

A second characteristic differentiating the transformed from the native state of the estradiol-receptor complex is its ability to bind to isolated nuclei (20) or to preparations of chromatin (21, 22). As shown in Fig. 3--left, when incubated with estradiol-cytosol mixtures, sucrose-purified uterine nuclei show little incorporation of the native complex at 0°C, in contrast to a large uptake at 25°C where transformation to the 5S form accompanies nuclear incorporation. But if the estradiol-cytosol mixture is first warmed to 25°C to effect receptor transformation, subsequent incubation with nuclei at either 0° or 25°C results in a substantial uptake of the transformed complex. Thus, receptor transformation appears to be a prerequisite for binding of estrophilin to isolated nuclei and probably represents the temperature-dependent process associated with nuclear incorporation in whole uterine tissue.

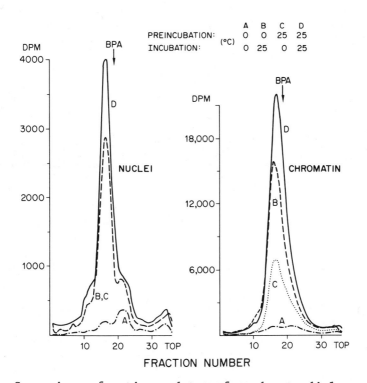

Fig. 3. Comparison of native and transformed estradiol-receptor complexes in their binding to calf endometrial nuclei (left) or to chromatin (right). Nuclei, isolated from 2.2 M sucrose, were incubated for 60 minutes at either 0°C (A,C) or 25°C (B,D) with endometrial cytosol (from a 20% homogenate in 0.32 M sucrose) that previously had been incubated with 5.6 nM estradiol-^{3}H for 45 minutes at either 0°C (A,B) or 25°C (C,D). Endometrial chromatin was resuspended with 1.5 mM NaCl in 0.15 mM citrate buffer, pH 7.0, and 1.0 ml portions of the suspension, containing 2 mg DNA, were incubated with endometrial cytosol (in 10 mM tris, pH 7.4) containing 10 nM estradiol-^{3}H in the manner described for nuclei except that preincubation at 25°C to effect receptor transformation (C,D) was for 60 minutes. After incubation with cytosol, the nuclei or chromatin were separated by centrifugation, washed (in 0.32 M sucrose for nuclei, 150 mM NaCl for chromatin), and extracted with 0.4 M KCl. A 200-μl portion of each extract was centrifuged in a 5 to 20 percent sucrose gradient containing 0.4 M KCl, and 100-μl fractions collected from the bottom. BPA indicates position of bovine plasma albumin (4.6S) marker. Reproduced from (22).

An analogous difference between the native and transformed complexes is seen in their binding to uterine chromatin (Fig. 3--right). As was also observed by McGuire et al. (21), temperature appears to exert a greater effect on the binding of pretransformed estradiol-receptor complex to isolated chromatin than to whole nuclei. When it was found that comparable transformation of the glucocorticoid-receptor complex of liver cytosol enhances its affinity, not only for nuclei, chromatin and DNA, but also for synthetic polyanions, such as carboxymethyl-Sephadex or sulfopropyl-Sephadex, it was suggested (23) that receptor transformation may involve the addition or exposure of acidophilic groups in the protein molecule. The reported (9) change from 5.8 to 6.5 in the isoelectric point of the androgen-receptor complex of rat prostatic cytosol as it acquires nuclear binding properties on warming at 30°C is consistent with this possibility.

Influence on RNA Polymerase

A third property of the transformed estradiol-receptor complex is its ability to increase the RNA polymerase activity of isolated uterine nuclei. As has been considered in more detail in several review articles (24-28), an early response of the rat uterus to estrogenic stimulation is enhanced biosynthesis of RNA, indicated by increased incorporation of labeled precursors, both in vivo and in surviving uterine tissue excised from estrogen-treated animals. It was also found that nuclei prepared from uteri excised one to four hours after the administration of estradiol to immature or castrate rats show RNA polymerase activity that is two to three times that of uterine nuclei from control animals (29-33). The stimulation appears to be predominantly of the nucleolar enzyme (polymerase I), inasmuch as the effect of estrogen is seen when the polymerase assay is carried out in the presence of α-amanitin or with Mg^{++} in low salt but not with Mn^{++} in high salt. The composition of the RNA produced by the stimulated nuclei is of a ribosomal type in that its nearest neighbor frequency spectrum shows an increased number of cytidine and guanosine linkages as compared to adenosine and uridine (34-36). More recently it was found that the nucleoplasmic enzyme (polymerase II) also is stimulated but at an earlier time, 15 to 30 minutes after the administration of estrogen (37).

After attempts to influence RNA synthesis by direct treatment of uterine nuclei with estradiol proved unsuccessful (31), an important advance was provided by the discovery by Raynaud-Jammet and Baulieu (38) that the RNA polymerase activity of isolated heifer endometrium nuclei, though not affected by estradiol alone, is increased two- to three-fold when the nuclei are incubated with estradiol in the presence of endometrial cytosol containing the extranuclear receptor. Further studies (39,40) showed that the

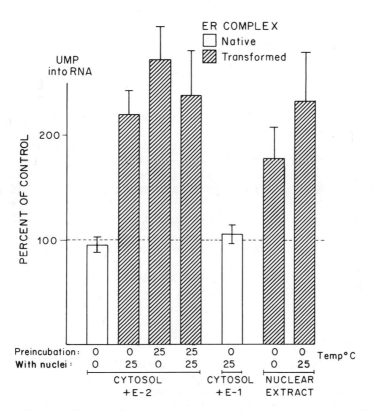

Fig. 4. Effect of transformed as compared to native estradiol-receptor complex on RNA synthesis in calf endometrium nuclei. Purified nuclei, from a homogenate of calf endometrium in 2.2 M sucrose containing 1 mM $MgCl_2$, were incubated for 30 minutes at either 0°C or 25°C with the cytosol fraction made 10 nM in estradiol (E-2) or estrone (E-1), or for 45 minutes with an extract of crude calf uterine nuclei previously exposed to 10 nM estradiol in uterine cytosol for 1 hour at 25°C. In two experiments with E-2, receptor transformation was effected prior to incubation with nuclei by warming the hormone-cytosol mixture at 25°C for 30 minutes. Before incubation with nuclei, the nuclear extract in 0.4 M KCl was concentrated (Diaflo XM-50 membrane) and then diluted with 2.2 M sucrose to an estrophilin concentration equivalent to that of the cytosol. After incubation, the nuclei were separated and resuspended in 0.32 M sucrose for assay of magnesium-dependent RNA polymerase, using UTP-^{3}H as the radioactive nucleotide. Control values were obtained without steroid, except for nuclear extract where heat-inactivated (50°C, 15 minutes) extract was used as a control. Results are based on seven replicate determinations. Reproduced from (22).

RNA polymerase of heifer endometrium nuclei, or of the enzyme extracted from these nuclei, could be enhanced by adding estradiol and certain uterine extracts directly to the polymerase assay system. As is the case when estradiol is administered in vivo (29), the susceptibility of RNA synthesis to stimulation by exposure to the estradiol-receptor complex is a specific characteristic of nuclei from hormone-dependent tissues such as uterus (41).

To produce the activating effect on RNA polymerase, the estrophilin complex must be in its transformed state (41). As shown in Fig. 4, enhancement is seen when incubation of calf endometrial nuclei with estradiol and endometrial cytosol is carried out at 25°C where conversion of the complex to the 5S form takes place but not at 0°C where the receptor remains in the 4S form. If the estradiol-cytosol mixture is first warmed to 25°C to transform the complex, subsequent incubation with nuclei at either 0° or 25°C leads to polymerase stimulation. Under conditions where estradiol induces receptor transformation but estrone does not, only estradiol causes increased activity of nuclear polymerase. Fig. 4 also shows that the 5S estradiol-receptor complex extracted from uterine nuclei is active in stimulating fresh nuclei on incubation at either 0° or 25°C.

The enhancement by estrogen of RNA synthesis in uterine nuclei appeared to involve effects both on chromatin template activity and on the enzyme itself (28). When calf endometrial chromatin is incubated with estrophilin complex and its template function then assayed using solubilized polymerase enzyme prepared from endometrium nuclei, there is an enhancement of RNA synthesis comparable to that seen when whole nuclei are exposed to transformed complex. However, increased chromatin template activity is not the sole explanation for the stimulatory effect of estradiol either in vivo or in vitro. When assayed with calf thymus DNA as template, the activity of the polymerase enzyme solubilized from uterine nuclei of estradiol-injected rats is found to be significantly enhanced. Similar stimulation is observed in the soluble portion of the polymerase enzyme of calf uterine nuclei after they have been exposed to transformed estrophilin complex in vitro (41).

The effect on RNA polymerase by treatment of isolated uterine nuclei with transformed estradiol-receptor complex resembles that produced by hormone administration in vivo in qualitative as well as in quantitative aspects. As seen in Fig. 5, nearest neighbor frequency analysis shows that the RNA synthesized by isolated uterine nuclei after exposure to transformed complex contains an increased proportion of cytidine and guanosine linkages (22), resembling the pattern observed by others after injecting hormone in vivo (34-36). Thus, the tissue-specific effects of the hormone given in vivo on the RNA synthesizing system of uterine nuclei are remarkably similar to those elicited by treatment of the nuclei

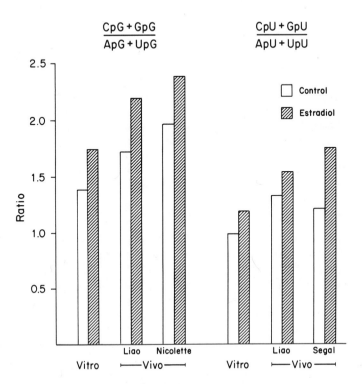

Fig. 5. Effect of estradiol-receptor complex on nearest neighbor frequency in RNA synthesized by calf endometrial nuclei using either α-^{32}P-labeled GTP or UTP as the reference nucleotide. A 20 percent homogenate of calf endometrium in 2.2 M sucrose containing 10 mM $MgCl_2$ was incubated for 60 minutes at 25°C in the presence or absence of 10 nM estradiol, after which the nuclei were separated and taken up in 0.32 M sucrose for synthesis of RNA in a polymerase assay system and determination of base linkages in the product by the procedure of Liao et al. (47). Changes in base-pair ratios are compared with those observed by Barton and Liao (34), Nicolette and Babler (36) and Trachewsky and Segal (35) with uterine nuclei obtained from rats given estradiol in vivo. Reproduced from (22).

with transformed estrogen-receptor complex in vitro. These findings suggest that the stimulation of isolated nuclei with the estrophilin complex has physiologic relevance and that estrogen-induced transformation of the receptor protein to a form that can interact with the genome is an important step in the biologic action of the hormone.

PURIFICATION OF ESTROPHILIN

A clearer understanding of the biochemical role of estrophilin in estrogen action and the chemical basis of the 4S to 5S transformation should be possible when the various forms of the receptor substance are available in pure state in amounts sufficient to permit determination of chemical composition and biochemical properties. We have now succeeded in isolating, in highly purified form, microgram amounts of both a calcium-stabilized modification of the estradiol-receptor complex of calf uterine cytosol and the nuclear complex obtained by incubating calf uterine nuclei with estradiol in uterine cytosol.

Because of their instability and tendency to aggregate, isolation of receptor proteins has proved difficult. Purification of the binding subunit of the extranuclear protein was facilitated by the finding (42,43) that addition of calcium ions to the salt-dissociated complex of uterine cytosol, prepared in the presence of EDTA, converts the binding unit to a stabilized 4.5S form that does not revert to the 8S form when salt is removed and which is resistant to aggregation. Although the 4.5S complex no longer undergoes transformation to the nuclear form, it provides the basic binding unit of the cytosol receptor in a form that can be purified by conventional techniques of protein chemistry. The 5S complex extracted from the nucleus, which aggregates to an 8S to 9S form when salt is removed, is not influenced by calcium ions, possibly because the calcium-sensitive factor, responsible for the stabilization of the cytosol complex (44), is not present in the nuclear extract. However, during the course of purification the nuclear complex loses its tendency to aggregate in low salt, a change that is accompanied by a decrease in its sedimentation coefficient to 4.8S. The molecular relation of these stabilized complexes to their aggregating counterparts, represented schematically in Fig. 1, is not yet understood. As indicated in Fig. 6, the stabilized forms of the cytosol and nuclear complexes can be readily differentiated, both by careful sedimentation in sucrose gradients and by acrylamide gel electrophoresis where the nuclear complex migrates nearly twice as fast as the cytosol complex (20).

By a modification of the previously described purification sequence (42,45), the calcium-stabilized (4.5S) estradiol-receptor

complex of calf uterine cytosol has been obtained in a form that
shows a single radioactive protein band on gel electrophoresis at
polyacrylamide concentrations ranging from 3 to 20 percent. By
isoelectric focusing on acrylamide gel, this product shows an isoelectric point of 6.8. On SDS gel electrophoresis, several proteins
bands are evident, the significance of which is not yet clear.

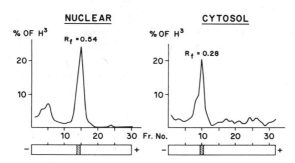

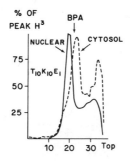

Fig. 6. Comparison of electrophoretic and sedimentation properties
of purified estradiol-estrophilin complexes from calf uterus. (A):
Preparations of calcium-stabilized (4.5S) and nuclear (4.8S) complexes, purified by salt precipitation, gel filtration, ion-exchange
chromatography and polyacrylamide gel electrophoresis, as described
previously (20), were subjected to polyacrylamide gel electrophoresis in duplicate; one gel was sliced, dried and combusted to give
the tritium distribution, and the other stained with amido black
to give the protein distribution as shown directly below. (B): A
third portion (200 μl) of each preparation was layered on a 10 to
30 percent sucrose gradient containing 10 mM KCl; after centrifugation for 16 hours at 300,000 g (nuclear) or 308,000 g (cytosol)
the sedimentation patterns were compared with that of a bovine plasma albumin (BPA) marker. Reproduced from (20).

More success has been achieved in the purification of the nuclear estradiol-receptor complex (28,46). By a sequence of ammonium sulfate precipitation, gel filtration through Sephadex G-200 in 0.4 M KCl (during which the complex loses its aggregation properties), electrophoresis of the desalted product in polyacrylamide gel at pH 9.2, and filtration through Sephadex G-200 in 0.03 M KCl, we have obtained a few hundred micrograms of highly purified preparations of the 4.8S form of estrophilin, extracted from calf uterine nuclei after incubation with tritiated estradiol in calf uterine cytosol. This, product, with isoelectric point 5.8, shows a single stained protein band on acrylamide gel electrophoresis, and analytical utracentrifugation indicates a single, ultraviolet absorbing macromolecular species (Fig. 7). Elution from a calibrated Sephadex G-200 column indicates a Stokes radius of 36.5, corresponding to a molecular weight of 66,000 daltons. The tentative aminoacid composition of this product is shown in Table 1. Rabbits have been immunized with the purified estrophilin complex in the hope that antibodies thus obtained will prove useful both in evaluating the homogeneity of the purified product and for the purification of aggregating forms of the receptor protein.

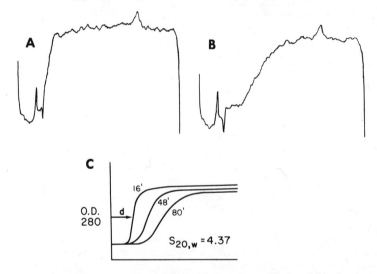

Fig. 7. Homogeneity on ultracentrifugation of purified estradiol-estrophilin complex from calf uterine nuclei. Scanner traces are of absorption at 280 nm of a solution of the complex (230 µg/ml in 50 mM tris buffer, pH 7.4, containing 0.3 M KCl), taken after (A) 16 minutes and (B) 64 minutes of sedimentation in a Spinco Model E analytical ultracentrifuge at 60,000 rpm at 11.4°C. Sedimentation is to the right. (C) Idealized curves of scanner traces after 16, 48, and 80 minutes to obtain the distance of migration, d, of the sample meniscus. A sedimentation coefficient, corrected to that in water at 20°C, was calculated to be 4.37S. Under similar conditions, bovine plasma albumin gave an $s_{20,w}$ of 4.3S. Reproduced from (28).

TABLE 1. AMINOACID COMPOSITION OF PURIFIED, NON-AGGREGATING ESTRADIOL-RECEPTOR COMPLEX FROM CALF UTERINE NUCLEI

CMCYS	6	MET	8
ASP	56	ILE	22
THR	30	LEU	46
SER	50	TYR	15
GLU	89	PHE	16
PRO	39	HIS	11
GLY	67	LYS	28
ALA	42	ARG	23
VAL	34	TRY	Not done

Determined on a Durrum D-500 aminoacid analyzer after reduction and carboxymethylation of cystine residues and acid hydrolysis with 0.1% phenol to protect tyrosine residues.

SUMMARY

The interaction of estradiol with uterine cells involves the association of the hormone with an extranuclear receptor protein known as estrophilin, followed by temperature-dependent translocation of the resulting complex to the nucleus. During this process, the steroid binding unit of estrophilin undergoes an alteration, called receptor transformation, that can be recognized by an increase in its sedimentation rate and by its acquisition of an ability to bind to isolated uterine nuclei and to enhance the RNA synthesizing capacity of such nuclei. Non-aggregating modifications of both the extranuclear and nuclear forms of estrophilin have been extensively purified, with the nuclear form obtained in what appears to be a homogeneous state.

ACKNOWLEDGMENTS

Investigations in our laboratory were supported by research grants or contracts from the Ford Foundation (690-0109), the American Cancer Society (BC-86), and the U.S. Public Health Service (CA-02897, HD 9-2109, and HD-07110).

REFERENCES

1. Raspé, G. (ed.): Schering Workshop on Steroid Hormone Receptors. Advances in the Biosciences, Vol. 7, Pergamon-Vieweg, Braunschweig, 1971.
2. Jensen, E.V., and DeSombre, E.R., Mechanism of Action of the Female Sex Hormones, Ann. Rev. Biochem., 41: 203-230, 1972.
3. Gorski, J., Toft, D., Shyamala, G., Smith, D., and Notides, A., Hormone Receptors: Studies on the Interaction of Estrogen with the Uterus, Recent Progr. Hormone Res., 24: 45-80, 1968.
4. Jensen, E.V., Suzuki, T., Kawashima, T., Stumpf, W.E., Jungblut, P.W., and DeSombre, E.R., A Two-Step Mechanism for the Interaction of Estradiol with Rat Uterus, Proc. Nat. Acad. Sci. U.S.A., 59: 632-638, 1968.
5. Jensen, E.V., Numata, M., Brecher, P.I., and DeSombre, E.R., Hormone-Receptor Interaction as a Guide to Biochemical Mechanism, In Smellie, R.M.S. (ed.): The Biochemistry of Steroid Hormone Action, Biochemical Society Symposium No. 32, Academic Press, London, pp. 133-159, 1971.
6. Munck, A., Wira, C., Mosher, K.M., Hallahan, C., and Bell, P.A., Glucocorticoid-Receptor Complexes and the Earliest Steps in the Action of Glucocorticoids on Thymus Cells, J. Steroid Biochem., 3: 567-578, 1972.
7. Jensen, E.V., and DeSombre, E.R., Estrogen-Receptor Interaction, Science, 182: 126-134, 1973.
8. Higgins, S.J., Rousseau, G.G., Baxter, J.D., and Tomkins, G.M., Early Events in Glucocorticoid Action. Activation of the Steroid Receptor and Its Subsequent Specific Nuclear Binding Studied in a Cell-Free System. J. Biol. Chem. 248: 5866-5872, 1973.
9. Mainwaring, W.I.P., and Irving, R., The Use of Deoxyribonucleic Acid-Cellulose Chromatography and Isoelectric Focusing for the Characterization and Partial Purification of Steroid-Receptor Complexes, Biochem. J., 134: 113-127, 1973.
10. Erdos, T., Properties of a Uterine Oestradiol Receptor, Biochem. Biophys. Res. Commun., 32: 338-343, 1968.
11. Korenman, S.G., and Rao, B.R., Reversible Disaggregation of the Cytosol-Estrogen Binding Protein of Uterine Cytosol, Proc. Nat. Acad. Sci. U.S.A., 61: 1028-1033, 1968.
12. Toft, D., and Gorski, J., A Receptor Molecule for Estrogens: Isolation from the Rat Uterus and Preliminary Characterization, Proc. Nat. Acad. Sci. U.S.A., 55: 1574-1581, 1966.
13. Jensen, E.V., DeSombre, E.R., Hurst, D.J., Kawashima, T., and Jungblut, P.W., Estrogen-Receptor Interactions in Target Tissues, Arch. Anat. Micr. Morph. Exp., 56 (Suppl): 547-569, 1967.
14. Puca, G., and Bresciani, F., Receptor Molecule for Oestrogens from Rat Uterus, Nature, 218: 967-969, 1968.
15. Jensen, E.V., Suzuki, T., Numata, M., Smith, S., and DeSombre, E.R., Estrogen-Binding Substances of Target Tissues, Steroids, 13: 417-427, 1969.

16. Gschwendt, M., and Hamilton, T.H., The Transformation of the Cytoplasmic Oestradiol-Receptor Complex into the Nuclear Complex in a Uterine Cell-Free System, Biochem. J. 128: 611-616, 1972.
17. Yamamoto, K.R., and Alberts, B.M., In Vitro Conversion of Estradiol-Receptor Protein to its Nuclear Form: Dependence on Hormone and DNA, Proc. Nat. Acad. Sci. U.S.A., 69: 2105-2109, 1972.
18. Notides, A.C., and Nielsen, S., The Molecular Mechanism of the in Vitro 4S to 5S Transformation of the Uterine Estrogen Receptor, J. Biol. Chem., 249: 1866-1873, 1974.
19. Little, M., Szendro, P.I., and Jungblut, P.W., Hormone-Mediated Dimerization of Microsomal Estradiol Receptor, Hoppe-Seyler's Z. Physiol. Chem., 354: 1599-1610, 1973.
20. Jensen, E.V., Mohla, S., Gorell, T., Tanaka, S., and DeSombre E.R., Estrophile to Nucleophile in Two Easy Steps, J. Steroid Biochem., 3: 445-458, 1972.
21. McGuire, W.L., Huff, K., and Chamness, G.C., Temperature-Dependent Binding of Estrogen Receptor to Chromatin, Biochemistry, 11: 4562-4565, 1972.
22. Jensen, E.V., Brecher, P.I., Mohla, S., and DeSombre, E.R., Receptor Transformation in Estrogen Action, Acta Endocrinol., 191(Suppl): 159-171, 1974.
23. Milgrom, E., Atger, M., and Baulieu, E.E., Acidophilic Activation of Steroid Hormone Receptors, Biochemistry, 12: 5198-5205, 1973.
24. Hamilton, T.H., Teng, C.S., Means, A.R., and Luck, D.N., Estrogen Regulation of Genetic Transcription and Translation in the Uterus, in McKerns, K.W. (ed.): The Sex Steroids Molecular Mechanisms, Appleton-Century-Crofts, New York, pp. 197-240, 1971.
25. Williams-Ashman, H.G., and Reddi, A.H., Actions of Vertebrate Sex Hormones, Ann. Rev. Physiol. 33: 31-82, 1971.
26. Mueller, G.C., Vonderhaar, B., Kim, U.H., and LeMahieu, M., Estrogen Action: An Inroad to Cell Biology, Recent Progr. Hormone Res. 28: 1-49, 1972.
27. O'Malley, B.W., and Means, A.R. Female Steroid Hormones and Target Cell Nuclei, Science 183: 610-620, 1974.
28. Jensen, E.V., Mohla, S., Gorell, T.A., and DeSombre, E.R., The Role of Estrophilin in Estrogen Action, Vitamins and Hormones, 32: 89-127, 1974.
29. Gorski, J., Noteboom, W.D., and Nicolette, J.A., Estrogen Control of the Synthesis of RNA and Protein in the Uterus, J. Cell. Comp. Physiol., 66: 91-109, 1965.
30. Nicolette, J.A., LeMahieu, M.A., and Mueller, G.C., A Role of Estrogens in the Regulation of RNA Polymerase in Surviving Rat Uteri, Biochim, Biophys. Acta, 166: 403-409, 1968.
31. Hamilton, T.H., Widnell, C.C., and Tata, J.R., Synthesis of Ribonucleic Acid during Early Estrogen Action, J. Biol. Chem. 243: 408-417, 1968.

32. Barry, J., and Gorski, J., Uterine Ribonucleic Acid Polymerase, Effect of Estrogen and Nucleotide Incorporation into 3' Chain Termini, Biochemistry, 10: 2384-2390, 1971.
33. Raynaud-Jammet, C., Biéri, F., and Baulieu, E.E., Effects of Oestradiol, α-Amanitin and Ionic Strength on the in Vitro Synthesis of RNA by Uterus Nuclei, Biochim. Biophys. Acta, 247: 355-360, 1971.
34. Barton, R.W., and Liao, S., A Similarity in the Effect of Estrogen and Androgen on the Synthesis of Ribonucleic Acid in the Cell Nuclei of Gonadohormone-Sensitive Tissues, Endocrinology, 81: 409-412, 1967.
35. Trachewsky, D., and Segal, S., Differential Synthesis of Ribonucleic Acid in Uterine Nuclei: Evidence for Selective Gene Transcription Induced by Estrogens, Europ. J. Biochem., 4: 279-285, 1968.
36. Nicolette, J.A., and Babler, M., The Regulatory Role of Protein in the Estrogen-Stimulated Change in the Nearest-Neighbor Frequency of Rat Uterine RNA Synthesized in Vitro, Arch. Biochem. Biophys., 149: 183-188, 1972.
37. Glasser, S.R., Chytil, F., and Spelsberg, T.C., Early Effects of Oestradiol-17β on the Chromatin and Activity of the Deoxyribonucleic Acid-Dependent Ribonucleic Acid Polymerases (I and II) of the Rat Uterus, Biochem. J., 130: 947-957, 1972.
38. Raynaud-Jammet, C., and Baulieu, E.E., Action de l'Oestradiol in Vitro: Augmentation de la Biosynthèse d'Acide Ribonucléique dans les Noyaux Utérins, C.R. Acad. Sci.[D](Paris), 268: 3211-3214, 1969.
39. Beziat, Y., Guilleux, J.C., and Mousseron-Canet, M., Effet de l'Oestradiol et ses Récepteurs sur la Biosynthèse du RNA dans les Noyaux Isoles de l'Utérus de Génisse. C.R. Acad. Sci.[D] (Paris), 270: 1620-1623, 1970.
40. Arnaud, M., Beziat, Y., Guilleux, J.C., Hough, A., Hough, D., and Mousseron-Canet, M., Les Récepteurs de l'Oestradiol dans l'Utérus de Génisse. Stimulation de la Biosynthèse de RNA in Vitro, Biochim. Biophys. Acta, 232: 117-131, 1971.
41. Mohla, S., DeSombre, E.R., and Jensen, E.V., Tissue-Specific Stimulation of RNA Synthesis by Transformed Estradiol-Receptor Complex. Biochem. Biophys. Res. Commun., 46: 661-667, 1972.
42. DeSombre, E.R., Puca, G.A., and Jensen, E.V., Purification of an Estrophilic Protein from Calf Uterus, Proc. Nat. Acad. Sci. U.S.A., 64: 148-154, 1969.
43. Puca, G.A., Nola, E., Sica, V., and Bresciani, F., Estrogen-Binding Proteins of Calf Uterus. Partial Purification and Preliminary Characterization of Two Cytoplasmic Proteins, Biochemistry, 10: 3769-3780, 1971.
44. Puca, G.A., Nola, E., Sica, V., and Bresciani, F., Estrogen-Binding Proteins of Calf Uterus. Interrelationship between Various Forms and Identification of a Receptor-Transforming Factor, Biochemistry, 11: 4157-4165, 1972.

45. DeSombre, E.R., Chabaud, J.P., Puca, G.A., and Jensen, E.V., Purification and Properties of an Estrogen-Binding Protein from Calf Uterus. J. Steroid Biochem., 2: 95-103, 1971.
46. Gorell, T.A., DeSombre, E.R., and Jensen, E.V., Purification of Estrophilin from Calf Uterus, Fed. Proc. 33: 1511, 1974.
47. Liao, S., Lin, A.H., and Barton, R. W., Selective Stimulation of Ribonucleic Acid Synthesis in Prostatic Nuclei by Testosterone, J. Biol. Chem., 241: 3869-3871, 1966.

TESTICULAR FEMINIZATION SYNDROME:

A MODEL OF CHEMICAL INFORMATION NON-TRANSFER

Robert C. Northcutt
David O. Toft

Division of Endocrinology and Department of Molecular Medicine, Mayo Clinic, Rochester, Minnesota, USA

In 1950, Lawson Wilkins[1] of Johns Hopkins Hospital reported a patient which he described as a "hairless women with testes" who failed to develop any signs of virilization on prolonged administration of large doses of testosterone and methyltestosterone. From this observation, he drew the inference that this syndrome might result from a "genetic end-organ unresponsiveness" to the action of testosterone and not to a deficiency of testosterone production. In 1953, Morris[2] is credited with the definitive clinical description of the testicular feminization syndrome. Since that time, many other investigators have contributed to our understanding of the clinical and pathophysiologic features of the disorder. The clinical features of the syndrome of testicular feminization are summarized in Table 1.

Table 1

Clinical Features of Testicular Feminization

1) Female habitus, breast development, and body contour.
2) Scanty or absent axillary and pubic hair.
3) Female external genitalia with a blind ending vagina.
4) Undescended testes.
5) Testes have a high incidence of neoplasia.
6) Rudimentary male duct system, epididymis, and vasa.
7) Absence of a female duct system, uterus, and fallopian tubes.
8) Negative sex chromatin and 46/XY karyotype.
9) Unresponsiveness to virilizing and metabolic effects of testosterone.

Clinical Features

In the complete syndrome, body habitus, changes in contour, and breast developement occur at the usual age of puberty and are characteristically normal for a woman. Menarche does not occur and pubic and axillary hair growth is scanty or absent. The external genitalia are normal for a woman but on examination, the vagina ends blindly and no cervix is seen. Rarely the vaginal canal is absent with only a superficial depression being present. The testes are undescended and usually are located in the labia majora, inguinal canals, or in the lower pelvis, but may be found as high as the inferior poles of the kidney. Half of such patients may have an associated inguinal hernia. The testes have been estimated by some authors to show as high as 30% incidence of neoplasia, particularly in older individuals. Interstitial cell and seminiferous tubule adenomata are not infrequently seen. Some of the tumors arising in these testes are malignant and are similar in type to those occuring in undescended testes of otherwise normal men. Histologically, the testes differ from the usual cryptorchid testes in having a smaller tubular diameter with decreased elastic and collagen fibers in the tubular walls and absent germinal elements. A rudimentary epididymis and vas deferens are usually present but end blindly within the pelvis and, consequently, cysts of these structures may be encountered. The uterus and fallopian tubes are absent.

Psychological orientation regarding gender identity and maternalism is essentially that of a normal woman. Many of these individuals have had successful marriages and have had normal female sexual responses.

The karyotype of the affected individuals is that of a 46/XY normal male. The familial nature of this order is well established but the precise form of inheritance is not clearly delineated. At the present time it seems that it is an X-linked disorder and which perhaps is reflected in mothers and sisters of the affected males as variable scantiness of their pubic and axillary hair, consistent with the variable expression of the mutant X-chromosome in these relatives.

Endocrine Features

From a classical endocrine standpoint, the affected individuals with this syndrome are near normal in that their secretion and metabolism of testosterone appears to conform to that of normal males. Slight modifications of this have been seen, but certainly, the abnormalities observed are quantitatively inadequate to account for the very dramatic absence of normal male virilization and striking feminization that these patients exhibit. In young adults with this syndrome, the levels of testosterone in plasma, urine, and spermatic venous blood are well within the range for normal adult males and much higher than those values observed in females.[3,4,5,6] Peripheral conversion of testosterone to estrogen is not excessive and fails to explain the pathogenesis of this syndrome.[4,5] The concentrations of circulating luteinizing hormone is slightly increased possibly on the basis that circulating testosterone does not adequately suppress the secretion of luteinizing hormone.[5] Following castration, the excretion of both FSH and LH rises to menopausal levels and hot flushes are experienced by the patient. Correction of the hot flushes is not usually successful by the administration of testosterone only. However, physiologic replacement doses of estrogen control these symptoms and suppress gonadotropins.

Pathogenesis

Returning to the Wilkins proposition that this syndrome is due to inability of the target tissues to respond to testosterone, a number of observations confirming this notion have been made. Furthermore, all of the clinical features in the syndrome of testicular feminization can be explained on the basis of the inability of the individual to respond to the action of testosterone.

As our understanding of the cellular action of testosterone has been expanded, an approach to the understanding of this syndrome has been undertaken. Significant are the observations of Bruchovsky and Wilson who adducted that testosterone (T) itself is not the biologically active form of the hormone but it is the 5α-reduction product, 5α-dihydrotestosterone (DHT) which mediates the androgen response in target tissues.[7] Following these observations this author reported in 1969[8] that the maximum conversion of T to DHT in the skin of these subjects was markedly depressed when compared to normal males. The decreased 5α-reductase activity in skin of patients with testicular feminization has been confirmed by several authors,[5] although other authors[9] who used in vitro incubation systems not providing maximum cofactors for this enzyme, have not observed such differences. The conclusion may be drawn from comparisons of these studies to indicate that 5α-reduction of T occurs at a reasonably normal rate but the maximum ability to convert T to DHT is reduced in patients with the testicular feminization syndrome.

Since the ability to activate T does not appear to be the causative factor in the pathogenesis of the testicular feminization syndrome, other sites of the cellular action of T must be examined in order to find an explanation of this syndrome. T must first enter the cell from the extracellular fluid and then be converted to DHT. The DHT then binds with a high affinity specific cytosol receptor protein which is transported into the nucleus, becomes slightly modified, and subsequently binds to acceptor sites on the genome. Transcription of new messenger RNA is initiated which then transfers the genetic information to ribosomal-RNA, resulting ultimately in the synthesis of the specific proteins of that particular target tissue. In subsequent studies the author was unable to demonstrate nitrogen retention in response to the administration of large doses of DHT.[10] This lack of response was confirmed and reported by Strickland and French.[11] It was suggested by these authors that the decreased 5α-reductase activity was a consequence of the lack of T/DHT action rather than its cause. This suggestion is attractive though not proven.

We have recently had the opportunity to study a case of the complete testicular feminization syndrome at Mayo Clinic and the following is a brief clinical summary of the patient's findings.

Case Report

The patient (MC# 3-019-817) was a 21-year-old music student who presented medically with bilateral inguinal hernia and amenorrhea. Past medical history was one of excellent health. Childhood growth and development were normal. Breast and body contour changes began at age 13. Menarche did not occur. Slight pubic hair growth occurred but no axillary hair developed. Asymptomatic bilateral inguinal herniae were discovered at age 14 years.

The patient's only sibling is a normally menstruating sister. A maternal aunt and two maternal cousins have bilateral inguinal hernias and amenorrhea and are presumed to be affected.

Physical examinations revealed her height to be 175 cm. and weight 98.5 kg. No axillary hair was present and only scant pubic hair was noted. Pelvic examination showed clitoris and perineum to be that of a normal female. The vagina was of normal depth and the mucosa was rugated but no cervix could be seen or palpated. Bilateral direct inguinal herniae were present.

Laboratory summary:

	Normal range, male
Plasma estrogen (E & E) - 104 ng/dl	(5-60)
Plasma testosterone - 991 ng/dl	(300-1200)
Serum LH - 20 mcgm/dl LER 907	(4-20)
Serum FSH - 29 mcgm/dl LER 907	(10-60)
Buccal smear - "X" body absent; "Y" body present	
Karyotype - 46/XY	

Pelvic exploration was performed. The uterus and fallopian tubes were not present. The testes were located at the site of the internal inguinal rings bilaterally each being associated with a rudimentary epididymis and vas deferens which ended blindly into the pelvic floor. Excision of the gonads and accessory structures was performed and the herniae were repaired. The testes were small weighing only 10 gm. each. The interstitial cells were hyperplastic and multiple small seminiferous tubule adenomata were present bilaterally.

In Vitro Study

The surgical specimens were immediately refrigerated at $0°C$. The rudimentary vas deferens and epididymis were dissected free from each testis and the fibrous tissue removed in the cold at $3°C$. The tissue was homogenized in Tris, EDTA, monothioglycerol buffer at pH 7.5, 1.5 ml. buffer per gram of tissue. The total tissue weight was 6.1 gm. The 100,000 g supernatant was prepared and 0.5 ml. aliquots were incubated with at $3°C$ for 2 hours with tritium labeled 5α-dihydrotestosterone ($3H-5\alpha-DHT-1\alpha^{3H}$, $2\alpha^{3H}$; 47Ci/mMol). A 0.2 ml. aliquot of this material was layered on sucrose density gradients (5-20%) containing 10% glycerol and centrifuged for 16 hours at 100,000 g at $3°C$. Fractions were collected and counted. Parallel competitive binding experiments using the charcoal/dextran method of

Korenman[12] were performed. No saturable specific binding of the ^{3}H-DHT was detectable.

Discussion

Deficient DHT cytosol receptor protein binding in the rat testicular feminization mutant has been reported[13,14] and our observation is consistent with this animal model. It would be presumptuous to state that the absence of a cytosol receptor is the primary molecular defect in the syndrome of testicular feminization inasmuch as the presence of this receptor may very well be an expression of the action of dihydrotestosterone rather than the ultimate cause of DHT inaction. Other sites in the mechanism of action of T in the target cell distal to the cytosol receptor could be at fault. These could include defects in the steroid receptor transport into the nucleus, modification of the cytosol receptor, nuclear receptor complex binding to the genome, the transcription of processing of messenger RNA, genetic information transfer to the ribosomes, or increased degradation of the new proteins synthesized at the ribosome.

Furthermore, there is no reason at the present time to believe that the syndrome of testicular feminization is a single genotypic entity. Multiple steps in the cellular action of T are potentially affected by genetic mutation. In the future, the syndrome of testicular feminization as a model of chemical information non-transfer may be found to have several genotypes contributing to the make-up of the so-called clinical entity.

In summary, a case of complete testicular feminization has been presented in which no cytosol receptor for DHT could be detected in the excised sexual accessory structures. A brief outline of the clinical syndrome is discussed.

REFERENCES

1) Wilkins, L.M.: The Diagnosis and Treatment of Endocrine Disorders in Childhood and Adolescence. Charles C. Thomas, Springfield, Ill. 1950, p. 271.

2) Morris, J.M.: The Syndrome of Testicular Feminization in Male Pseudohermaphrodites. Am. J. Obstet. & Gynec. 65:1192, 1953.

3) Rosenfield, R. L., Lawrence, A. M., Liao, Landau, R. L.: Androgens and Androgen Responsiveness in the Feminizing Testis Syndrome. Comparison of Complete and "Incoplete" Forms. J. Clin. Endocrinol. & Metab. 32:625, 1971

4) Southren, A. L., Sharma, D. C., Ross, H., Sherman, D., Gordon, G.: Plasma Concentration and Biosynthesis of Testosterone in the Syndrome of Feminizing Testes. Bull. N. Y. Acad. Med. 40:86, 1964.

5) Judd, H. L., Hamilton, C. F., Barlow, J. J., Yen, S. S. C., Kliman, B.: Androgen and Gonadotropin Dynamics in Testicular Feminization Syndrome. J. Clin. Endocrol. & Metab. 34:229, 1972.

6) Morris, J. M., Mahesh, V. B.: Further Observations on the Syndrome, Testicular Feminization. Am. J. Obstet. & Gynec. 87:731, 1963.

7) Bruchovsky, N., Wilson, J. D.: The Conversion of Testosterone to 5 -Androstan-17 -ol-3-one by Rat Prostate *in Vivo* and *in Vitro*. J. Biol. Chem. 243:2012, 1968.

8) Northcutt, R. C., Island, D. P., Liddle, G. W.: An Explanation for the Target Organ Unresponsiveness to Testosterone in the Testicular Feminization Syndrome. J. Clin. Endocrinol. & Metab. 29:422, 1969.

9) Schindler, A. E., Keller, E., Friedrich, E., Jaeger-Whitegiver, E. R.: *In Vivo* and *in Vitro* Studies in Patients with Testicular Feminization. J. Steroid Biochem. 5:371, 1974. (Abstract 317)

10) Northcutt, R. C., Kaminsky, N. I.: Unpublished Observations.

11) Strickland, A. L., French, F. S.,: Absence of Response to Dihydrotestosterone in the Syndrome of Testicular Feminization. J. Clin. Endocrinol. & Metab. 29:1284, 1969.

12) Korenman, S. G.: Relation between Estrogen Inhibitory Activity and Binding to Cytosol of Rabbit and Human Uterus. Endocrinology. 87:1119, 1970.

13) Bardin, C. W., Bullock, L. P., Sherins, R. J., Mowszowicz, Blackburn, W. R.: Androgen Metabolism and Mechanism of Action in Male Pseudohermaphroditism: A study of Testicular Feminization. Rec. Progr. Hormone Res. 29:65, 1973.

14) McLean, W.S., Smith, A.A., Hansson, V., Nayfeh, S.N., French, F.S.: Androgen Receptor in Seminiferous Tubules of Immature Rat Testes: Absence of the Receptor in the Androgen Insensitive Stanley-Gumbreck (Tfm) Rat. Program. 56th Annual Meeting Endocrine Society, 1974, p. A-198, Abstract 286.

EFFECT OF THE SYNTHETIC ESTROGEN DIOXYDIPHENYLHEXANE ON THE CONTENT OF NUCLEIC ACIDS AND PROTEINS DURING DEVELOPMENT

G. Uzunov

Institute of Meat Technology

Sofia, Bulgaria

In the problems concerning development and differentiation the specific protein synthesis and the study of the conditions influencing this process have become of considerable interest. One of the ways to study biosynthesis is to use biologically active substances influencing the process between genetic material and specific proteins.

The selective action of hormones on individual cell structures and their functions, including the genetic apparatus of cells, has been studied (Lezzi, 1970; Mosebach et al., 1972). In this direction, the biological activity of estrogens remains an interesting topic. The information on it having been enlarged only in respect of their many-sided action on metabolic processes (O'Malley, 1967; Palmitter et al., 1970; Voytovich et al., 1969) and the genic-hormonal phenomena (Galand and Dupont-Mairesse, 1972).

The experimental works of Bullough (1955), Hollander (1960) and others cosider estrogens as regulators and activators of mitosis in definite tissues, and in the studies of Swann (1958), estrogens are found to switch over the type of metabolism in cells to the synthesis of the basic biopolymer of the mitotic apparatus. When studying early changes in the metabolism of target organs stimulated by estrogen, Mueller (1960) proved estrogens to influence the metabolic activity of enzyme systems in the biosynthesis of nucleic acids and protein, the altered enzyme activity being due to the synthesis of molecules de novo, and not to enzymes or proenzymes existing before.

On the other hand, according to the same author, the action of estrogens is related also to the processes of including activated amino acids, by t-RNA, to the ribosomes. In the works of Hamilton (1963, 1965) and Lezzi (1970) estrogens are found to activate DNA-dependent RNA-polymerase which results in an enhancement of RNA synthesis.

In the literature there is a lack of studies on the change of the intracellular biochemical activity realized by the inductive function of estrogens, by way of following the characteristics of soluble cell (tissue) proteins and the synthesis of nucleic acids, which is the object of the present investigation carried out on poultry.

MATERIAL AND METHODS

The study was effected on chicken and hens (Rhode Island breed), wit a maximum standartization of material by origin, age, weight and rearing conditions. The effect of dioxydiphenylhexane on the concentrations of nucleic acids and proteins in the hen organism was followed in four experiments: (a) one-month-old chicken treated with 2,000, 4,000, and 10,000 IU of dioxydiphenylhexane per kg of live weight; (b) two-month-old chicken treated with 4,000, and 10,000 IU; (c) three-month-old chicken treated with 3,000, and 10,000 IU, and (d) yearling hens trested with 8,000, and 100,000 IU of dioxydiphenylhexane. Experimental and control groups consisted of 8 birds each. Estrogen was injected in breast muscles four times at 72 hour intervals.

A study was made of the concentrations of RNA, DNA, proteinaceous nitrogen, the electrophoretic spectrum of the soluble proteins of the liver, oviduct and serum proteins. Nucleic acids were determined by the two-wave spectrophotometric method of Tsanev and Markov (1960). The coefficient for RNA values was obtained by determining the nucleotide composition by the method of Flack and Munro, as modified by Khadzhiolov et al. (1964). The protein nitrogen concentration in the organs and serum was examined by the micro-Kjeldal method and parallel whith that an electrophoretic study was made of the soluble proteins of liver, oviduct and serum proteins. Soluble tissue proteins were obtained by the method of Kaplanskiy (1956).

RESULTS

The introduction of estrogen resulted with all experimental animals in a rapid rise of total RNA in the liver, but the degree of increasing RNA concentration depended not only on the dose applied, but also on age (Fig.1). The figure indicates that, in one-month-old chicken treated with 2,000 IU of estrogen, the P-RNA concentration increases by 110 mg%, with the 4,000 IU dose by 164,9 mg%, and on tretment with 10,000 IU by 194 mg%, whereas in two-month-old chicken treated with 4,000 IU of dioxydiphenylhexane the P-RNA concentration increases by only 96 mg%, and with 10,000 IU by 154 mg%. Apparently, three-month-old chicken and yearling hens show a weaker response to estrogen. With introducing 3,000 IU in three-month-old chicken, the P-RNA concentration increases by only 62 mg%, and with 10,000 IU by 121 mg%. In yearling hens, the application of 8,000 IU of estrogen increases the P-RNA concentration by 94 mg%, and only with administering 100,000 IU, the concentration rises by 151 mg%. The DNA concentration related to dry substance decreases, but when related to fresh tissue the values remain almost constant in the individual groups of birds. The RNA/DNA coefficient acquires very high values, resulting mainly from the increase in the RNA concentration (Fig.1).

In comparison with liver the oviduct reacts in a different way to the action of estrogens, (Fig.2). The effect of dioxydiphenylhexane on the RNA concentration is much greater with the lower doses. The increase in estrogen level is not accompanied by a proportional rise in the RNA concentration. The maximum doses in one- and two-month old chicken have a weaker effect, and in hens, the massive dose of 100,000 inhibits the synthesis of RNA. It is only in three-month-old chicken that the high estrogen doses encourage RNA synthesis.

The DNA concentration is relatively reduced with the application of higher estrogen levels (Fig.1). The RNA/DNA coefficient reaches its highest values under the action of low estrogen doses (Fig.1).

Compared to liver and oviduct, the ovary reacts in a rather different way to the action of estrogen. As a general tendency in following its action with age on the RNA and DNA concentrations in the ovary, it could be noted that, regardless of its level, estrogen inhibits the RNA synthesis and reduces the relative content of DNA (Fig.1).

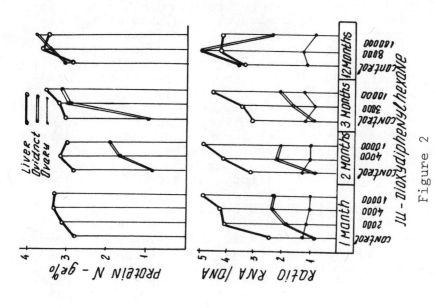

Figure 1

Changes in P-RNA and P-DNA concentrations in the liver, ovary and oviduct of chicken and hens, depending on the dose of dioxydiphenylhexane

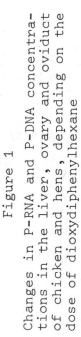

Figure 2

Effect of dioxydiphenylhexane on protein nitrogen concentration and the P-RNA/P-DNA coefficient in the liver, ovary and oviduct of chicken and hens

CONTENT OF NUCLEIC ACIDS AND PROTEINS DURING DEVELOPMENT 241

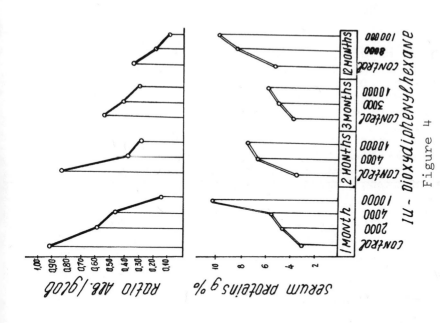

Figure 4

Effect of different dioxydiphenylhexane doses on total protein concentration and the albumin/globulin coefficient of the serum of chicken and hens.

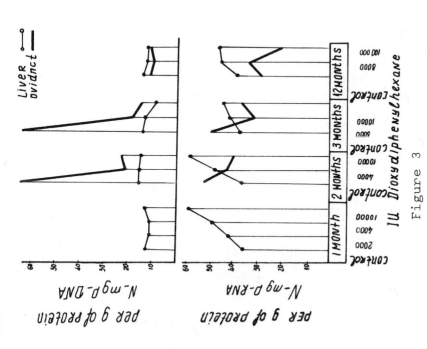

Figure 3

Effect of dioxydiphenylhexane on the coefficient of protein:P-RNA and protein:P-DNA in the liver and oviduct of chicken and hens.

Comparatively distinct changes occur also in the degree of protein synthesis. No great divertion in the relative content of protein nitrogen corresponds to the considerable changes in the RNA concentration in liver under the influence of dioxydiphenylhexane (Fig.1). RNA synthesis outstrips protein synthesis and the coefficient expressing the quantitative relations of P-RNA to a unit of protein nitrogen is very high in the experimental groups (Fig.3). The application of the massive estrogen doses in one- and two-month-old chicken was followed by even lower protein concentrations in liver. This is probably a consequence of the more intensive emanation of soluble proteins to the serum, whose protein content increases 2 to 3-fold in chicken (Fig.4).

In the oviduct of treated birds protein synthesis also gains strenght with the increase in the RNA concentration and accordingly the values of protein nitrogen exceed up to three times those of the control (Fig.3). These diversions are observed in a smaller degree with one year-old hens. The amount of P-RNA per g of protein nitrogen diminishes in the experimental groups of birds. This trend is clearly marked especially in the coefficient of P-DNA to a unit of protein nitrogen in the oviduct of estrogen-treated chicken.

The effect of dioxydiphenylhexane on the synthesis of soluble proteins and their respective spectra in liver, is similar in all ages of the birds, the changes in the proteinogrammes being more definite with the high estrogen doses. The fast-moving (to the anode) fractions, 1, 2, 3, and 4, decrease considerably in relative protein content, and fractions 6, 7, and 8, which have α_2-, α_3- and β-globulins as analogues in serum (Fig.5), acquire higher values. In all ages, the displacement is directed to fraction 7 (Fig.6).

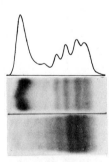

Fig. 5. Electrophoregrammes of the soluble proteins of hen liver (a) and Serum (b): a comparison

CONTENT OF NUCLEIC ACIDS AND PROTEINS DURING DEVELOPMENT 243

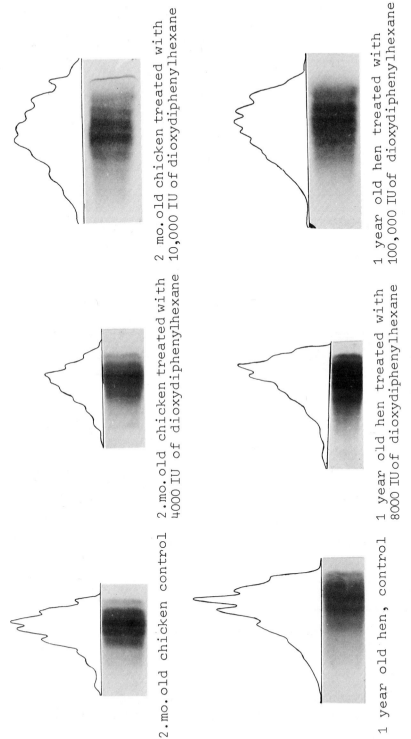

Fig. 6. Electrofphoregrammes of the soluble liver proteins of chicken and hens treated with different dioxydiphenylhexane doses.

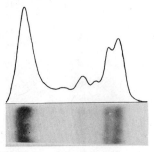
2 mo. old chicken, control

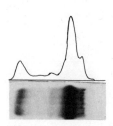
2 mo. old chicken treated with 10,000 IU of dioxydiphenylhexane

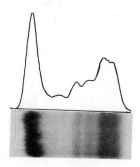
3 mo. old chicken, control

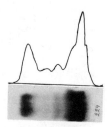
3 mo. old chicken treated with 10,000 IU of dioxydiphenylhexane

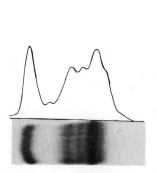
1 year old hen, control

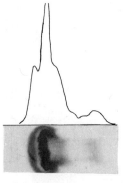
1 year old hen treated with 100,000 IU of dioxydiphenylhexane

Figure 7

Electrophoregrammes of the blood serum of chicken and hens treated with different dioxydiphenylhexane doses.

CONTENT OF NUCLEIC ACIDS AND PROTEINS DURING DEVELOPMENT 245

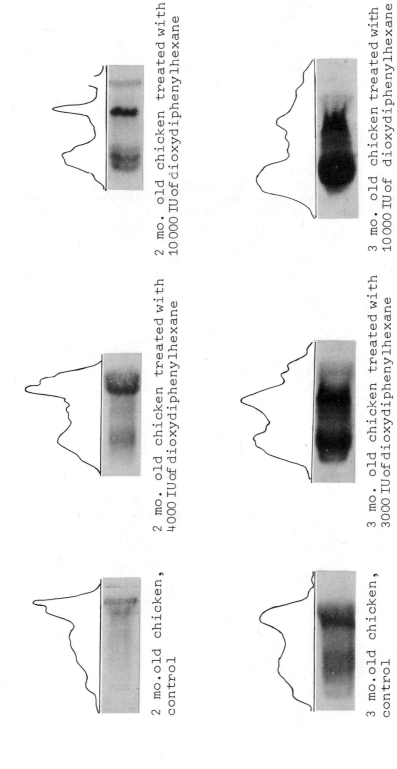

Fig. 8. Electrophoregramme of the soluble oviduct proteins of chicken treated with different dioxydiphenylhexane doses.

Total protein content in serum increases in gradation with the level of estrogen applied to the animals. The hyperproteinemia observed is at the expense of the increase in globulins (Fig.7). It is characteristic of this hyperproteinemia that, in the zone of globulins, a specific redistribution of protein among the individual fractions takes place, depending on the dose of estrogen. With the high estrogen doses, albumin and α_1-globulin decrease, while β-globulin acquires higher values. Massive estrogen doses affect all fractions; the protein content of albumin decreases most (from 47 to 14%) in one-month-old chicken, and from 25.5 down to 9.14 in the hens, after the administration of 100,000 IU of dioxydiphenylhexane. α_1- and α_2-globulins also decrease in percentage, and α_3- and β-globulins increase in absolute and relative protein content. Proteinogrammes are deformed so that the highest peak is obtained with β-globulin (Fig.8).

Significant differences were found also in the profile of soluble proteins of oviduct mucosa under the influence of dioxydiphenylhexane. A greater number of fractions, six, are obtained in the experimental groups of two-month-old chicken, in comparison with controls (where four fractions are shown) and the proteinogrammes of two- and three-month-old chicken approach in character those of sexually mature birds (8). Albumin content increases and the individual ovi albumin components are clearly outlined. In the hens treated with low estrogen doses, it is mainly the albumin fraction that increases (especially the A-1 component). The effect of massive estrogen doses on soluble oviduct proteins in hens is weak.

DISCUSSION

The increase in the concentration of RNA and protein in the hen organs studied depending solely on the introduction of dioxydiphenylhexane, with all other conditions equal, indicates that estrogen achieves that level of synthesis by way of regulating the activity of RNA-synthesizing systems, and that such a concentration of ribosomes in the cell is maintained, as will accomplish the intensive protein synthesis. Since the velocity of protein synthesis related to one ribosome is constant and the overall level of protein synthesis is determined by the number of ribosomes in the cell (Shantren, 1966), it could be assumed that, after the effect of estrogen, only ribosomal RNA increases. But, on the other hand, ribosomal RNA does not carry information to the synthesis of proteins and, if it is assumed that only the ribosomal fraction increases upon the introduction of estrogen, then just a simple quantitative change in protein content would result - without the appearance of any qualitative differences in the spectrum of the soluble proteins in the organs studied. But the appearance, after estrogen application, of non-specific (for the age) protein fractions in the liver of 5-day old chicks (Uzunov, 1968), which are characteristic of mature specimen, as also fractions 2 and 6 in the soluble oviduct proteins in chicken, non-specific for the intact organism, indicates a certain degree of change in the information to protein-synthesizing structures also. This makes it possible to presume that, informational RNA increases also, together with the increase in ribosomal RNA. But only the cistrons that are not repressed by histones at a given stage of cell development, are actively synthesizing RNA. Entering the chain of events in protein biosynthesis, estrogen probably activates accordingly other loci of the genome by way of changing the connections of histones with DNA (Agrell and Persson, 1956) and imposes other information followed further by a change in the characteristics of proteins in tissues affecting the spectrum of serum proteins.

These changes in protein synthesis can explain, to a great extent, the processes in the intact organism also, where the gradual turn of DNA matrix function under the influence of endogenous factors (hormones) and their action results in the age specificity of tissue proteins.

It is also interesting to note that, in liver, ovary and oviduct, estrogen effect does not meet with an equivalent response in respect of the synthesis of RNA and protein with the different doses.

While liver and oviduct react by a stimulated synthesis of RNA and protein with the increase of estrogen level, though to a certain degree and disproportionately, an inverse effect is observed in most cases in the ovary of chicken and hens. Probably the higher estrogen doses exert a restraining effect on cell-division. It can be assumed that the smaller ability of ovary tissues to metabolize estrogen to weaker biologically active substances is of significance for this effect. We are inclined to accept the view of Geschicter (1961), according to whom, ovaries respond to the intensive introduction of estrogens by follicles coming to rest and a subsequent atrophy.

Another important point which is a logical consequence of the data analysis, is that the estrogen-induced synthesis of RNA and proteins and the increase in their concentrations underlies encouraged organ growth (Uzunov, 1968), which speaks, on the one hand, of stimulated cell division and enhanced metabolic activity, but in most cases it is a matter of cell hypertrophy, and that can be judged by the increased P-RNA/P-DNA coefficient and the accumulation of a greater amount of proteins expressed also by the quantitative relations of protein nitrogen to DNA.

In conclusion, it could be assumed that the physiological function of dioxydiphenylhexane is addressed to the genetic apparatus of cells and this is demonstrated by the synthesis of protein fractions non-specific for the intact organism, and intensive RNA synthesis.

REFERENCES

Agrell J., H. Persson, Biochem. Biophys. Acta, 1956, 20, 3, 543-546

Bullough W. S., in: Vitamins and Hormones, 1955, 13, 261-292

Galland P. A. and N. Dupont-Mairesse, FEBS-Congress, Varna, 1972, Abstr., 700, 254.

Geschicter F. C., in: Diseases of the Braget, 2nd ed., Lippicot-Philadelphia, 1955, 836-849.

Hamilton T. H., Proc. Nat. Acad. Sci. U.S.A., 1963, 49, 3, 373-379

Hamilton T.H., C.C. Widnel, J.R. Tata, Biochem. Biophys. Acta, 1965, 108, 1, 168-172.

Hollander V. P., M. S. Stephens, T. E. Adamson, Endocrinology, 1960, 66, 39-47.

Kaplanskiy S. Ia. et al., Biokhimia, 1956, 21, 4, 469-477 (in Russian)

Khadzhiolov A.A. et al., Bulletin of the Central Lab. Biochemistry (Bulg. Acad. Sci.), 1964, 2, 31-37, Sofia.

Lezzi M., 8th Int. Congress of Biochem. Switzerland, 1970

Mozelbach K.O., A. Scheuer and H.G. Dahnke, FEBS-Congress, Varna, 1972, Abstr., 701, 254

Mueller G. C., in: Biological Activities of Steroides in Relation to Cancer., Eds. Pincus & Vollumer, N.Y.-London, 1960, 129-142.

O'malley B. W., Biochemistry, 1967, 6, 2564.

Palmitter R.A., Christensen A.K., Schimke R.T., J. Biol. Chem., 1970, 245, 833.

Shantren Iu., in: Protein Biosynthesis, Ed. Sbarskij I.B., Moscow, 1963 (in Russian)

Swann M.M. Cancer Res., 1958, 18, 10, 1118-1160.

Tsanev R., G.Markov, Biokhimia, 1960, 25, I, 151-159 (in Russian)

Uzunov G., Thesis 1968, Sofia (in Bulgarian)

Voytovich A.E., I.S. Owens and Y.J.Topper, Proc. Natl. Acad. Sci. USA, 1969, 63, 312.

POSSIBILITY FOR A DIRECT HORMONE-HORMONE INTERACTION

Julia G. Vassileva-Popova and Dimitur Dimitrov

Central Laboratory of Biophysics, Bulgarian

Academy of Sciences, 1113 Sofia, Bulgaria

The present <u>in vitro</u> study is an attempt for the further understanding of the hormonal regulatory behaviour and especialy of the hormone-hormone interrelations. The possible interaction between protein hormone (PH) and sex steroid (SS) is in the focus of this biophysical level experiment. Because of lack of enough data the biological significance of the present experiment will not be discussed here. The possible application of the results has to be found in our work on mathematical modeling of the hormonal regulation.

MATERIALS AND METHODS

The method of quenching the tryptophane (Trp) fluorescence was applied. The estimation of the Trp fluorescence of the PH was made with and without presence of Δ^4-androstene-17β-ol-3-one and Δ^4-androstene-3,11,17-trione. The comparison of the Trp fluorescence in both cases became a criterion for the conformational state of the protein molecule. Some earlier studies (1-5) indicate that the iodine in a low concentration produce a diffusional quenching of the Trp fluorescence. This effect is also shown by an oxygen application (6).

Symbols used: Luteotropic hormone (LTH) or Prolactin (PR); Luteinizing hormone (LH); Protein hormone (PH); Sex steroid (SS); Tryptophan (Trp); Concentration (Q).

The Trp quenching method is quite well established for the protein conformation study (7-10), but not so intensive in the investigations on the hormones containing Trp (11) and espetialy for Prolactin (Pr). In our study the classical principle of Stern-Volmer (7) is well observed when as a quencher I⁻ was used

$$\varphi_0 = \varphi[1 + K_Q(X)] \qquad /1/$$

where:
  φ_0 - the quantum yield without a quencher,
  φ - the quantum yield with a quencher,
  K_Q - the Stern-Volmer constant

An extempore made solution of sodium sulphite was added in order to prevent the possible formation of I_3^- and a preliminary test for ion capacity was made.

The PH was kindly supplied by NIAMD, NIH, Bethesda. As protein hormones Pr and LH were used after further purification. Because of a possibility of albumin traces in the purified samples the mixture of albumin (BSA grade) and SS was also examined and no significant differences were observed in presence of SS (Fig.1).

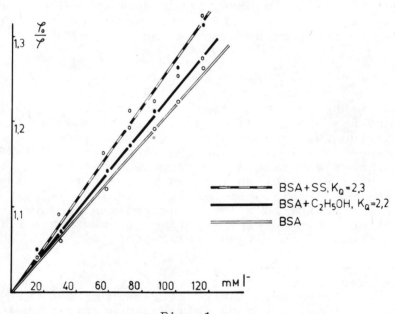

Fig. 1

One of the time consuming procedures of this assay was the tuning of optimal conditions for the simultaneous preservation of PH and SS in solution. The approach used was the preparation of a water solution of the PH and an organic (ether) solvent for the SS. After mixing of the PH and SS solutions an evaporation was made up to a determinate volume of the total mixture. It was also proved that the organic solvent did not interfere with the results. For this reason fluorescence estimations of the PH with and without ethanol and ether were made. Simultaneously the experiment was carried out in the absence of organic solvent for Δ^4 -Androstene-3,11,17-trione. In these conditions because of the slight water solublity of this SS it appears to be in a very low concentration in comparison with the PH. The PH concentrations used were 10^{-5}M and 10^{-6}, and that the SS was 10^{-5}M respectively. The quantum yield of Trp in the PH was determined by comparison with the Trp quantum yield in water solution with pH = 7.0 at 23°C. The fluorescence excitation was achieved at 297 nm and the measurement was carried out at 340 nm.

Because of lack of Trp residues in the LH molecule and the relatively small contribution to the fluorescence of other aromatic residues a not yet fully completed isotope assay was carried out.

RESULTS AND DISCUSSION

The results indicate that the Pr molecule is not indifferent to the SS presence. The data are shown in the Stern-Volmer coordinates with some modifications (12). On the ordinate the ratio $\varphi/(\varphi_0-\varphi)$ is plotted and on the abcissa the concentration increase of the quencher indicated

$$\frac{\varphi_0}{\varphi_0 - \varphi} = \frac{1}{X f_a K_Q} + \frac{1}{f_a} \qquad /2/$$

where:

φ and φ_0 were explained previously,

K_Q - the Stern'volmer constant,

f_a - % of exposed Trp residues,

X - Q of the quencher ions.

The difference between the PH molecule alone and with SS presence is significantly evident (Fig.2).

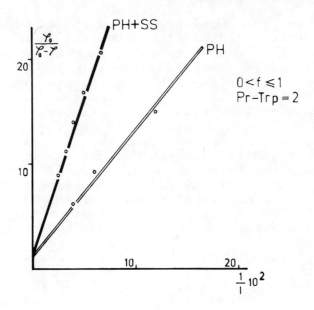

Fig. 2

In fact f has a value in the interval 0÷1. f=1, only in the cases where all the Trp residues are exposed to the buffer and in the interval 0÷1 in all the other combinations of Trp residues exposures.

According to the established protein sequences for Pr (13), Trp = 2. When f ≈ 1 the two Trp residues are on the molecule surface and are exposed to the surrounding medium. In the presence of the SS there is not a significant change of the Trp position in the Pr molecule, but the constant K_a is considerably altered (Fig.3). This observation indicates that Pr interacts with the SS without the direct involvement of the Trp residues. The interaction between the two hormones appears to take place in the close vicinity of the Trp residues. It looks possible the hydrophobic "pockets" in the Pr molecule to be at 112 and at 163 places. In a comparative way LH and Pr have more than five hydrophobic places for the possible SS interaction (14). The effect is also accompanied by a conformational change of the PH and a limited approach of I^- ions to the PH. All the changes observed are an indication for a complex formation between the PH and the SS. Probably it is not unexpected that the protein-ste-

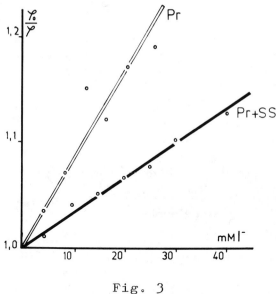

Fig. 3

roid interactions have some kind of support by the presence of the hydrophobic centre in the protein molecule (15).

The difficulties we had in proving the existence of the hormone-hormone complex formation were probably due to a very weak interaction and quite an unstable PH - SS complex.

The discussion on the possible biological significance of this finding will be limited in this paper. The SS-dependent regulation of LH and the steroidogenetic function of LH, resp. ICSH, were established a long time ago and this was in relation with the idea of this study. Relatively new is the topic of the Pr and SS synergism and especially with androsteron (16-22).

The further discussion of the possible significance of the in vitro finding of a non-covalent weak hormone-hormone interaction should be postponed until more data are available.

ACKNOWLEDGEMENT

We would like to thank Dr J.R.Tata for the materials supplied and for the useful discussions. Some of the experiments were carried out with the facilities kindly provided by the Institut für Biochemie und Biophysik, 1 Berlin 33, BRD.

REFERENCES

1. Lehrer, S.S., Biochem. Biophys.Res.Commun., 29, 767, (1967).

2. Winkler, M.H., Biochemistry, 8, 2587, (1969).

3. Arrio, B., Rodier,F., Boisson, C. and Vernotte,C., Biochem. Biophys. Res. Commun., 39, 589, (1970).

4. Teale,F.W.J. and Badley,R.A., Biochem. J., 116, 341, (1970).

5. Vaughan, W.M. and Weber, G., Biochemistry, 9, 464, (1970).

6. Burstein, E.A., Biofizika, 13, 433, (1968), (in Russian).

7. Stern, O. and Volmer, M., Physiol. Z., 20, 183, (1919).

8. Badley, R.A. and Teale, F.W.J., J. Mol. Biol., 44, 71, (1969).

9. Lehrer, S.S. and Fasman, G.D., Biochem. Biophys.Res. Commun., 23, 133, (1966).

10. Teale, F.W.J., Biochem.J., 76, 381, (1960).

11. Dixon, J.S., in: Peptide Hormones, Eds., S.A. Bernson and R.S.Yalow, North-Holland Publ.Co., Amsterdam-London, 1973.

12. Lehrer, S.S., Biochemistry, 10, 17, (1971).

13. Li, C.H., Dixon, J.S., Lo, T.B., Schmidt, K.D. and Pankov, Y.A., Arch.Biochem.Biophys., 141, 705, (1970).

14. Pierce, J.G., in: Peptide Hormones, Eds., S.A. Bernson and R.S.Yalow, North-Holland Publ.Co., Amsterdam-London, 1973.

15. Varshavsky, Ja.M., in this volume, 1975.

16. Segaloff, A., Steelmann, S.L. and Flores, A., Endocrinology, 59, 233, (1956).

17. Chase, M.D., Geschwind, I.I. and Bern, H.A., Proc. Soc.Exp.Biol.Med., 84, 680, (1957).

18. Antliff, H.R., Prasad, M.R.N. and Meyer, R.K., Proc. Soc.Exp.Biol.Med., 103, 77, (1960).

19. Grayhack, J.T., Nat. Cancer Inst.Monogr., 12, 189, (1963).

20. Grayhack, J.T. and Lebowitz, J.M., Invest.Urol.,5, 87, (1967).

21. Peyre, A., Revoult, J.P. and Laporte, P., C.R.Soc. Biol., 162, 1592, (1968).

Cyclic AMP, Enzyme, and Hormone Networks in Bio-Development

GROWTH AND DEVELOPMENTAL HORMONES AS CHEMICAL MESSENGERS

J.R. Tata

National Institute for Medical Research

Mill Hill, London NW7 1AA, England

INTRODUCTION

Hormones are a most appropriate topic for a meeting dealing with physical and chemical means of information transfer in living systems. The word HORMONE is often used to denote circulating chemical messengers, although we realise that not all chemical messengers need be circulating. For example, local hormones or neurotransmitter substances, such as acetylcholine, serotonin, noradrenaline, etc., act very near the site where they are produced without having to be transported via the circulatory system. Similarly, a very important group of most recently discovered substances called pheromones, which are used for communication between individual organisms, are transmitted through the air and impinge on the brain via the olfactory system. Most plant hormones, such as auxins, gibberellic acid, dormins, etc., do not need to be transported from one part of the plant to another. It is only in higher animals that hormones are made in very specialized organs, called endocrine glands, and these hormones can only reach their target tissues if appropriately carried to them in blood or other extracellular fluids.

Perhaps the most important common feature of all hormones, that needs to be highlighted for the purpose of this article, is that they are chemical intermediaries in the transfer of information from the environment to the organism and then from one group of cells within the organism to another; the overall effect of such information transfer being the co-ordination of multiple activities in different cells and eventually allowing the organism to adapt to environmental demands. Figure 1 represents the various levels at which hormones can convey information. The

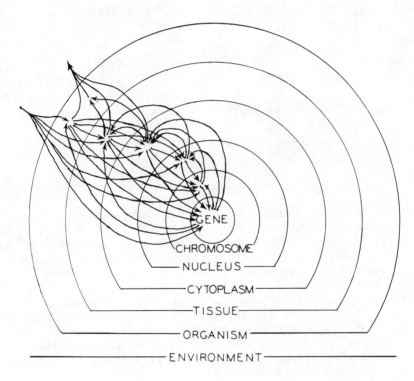

Figure 1: Schematic representation by P. Weiss of the different levels at which the environment can influence an organism, its cellular constituents, and the function of intracellular elements. Hormones are major chemical intermediaries in information transfer across these different compartments as indicated by the arrows.

other important feature to bear in mind is that in bringing about such information transfer, the physical and chemical forms of information are interconvertible. Thus, for example, a physical signal from the environment, such as change in outside temperature or in the duration or intensity of light, which are physical parameters of information, is first converted to electrical signals in the brain and eventually into chemical form in the hypothalamus in the form of hypothalamic release hormones, about which a lot has been said in this colloquium. The hypothalamic release hormones can then set in motion the release of several hormones by a series of interactions shown in Figure 2. It should also be noted that chemical information can be converted back into physical information as, for example, the action of neurotransmitters on nervous tissue or that of pheromones on the olfactory apparatus of the brain.

It is important, in considering hormones as chemical signals, to take into account the importance of evolution of hormones and their actions. Quite simply, this consideration becomes a question of whether in evolution the hormones or their physiological actions came first. This question is not merely one of some trivial academic value but its realisation can often simplify some of the confusion regarding hormone action. In general, when one considers this question, it turns out that most often hormones appeared before the very sophisticated actions we attribute to them in higher organisms. For example, the sex steroids, oestrogen and androgen, which have such important functions in controlling the maturation and differentiation of many tissues in higher animals, are extremely primitive substances and have been identified in plants and even micro-organisms. On the other hand, their effects on behaviour, maturation of cells, such as mammary cells, and sexual tissues can only have evolved more recently in higher organisms. Similarly, the two thyroid hormones, L-thyroxine and 3,3'-,5-triiodo-L-thyronine, are found in very primitive organisms such as marine invertebrates and sponges, but it is only in higher vertebrates that they have acquired a very important physiological action.

RAPID AND SLOW RESPONSES TO HORMONES

It is impossible, in such a brief article, to catalogue all the different types of plant and animal hormones and the wide variety of physiological functions that they regulate. Suffice it to say that there is virtually no metabolic or growth process in plants and animals which does not come under hormonal regulation in one organism or the other at some developmental stage of the organism. For the purpose of this article, it is important to distinguish between growth and developmental hormones and those that control rates of metabolic processes in differentiated or undiffer-

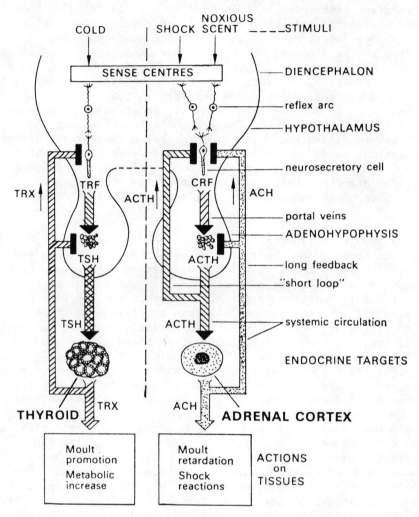

Figure 2: Two examples of the conversion of information from the environment as physical stimuli to chemical information as a series of hormones and leading to important physiological changes in vertebrates. The initial conversion takes place in the central nervous system to give rise to hypothalamic release hormones or factors (TRF and CRF, in this figure, to represent TSH and ACTH releasing factors). These hypothalamic hormones act on the anterior pituitory (hypophysis) to release TSH (thyroid stimulating hormone) and ACTH (adrenocorticotrophic hormone). The latter reaches the thyroid and adrenal glands via the circulation and stimulates the formation and release of thyroid and adrenocortical hormones, each of which regulate different physiological functions in various target tissues.

entiated cells. In an experimental situation, such a distinction amounts to whether or not the target cell exhibits a response to a hormone almost immediately upon their interaction or whether there is a long latent period involved in the response. Table 1 lists a few examples of the two classes of hormones according to the rapidity of response to them by the targets.

Thus, acording to Table 1, hormones which elicit a very rapid response, such as adrenaline and insulin, regulate the rate of the metabolic process, such as glycogenolysis and sugar uptake, almost as soon as they reach their target tissues (muscle or liver tissues) and without affecting the growth and maturation of such tissues.

The other group of hormones, such as thyroid hormones, ecdysone and sex steroids, all bring about their action after a substantial latent period between their interaction with cellular receptors and a physiological change. In all instances, such a change involves the regulation of protein and nucleic acid synthesis in their target cells. It is this type of hormonal control that we shall now examine below. Before considering the major biochemical mechanisms involved in the response of immature cells to their respective growth and developmental signals, it is worthwhile drawing attention to two important features concerning such hormones. Firstly, the same hormone can regulate metabolic activity rapidly in a given tissue or species and control growth and development less rapidly in another. Thus, in an immature rat, growth hormone (somatotrophin) has a profound effect on the rate of growth of virtually all tissues. However, in adults this hormone also plays a very important metabolic role in controlling fat mobilization.

SOME IMPORTANT CONTROL MECHANISMS INVOLVED IN RESPONSE TO GROWTH AND DEVELOPMENTAL HORMONES

Since growth and development can best be expressed in biochemical terms as a regulation of protein synthesis we shall consider here very briefly the possible levels at which growth and developmental hormones may provoke a response in their target cells which could trigger off a chain of events leading to the characteristic physiological action of these hormones. This is not to say that mechanisms other than those involved in the control of protein and nucleic acid synthesis are excluded from the multiple response to the hormones, but that such other responses cannot, at the present moment, be fitted into a cause-effect relationship which fits into current biophysical and biochemical dogmas.

Table 1: A few examples of hormones that rapidly control the metabolic activity of target cells and those that regulate growth and development after a latent period.

HORMONE	RELATIVE SPEED OF RESPONSE	PHYSIOLOGICAL AND BIOCHEMICAL ACTIONS
ADRENALINE	Rapid	Cardiac activity; glycogenolysis; regulation of cyclic AMP levels.
INSULIN	Rapid	Carbohydrate metabolism; uptake of sugars.
PARATHYROID HORMONE	Rapid	Ca and P metabolism in bone and kidney; uptake and intracellular levels of Ca^{++}.
VASOPRESSIN	Rapid	Water and ion metabolism in bladder; regulation of cyclic AMP level and Na^+ pump.
ECDYSONE	Slow	Insect metamorphosis; "gene" puffs; sclerotization and induction of enzymes in larval tissues; regulation of RNA and protein synthesis.
THYROXINE	Slow	Amphibian metamorphosis; induction of urea cycle enzymes and albumin in liver; regression of tail, gills and gut; development of CNS and regulation of basal metabolic rate in mammals; control of RNA and protein synthesis.
OESTROGEN	Slow	Growth of accessory sexual tissues (uterus, mammary gland) in mammals; induction of egg white and yolk proteins in birds, amphibia, reptiles; gene activation.

DNA Replication

It is well-known that DNA synthesis is an important event in cellular differentiation and tissue growth. Besides DNA synthesis as a part of the process of cell division, many workers have recently considered the role of gene amplification, i.e. limited DNA synthesis involving only a few genes and not leading to cell division, as an important facet of the possible way in which DNA synthesis may be involved in the regulation of development. The best documented case for this type of regulation is in the multiplication of genes coding for ribosomal RNA during oogenesis in insects and amphibia. Professor Arnstein has considered how cell division could play an important role in erythropoiesis, a process regulated by the hormone erythropoietin, and we shall not, therefore, deal with this particular regulatory system. However, if we consider erythropoietin to be a growth and developmental hormone, then its action, which requires some participation of DNA synthesis, stands out as an exception to the general pattern seen with all other classical growth and developmental hormones. For this reason, we shall not consider DNA synthesis any further in this review.

Transcription

Currently, the actions of most growth and developmental hormones could be best explained in terms of their regulation of gene expression or transcription in their target cells. Transcriptional control can be considered from three different angles:

(a) <u>Regulation of template activity</u>. It is known that the bulk of the information encoded in the genome of higher organisms is not expressed and that during major developmental events the transcription of certain genes is either "turned on" or preferentially enhanced. This has led to the currently popular concept of the regulation of template activity of chromatin (the operational term used for denoting isolated chromosomal content). It is well-known that the repression of most of the inactive chromosomal material of a cell is accomplished by histones, but more recently it has been found that a group of proteins, call acidic or non-histone nuclear proteins associated with chromatin, play an important role of positive control in determining actively transcribed genes. For this reason, a large number of papers have recently appeared on the possibility of growth and developmental hormones controlling the amount of different acidic nuclear proteins or their chemical modification.

(b) <u>RNA polymerases</u>. That plant and animal cells have multiple RNA polymerase involved in the transcription of different types of genes is now a well-established fact. Recent work on sigma factor in <u>E. coli</u> RNA polymerase has encouraged many workers to look for similar regulatory elements in the different animal RNA polymerases. So far, components of animal RNA polymerase with similar regulatory properties have not yet been discovered and, although both cytoplasmic and nuclear factors controlling the overall activity of RNA polymerases are known to exist, it seems that the amount of RNA polymerase is not rate-limiting for transcription in higher organisms.

(c) <u>Post-transcriptional processing of RNA</u>. It seems that in higher organisms much of ribosomal and messenger RNA is transcribed as large units (45S pre-ribosomal RNA and heterogeneous nuclear RNA or HnRNA, respectively) and these are then cleaved to smaller fragments (28S and 18S ribosomal RNA and messenger RNA) by specific cleavage enzymes within the nucleus. In addition to cleavage, the primary transcript is modified in a number of ways, particularly methylation and polyadenylation. The latter process has stirred up much interest recently among investigators looking at the importance of regulation of transport of RNA from the nucleus to the cytoplasm since most messenger RNA molecules are now thought to be polyadenylated in the same position (the 3'-end).

(d) <u>Nuclear structure and compartmentation of transcription and post-transcriptional processes</u>. The synthesis of ribosomal RNA and transfer RNA is compartmentalized and takes place in different parts of the nucleus. Furthermore, it is known that active and inactive genes can be separated by euchromatin and heterochromatin by virtue of their different densities and structures. These two factors underline the importance of nuclear structure in considering transcription. Yet, very little attention so far has been paid to the way in which the intimate association between nuclear structure and transcription may be involved during regulation of gene expression by hormones or other developmental stimuli.

Translation

Messenger RNA in animal cells is relatively stable, unlike in micro-organisms, whereas some animal proteins made on such stable messengers can have a very short life. This means that the rate of messenger RNA synthesis is not necessarily the only rate-limiting step in the regulation of induction of proteins but that the rate of translation can also be an important factor. The process of translation is a very complex one in animal cells and so far it has been difficult to pin-point any one of the numerous steps likely to be involved in the control of synthesis of key proteins during development. The best example of a possible translational control operating during enzyme induction is the system of induction of tyrosine aminotransferase by glucocorticoids in hepatoma cells. Work carried out over the last five years in Tomkins' laboratory had suggested that there may be a specific factor in the cytoplasm with which the hormone would combine in order to stabilize the messenger for tyrosine aminotransferase. However, in the light of more recent evidence from Tomkins' own laboratory, it now seems that even in this system the major role of the hormone is to control the transcription of the messenger RNA coding for the induced enzyme. For this reason, we shall not consider individual translational steps as sites of regulation by growth and developmental hormones, but we shall consider a special kind of translational control which involves cellular structure in so far as it concerns the deployment of ribosomes on intracellular membranes.

It is well-known that proteins destined for export from the cell are synthesized on polyribosomes attached to membranes of the endoplasmic reticulum and that some of the proteins retained for intracellular use may be synthesized on non-membrane-bound or free polyribosomes. However, it is well-known that many tissues that are not predominantly protein-secreting tissues can have a large proportion of their protein synthesized on membrane-bound ribosomes, particularly during rapid developmental change. This suggests that membrane-bound ribosomes must play some other role in the control of protein synthesis during development. What this role is has not yet been defined but it is well worth considering it in the discussion below.

SEQUENTIAL RESPONSES TO GROWTH AND DEVELOPMENTAL HORMONES

The manner in which information is transmitted through hormones and consequently expressed has been approached from two opposite directions. One direction is based on the study of "receptors" whereby the target tissue has the inherent capacity to recognise certain signals represented by specific growth and developmental hormones. A large amount of work has now been carried out on determining the nature of the components in a cell that interact with hormones when the cell first sees them and then to relate such an interacting component or receptor with a series of events at the molecular and cellular levels that would ultimately lead to the physiological action of the hormones. In this colloquium several investigators have referred to this approach, both for protein hormones that can trigger off a change within a cell from the outside, as well as steroid hormones which penetrate into cells and interact with intracellular elements. We shall, therefore, not deal with this approach, but discuss the second approach in which one analyses backwards from late physiological responses to a series of earlier and earlier events which are likely to bear a cause-effect relationship to one another.

Transcriptional Events

About ten years ago it became quite clear that the action of any growth or developmental hormones can be abolished by the simultaneous administration of inhibitors of RNA synthesis, such as Actinomycin D, which showed the key role played by gene transcription in the expression of the action of the hormone. Furthermore, changes in the rate or nature of RNA synthesized are among the earliest responses observed following the interaction between a developmental hormone and its target. Figure 3 gives a schematic account of the changes one is most likely to see in the nucleus of a responsive cell such as the oviduct or the uterus to the female sex hormone, oestradiol. This is the best example of changes in transcription within a nucleus in response to a developmental hormone and, although similar detailed information is not available for other hormones, this seems to be the general pattern of events.

The first nuclear response to a stimulation of development seems to be a rather subtle change in the template activity of the target cell chromatin which, in the case of the oviduct or uterus, can take place within 30 to 60 minutes following hormone administration. This alteration in template activity, which becomes obvious both when it is tested with an exogenous RNA polymerase enzyme or with the endogenous enzyme associated with the template in vivo, is thought to be brought about by an alteration in the amount or nature of one or more acidic chromosomal proteins. The modification itself may involve the phosphorylation or dephosphorylation of proteins and it is interesting that cyclic-AMP is a major regulatory metabolite in this process. The next event has only been deduced from negative results obtained with inhibitors of protein synthesis in that it is thought that a consequence of this initial change in transcriptional activity leads to the generation of one or a few key proteins which act on the nucleus. If this were to be finally proven, it would appear that such a protein regulates the synthesis of ribosomal RNA since the next event is a massive increase in the synthesis of ribosomal RNA in ribosomes. These responses can be observed both in vivo and in isolated preparations of nucleoli or RNA polymerase A, the enzyme responsible for the transcription of ribosomal genes. The massive increase of ribosomal RNA synthesis masks that of non-ribosomal RNA species, which would include messenger RNA. The next sequence of events takes place in the cytoplasm where new proteins may make their appearance or one may observe a preferential synthesis of one or two classes of proteins whose function would be associated with a particular developmental change. Indeed, messenger RNA for ovalbumin has been isolated from chick oviduct polyribosomes from newborn chicks treated with oestrogen and the appearance of this message in the cytoplasm immediately precedes the first detection of ovalbumin. Furthermore, the accumulation of messenger RNA for ovalbumin is anticipated by the detection of ovalbumin coding

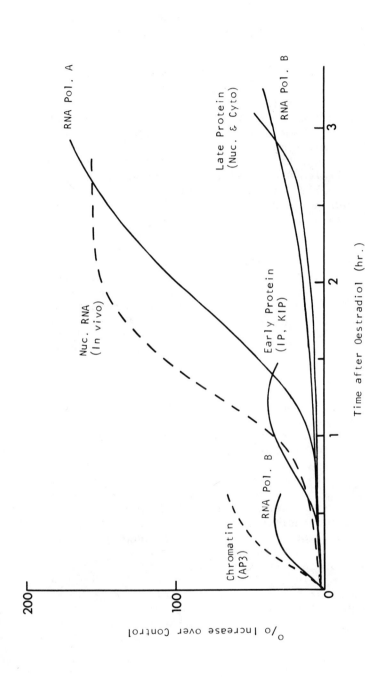

Figure 3: Schematic depiction of the sequential responses of the nuclei of target cells (oviduct, uterus, etc.) to the hormone oestrogen. The earliest transcriptional response is a small but significant change in chromatin template and RNA polymerase B (or II) activities, followed by an increase in amount of a short-lived protein ("early protein") of as yet unknown function. The next event is a massive increase in all classes of nuclear RNA, most marked for ribosomal RNA and the activity of RNA polymerase A (or I). The last step is the appearance of specific mRNA sequences in the nucleus and cytoplasm leading to the appearance of hormone-specific proteins ("late proteins").

sequences in the primary transcript generated in the nucleus. Before such a transcript is converted into the cytoplasmic messenger RNA, it has to undergo certain modifications, particularly polyadenylation, and the addition of specific proteins involved in transport of messenger RNA from the nucleus to the cytoplasm.

In my Laboratory, studying the induction of vitellogenin by oestradiol in male frogs, we have observed that the polyadenylated precursor to the putative vitellogenin messenger is highly compartmentalized within the nucleus. We find that high molecular weight poly(A)-rich RNA is found associated with euchromatin or with interchromatin granules which may, presumably, represent the state in which processed messenger RNA is packaged before release to the cytoplasm. The poly(A)-rich RNA found in the nuclear sap is of a molecular weight much smaller than would be expected for most animal protein messenger RNA's and, therefore, represents degradation products of the rapid turnover of non-messenger, non-ribosomal RNA within the nucleus. Such a compartmentation shows that the sites of polyadenylation and transcription may be very close to one another and also highlights the importance of the structural organization of the nucleus when studying the functional responses of this type. Perhaps the best example one can cite of a coupling between transcription and nuclear structure is that of the phenomenon of "gene puffing" observed in the development of some insect larval tissues triggered off by the insect metamorphic hormone, ecdysone.

Cytoplasmic Protein Synthesis

Soon after the burst in nuclear activity related to the synthesis or processing (cleavage, polyadenylation, ribonucleoprotein particle formation) one observes a change in the activity of the protein synthetic apparatus in the cytoplasm as a consequence of hormonal stimulation. Such a change may either involve the appearance of new proteins, as in the case of oestrogen-induced ovalbumin or vitellogenin synthesis, or an overall increase in the rate of amino acid incorporation per ribosome, as in the case of a general enhancement of protein synthesis during growth of accessory sexual tissues (uterus, prostate, etc.) by oestrogen and androgen. In either case, the pattern of changes accompanying those in protein synthesis in the cytoplasm are the same, except, of course, that in the former situation there is the induction of specific messengers. The increase in protein synthetic activity is not abrupt but relatively more gradual than that observed for the stimulation of nuclear RNA synthesis. The change in the protein synthetic rate per ribosome is accompanied by a very complex alteration at multiple levels of the translational process.

These co-ordinated changes in translational activity include:

(a) an increase in the number of newly-formed ribosomes per cell;

(b) an increase in the size of polysomes or a decrease in the monomeric or sub-unit forms of ribosomes;

(c) an alteration in the relative numbers of polysomes that are free or bound to membranes of the endoplasmic reticulum;

(d) an increase in the levels of several soluble components of protein synthesis, particularly tRNA and two or more elongation and initiation factors.

In fact, during the early stages of stimulation of protein synthesis the increase in the rate of amino acid incorporation is accompanied by a parallel increase in the rate at which newly-formed ribosomes accumulate in the cytoplasm. It almost seems that the hormone alters the ribosome content both qualitatively and quantitatively. It is difficult to say why a cell needs to double or treble its ribosomal population if it is only a question of induction of a few extra proteins, particularly in cells that already have a high complement of ribosomes, before their exposure to the growth or developmental hormone. A clue to this problem may lie in considering the changes that occur in the ultrastructural distribution of ribosomes during such a stimulation.

Cellular Structural Changes

As ribosomes accumulate in the cytoplasm during the onset of response to a growth or developmental hormone, one observes an increasing number of ribosomes that are bound to the membranes of the endoplasmic reticulum. This is not merely a redistribution of ribosomes on existing membranes, but represents a true proliferation of the rough endoplasmic reticulum. In other words, the increase in the number of ribosomes is accompanied by an enhanced synthesis and assembly of membranes to which these additional ribosomes are bound. This feature is illustrated in Figure 4 which summarizes some of the major events occurring in a tadpole liver following the induction of metamorphosis by thyroid hormones. In this developmental situation, there is both a preferential induction of certain proteins (enzymes of the urea cycle, mitochondrial proliferation, induction of serum albumin) and an overall increase in protein synthesis in the liver cell. Just prior to the appearance of new proteins, one observes the appearance of newly-formed ribosomes and membranes to which they are bound, as indicated by the measurement of labelled RNA and phospholipid in the rough endoplasmic reticulum by sub-cellular fractionation.

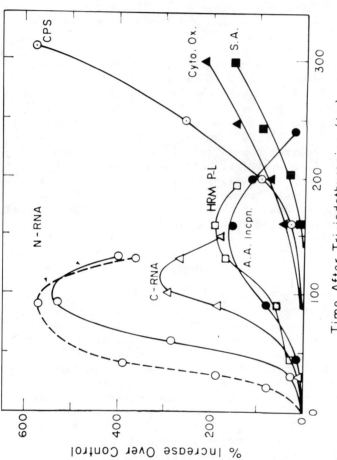

Figure 4: Schematic representation of the changes occurring in bullfrog tadpole liver upon induction of metamorphosis by thyroid hormone (tri-iodothyronine). The first major event is a burst of nuclear RNA (nRNA) synthesis followed by the appearance of additional RNA in the cytoplasm (cRNA). Some hours later, there is a gradual accumulation of ribosomes and membranes of the endoplasmic reticulum, the latter determined by labelling of membrane phospholipids (HRM-PL). A day later one can observe the preferential formation of urea cycle enzymes, particularly carbamyl phosphate synthetase (CPS), cytochrome oxidase and serum albumin (SA).

This type of co-ordinated proliferation of a major intracellular structure is typical of almost all situations whether hormonally or non-hormonally regulated and not restricted to just amphibian metamorphosis. Since the only major well-established function associated with the binding of ribosomes with membranes of the endoplasmic reticulum is that of facilitation of secretion of proteins, it is important to note that this type of proliferation is found during the development of most cells whether or not they are predominantly protein secretory. Thus, even during the accumulation of extra intracellular proteins, or the specific induction of the proteins that do not leave the cell, we have a proliferation of rough endoplasmic reticulum. This suggests that there may be some other functions, besides that of secretion, for the attachment of ribosomes to membranes of the endoplasmic reticulum and which may be critical during developmental or adaptational change involving the induction of specific proteins. What this other function may be is difficult to specify at the moment, but it can be hypothesized that the attachment of newly-formed and programmed polyribosomes to membranes may accomplish a kind of segregation within the cell of populations of polyribosomes engaged in the synthesis of different classes of proteins. This is schematically expressed in Figure 5 as a sequence of events occurring in the target cell between the time it first sees the hormone and the appearance of specific proteins which characterise a given developmental process. Finally, it is worthwhile mentioning that during the initial phase of proliferation, indeed, one sees in the electron microscope dense pockets of newly-formed rough endoplasmic reticulum in the perinuclear region and it would be most interesting to design experiments showing that it is in this part of the cytoplasm that the induced proteins are first synthesized.

RAPID AND SLOW RESPONSES TO GROWTH AND DEVELOPMENTAL HORMONES

The sequential events concerning protein synthetic activity of a cell modified by an appropriate hormonal stimulus, that we have considered above, are all relatively slow manifestations on a biochemical time-scale. It is known that most hormones reach their target sites very rapidly (within a minute or two) and yet there is a fairly long lag period (of the order of 0.5 - 6 hours) before the earliest response of any of the steps of the RNA synthesizing machinery can be detected. This raises the question of whether there are any other responses of the cell, which may or may not be relevant to the regulation of protein synthesis, which may occur very rapidly following the interaction between the hormone and the target cell.

The answer to the above question is that, indeed, the same growth and developmental hormone, which provokes changes in the

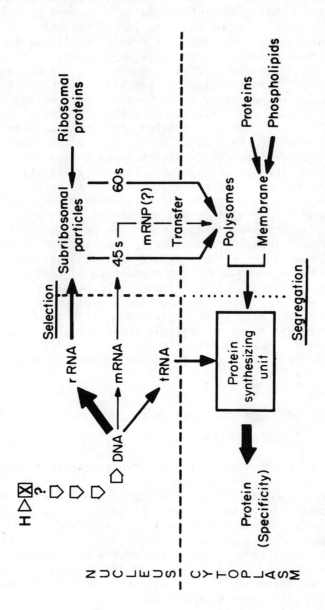

Figure 5: Scheme summarizing the sequential events in the nucleus and cytoplasm of a target cell following its stimulation by a growth and developmental hormone (H). The site of action of the hormone or its "receptor" can either be in the nucleus or the cytoplasm. The width of the arrow gives relative magnitude of the response.

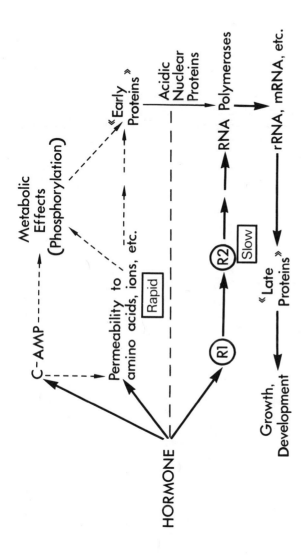

Figure 6: Scheme suggesting the rapid and slow responses to growth and developmental hormones. The major physiological actions would arise out of a regulation of protein synthesis in the target cell via a series of interactions depicted in Figure 5 and through specific receptors (R1, R2). However, the same hormone would elicit rapid responses of a general nature, common to all hormones and metabolic regulators, such as cyclic-AMP (cAMP) or Ca^{++} ions. These effects are not thought to be causal for the late responses but are necessary to alter the metabolic milieu of the cell to facilitate the action via protein synthetic mechanisms.

protein synthetic apparatus, also elicits very rapid changes in the metabolic properties of the responsive tissue. These include a raising or lowering of the intracellular levels of cyclic-AMP and cyclic-GMP, changes in the levels of metabolically critical intracellular ions, such as Ca^{++} and Na^+, and alteration in the permeability of the tissue towards small molecular weight substances, such as amino acids, sugars, nucleotides, etc. It is interesting that these rapid responses to growth and developmental hormones are the same that one observes for rapidly acting hormones which control metabolic functions and not growth and development, such as adrenalin, insulin, etc. (see Table 1). The next question that comes up is whether the mechanisms representing rapid responses are the cause for later responses involving protein synthesis. Whereas no direct experiments have so far been possible to answer this question, the conclusion that one can draw from a number of indirect observations is that there is no cause-effect relationship. On the other hand, it is most likely that mechanisms involving changes in cyclic-AMP levels or permeability properties of the cell towards nutrients play a very important facilitative role in sustaining the later events involving protein synthesis. Some of the important multiple responses of target cells to growth and developmental hormones (also known as "pleiotypic" responses) are linked together in a hypothetical scheme in Figure 6. The link between rapid and slow responses is not yet determined and it is only a suggestion that the formation or modification (say by phosphorylation catalysed by cyclic nucleotides) of "early proteins", of the type mentioned in Figure 3, may possibly act as a co-ordinating agent between a rapid adjustment of the metabolic milieu of the cell and more specific changes in the amounts of those proteins that underlie a particular developmental process. Whatever the mechanism for such a co-ordination may be, this scheme suggests that one has to consider more than one site of action for growth and developmental hormones: one operating through a receptor(s) associated with a non-specific type of rapid metabolic regulation and the other through a highly specific receptor associated with the expression of specific genes.

FURTHER READING

1. Barrington, E.J.W.: Hormones and Evolution. The English Universities Press, London. 1964.

2. Tata, J.R.: Regulation of Protein Synthesis by Growth and Developmental Hormones. In Biochemical Actions of Hormones, ed. by Litwack, G., Vol. 1, pp. 89-133. 1970.

AGE-DEPENDENT SENSITIVITY AND SPECIFIC ISOENZYME INHIBITION OF GLUCOSE-6-PHOSPHATE DEHYDROGENASE BY DEHYDROEPIANDROSTERONE

Milka S. Setchenska, E.M. Russanov,

Julia G. Vassileva-Popova

Central Laboratory of Biophysics, Bulgarian

Academy of Sciences, Sofia 1113, Bulgaria

INTRODUCTION

Glucose-6-phosphate dehydrogenase (EC 1.1.1.43) is the rate-limiting enzyme in the hexosemonophosphate pathway and plays an important role in lipogenesis and steroid biosynthesis. The enzyme is inhibited by steroid hormones, as was first reported by Marks and Banks /1/ and by McKerns et al /2/. Levy /3/ found that dehydroepiandrosterone inhibited the NADP - but not the NAD - linked G-6-PDH isolated from mammary glands of lactating rats. Urea, glycerol, high pH and increased temperature were found to reverse the inhibition by dehydroepiandrosterone, suggesting alterations in the three-dimensional structure of the enzyme and the removal of the steroid-binding site from the proximity of the active centre. Human testicular /4/, ovarian /5/ and placental /6/ G-6-PDH are also inhibited by different steroid hormones as well as by their derivatives, analogues and conjugates, dehydroepiandrosterone (DHEA) being the most potent inhibitor in vitro. With rat liver G-6-PDH, Lopez and Rene /9/ have shown that at $5 \cdot 10^{-5}$ M dehydroepiandrosterone uniformly inhibits all four isoenzymes with complete disappearance of band D at the higher concentration of 10^{-4} M.

The object of the present study was to investigate the age-dependent sensitivity of rat heart G-6-PDH to dehydroepiandrosterone (Δ^5-Androstane-3β-Ol-17-one) as well as to look for a pattern of specific isoenzyme inhibition.

MATERIALS AND METHODS

Two groups of six male 4 weeks-old (immature) and 3 months-old (adult) rats were used. A 10,000 x g supernatant obtained from heart homogenate was used as the enzyme preparation and the reaction mixture for determination of the total G-6-PDH activity consisted of 0.1M tris-HCl buffer, pH 7.6, 0.5mM NADP and the enzyme, the reaction being started by the addition of 3.5mM glucose-6-phosphate. Dehydroepiandrosterone was dissolved in dioxane and added just before the enzyme. Dioxane alone (0.7% of the final assay volume) had no effect on the enzyme activity. Changes in optical density at 340 nm were measured at 37° in a cell with 1 cm light path at 30 sec. intervals. Protein was determined by the Lowry method. The specific activity of the enzyme was expressed in micromoles NADP reduced per minute per mg protein. The results presented were obtained from six experiments and were statistically significant.

Isoenzyme patterns were determined by the standart Davis procedure /11/ for polyacrylamide disc electrophoresis with tris-glycine electrode buffer, pH 8.3, at a constant current of 4mA per tube. On each gel 500 μg protein were layered and, after electrophoresis, the gels were stained for enzyme activity at 37° for 60 min. in the following reaction mixture: 95 μmoles tris-HCl buffer, pH 7.6, 0.9 μmoles NADP, 1 mmol $MgCl_2$, 0.04 mg PMS, 1 mg nitrotetrazolium blue and 8 mmoles glucose-6-phosphate.

RESULTS AND DISCUSSION

While Table 1 shows that the total activity of heart G-6-PDH from immature rats is about 25% higher than that from mature ones which may reflect an in vivo inhibition of G-6-PDH by the increasing endogenous steroid hormones as the rats mature. The enzyme from immature rats is also more sensitive to dehydroepiandrosterone, a 50% inhibition being obtained at 10^{-6} M with the enzyme from young rats whereas a 50% inhibition of the enzyme from old rats required a concentration of 10^{-5} M dehydroepiandrosterone. Complete inhibition of the enzyme was obtained at 10^{-5} M and 10^{-4} M for the immature and adult rats, respectively. These results further showed that rat heart G-6-PDH is more sensitive to steroid inhibition than rat liver G-6-PDH, the latter being inhibited by 50% with 5.10^{-5} M dehydroepiandrosterone.

The addition of digitonin (0.02% final concentration) to the reaction mixture abolished dehydroepiandrosterone inhibition and at the same time increased G-6-PDH activity from mature rats, but not young rats, by about 30% (Table 1). It is known that digitonin can react with the 3-hydroxy group of steroids forming an insoluble complex as well as affect hydrophobic bonds. The increase in G-6-PDH activity from mature rats by the addition of digitonin can be explained by digitonin interacting with endogenous steroids in heart homogenates since it did not increase the activity of G-6-PDH from immature rats.

Raineri and Levy /7/ suggested that the inhibitory steroids were bound to a hydrophobic pocket in the G-6-PDH molecule which had the approximate dimensions of 15 x 8 x 6 Å, one side of which appeared to be virtually planar under optimum conditions of binding since the best inhibitors tested were characterised by a large planar surface on the α-side which played the dominant role in binding the steroid to the enzyme. Our results with digitonin support the idea of hydrophobic interactions involved in the loose binding of steroid to the enzyme. Levy /3/ has shown that urea, glycerol, high pH and temperature reversed dehydroepiandrosterone inhibition which was interpreted as a result of an alteration in the three-dimensional structure of the enzyme. Our experiments with digitonin suggest that no conformational change was necessary for recovering the enzyme activity inhibited, but that only the damaging of hydrophobic bonds between steroid hormone and G-6-PDH molecule sufficed to reverse the inhibition.

From the electrophoretic isoenzyme patterns shown in Fig. 1, it was evident that rat heart G-6-PDH had three NADP-linked isoenzymes: bands A, B and C with relative electrophoretic mobilities of 0.13, 0.21 and 0.33, respectively. The most pronounced difference between the zymograms of immature and mature rats is the higher activity of band C of G-6-PDH from immature rats. Before staining, some gels were pre-incubated at room temperature for 15 min. in 0.1M tris-HCl buffer, pH 7.6, containing dehydroepiandrosterone in that concentration which caused 50% inhibition of the total G-6-PDH activity, i.e. 10^{-5} M for mature and 10^{-6} M for immature rats.

Table 1

Inhibitory Effect of Dehydroepiandrosterone (DHEA) on the Total Activity of Heart Glucose-6-phosphate Dehydrogenase from Immature and Mature Male Rats

Addition to the Assay Mixture	Immature Rats		Mature Rats	
	Specific Activity	% Activity	Specific Activity	% Activity
None	0.018	100.0	0.015	100.0
+ 10⁻⁶M DHEA	0.010	55.5	0.012	80.0
+ 5·10⁻⁶M DHEA	0.006	33.3	-	-
+ 10⁻⁵M DHEA	no activity	0.0	0.008	53.3
+ 5·10⁻⁵M DHEA	-	-	0.005	33.3
+ 10⁻⁴M DHEA	-	-	no activity	0.0
+ 10⁻⁶M DHEA+Digitonin	0.018	100.0	-	-
+ 10⁻⁵M DHEA+Digitonin	-	-	0.020	133.3
+ Digitonin only	0.018	100.0	0.020	133.3

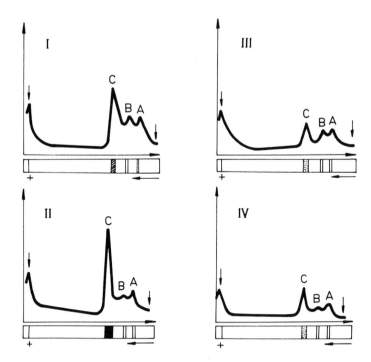

Fig. 1: Isoenzyme patterns for heart glucose-6-phosphate (G-6-PDH) dehydrogenase from mature and immature rats and the specific inhibition by dehydroepiandrosterone (DHEA) of isoenzyme band C.

I. G-6-P DH isoenzymes from mature rats
II. G-6-P DH isoenzymes from immature rats
III. G-6-P DH isoenzymes from mature rats + 10^{-5} M DHEA
IV. G-6-P DH isoenzymes from immature rats + 10^{-6} M DHEA

After such treatment the activity of band C isoenzyme was markedly inhibited while the activities of bands A and B were only slightly affected. Thus, the inhibitory effect of dehydroepiandrosterone on heart G-6-PDH is manifested mainly through the selective decrease of the band C isoenzyme activity. The lower activity of isoenzyme C from mature rats is probably due to its specific in vivo inhibition by the higher levels of endogenous steroid hormones. This selective inhibitory effect of dehydroepiandrosterone on G-6-PDH isoenzymes is a novel finding of special interest as little is known in general about the regulation of isoenzymes in mammalian tissues.

REFERENCES

1. Marks, P. & Banks, J. (1960). Proc.Natl.Acad.Sci. USA, 46, 447-452.

2. McKerns, K. & Bell, P. (1960). Recent Progr.Horm. Res., 16, 97-117.

   McKerns, K. & Kaleita, E. (1960). Biochem.Biophys. Res. Comm., 2, 344-348.

3. Levy, H.R. (1963), J.Biol.Chem., 238, 775-784.

4. Benes, P. & Oertel, G. (1971), Experientia, 27, 1157.

5. Oertel, G., Menzel, P. & Banks, D. (1971), Steroidologia, 1, 2-7.

6. Benes, P. & Oertel, G. (1972), J.Steroid.Biochem., 3, 799-801.

7. Raineri, R. & Levy, H.R. (1970), Biochemistry, 9, 2233-2243.

8. Lopez, S.A. & Rene, A. (1973), Proc.Soc.Exp.Biol.Med. 142, 258-261.

9. Davis, B. (1964), Ann.N.Y.Acad.Sci., 121, 404-427.

THE BIOSYNTHESIS OF HAEMOGLOBIN, ENZYMES AND NUCLEIC ACIDS DURING DIFFERENTIATION AND DEVELOPMENT OF ERYTHROID BONE MARROW CELLS

H.R.V. Arnstein

Department of Biochemistry, University of London

King's College, Strand, London WC2R 2LS, England

INTRODUCTION

In adult animals, mature red blood cells arise by the differentiation and development of stem cells in the bone marrow. This process is complex and takes place in a number of well-defined stages distinguishable on the basis of morphological, physiological and biochemical criteria. The underlying biochemical mechanisms are as yet not well understood, but their elucidation may provide useful information about the regulation of developmental systems in general.

The most primitive, undifferentiated cell, termed the multipotent or pluripotent stem cell, is capable of both self-renewal and differentiation into all other haematopoietic cell lines. When injected into lethally irradiated mice, multipotent stem cells are capable of forming colonies of haematopoietic cells in the spleen and hence this cell is also known as the colony forming unit (CFU).

A second stem cell, which responds to the hormone erythropoietin (Ep) by increased proliferation to recognizable erythroid cells, has been characterized on the basis of its differential sensitivity to irradiation and differences between the subsequent recovery pattern of the multipotent cell number and the erythropoietin response. For example, after chronic irradiation of mice, the number of multipotent stem cells was observed to fall to less than 1%, whereas the response to Ep was 20% of normal (Porteous and Lajtha, 1966). Moreover, suppression of erythropoiesis by hypertransfusion does not affect the number of spleen colony

forming units present in the marrow (Till et al., 1967) and
deficient production of erythroblasts has been shown to occur in
a genetically defective strain of anaemic mice even though the
content of multipotent cells was normal (Fowler et al., 1967).

Following stimulation by Ep, the ERC compartment gives rise
to proerythroblasts, which undergo a limited number of mitotic
cell divisions without much change in total mass so that the cell
volume is approximately halved at each division. On the basis of
this difference in cell volume, bone marrow erythroblasts have
been separated by velocity sedimentation into five fractions of
relatively homogeneous cell populations at different stages of
development (McCool et al., 1970; Denton and Arnstein, 1973).

THE RESPONSE OF EARLY ERYTHROID CELLS TO ERYTHROPOIETIN

The erythropoietin-responsive cell (ERC) compartment turns
over continuously with a cell generation time of approx. 9 hours
(Hodgson, 1967). Studies of the time of maximum and minimum
sensitivity of the Ep response to irradiation or to methotrexate,
a drug which blocks cells in S-phase, suggest that cells are
responsive to Ep immediately following mitosis, i.e. in G_1, after
which stimulated cells enter S-phase (Hodgson, 1967; Hodgson and
Eskuche, 1968). In addition to the effect of Ep on the
differentiation of the Ep-responsive cells, the hormone appears
to be required also for the further development and maturation of
proerythroblasts (Clissold, 1974).

At the molecular level, an early effect of erythropoietin
involves stimulation of RNA synthesis. In foetal liver, there is
also an increase in DNA synthesis, within 20 minutes after
addition of Ep, and this is followed by a period of enhanced
synthesis of both RNA and protein. The secondary response of RNA
synthesis is dependent on synthesis of DNA, since synthesis of
both RNA and haemoglobin are abolished by inhibitors of DNA
synthesis, such as fluorodeoxyuridine (Paul and Hunter, 1969),
hydroxyurea and cytosine arabinoside (Paul et al., 1972).

In cultured bone marrow cells, Ep stimulates the synthesis of
RNA within 5 - 15 minutes (Gross and Goldwasser, 1969; Goldwasser,
1972) whereas DNA synthesis is increased only later (Gross and
Goldwasser, 1970) and fluorodeoxyuridine does not affect the
response of RNA synthesis (Gross and Goldwasser, 1970; Goldwasser,
1972). The earliest effect of Ep on RNA synthesis appears to be an
increase of a very large (150 S) nuclear species, but subsequently
there is stimulation of the synthesis of other RNAs, including
ribosomal RNA as well as a 9 S RNA, which may be globin messenger
(Gross and Goldwasser, 1971).

The increase in protein synthesis by bone marrow cells after addition of Ep is not restricted to haemoglobin, but synthesis of δ-aminolaevulinic acid synthetase (Bottomley and Smithee, 1969) as well as of a protein concerned with the uptake of iron (Hrinda and Goldwasser, 1969) by marrow cells has also been shown to be stimulated. Furthermore, in cell cultures of goat erythroblasts, high levels of Ep induce a switch from haemoglobin A to haemoglobin C synthesis (Barker et al., 1973) which occurs also in anaemic animals.

Recent observations suggest that the effect of Ep on the synthesis of RNA is probably indirect since no stimulation of RNA synthesis by isolated bone marrow cell nuclei could be obtained with Ep, whereas a cytoplasmic protein factor present in extracts of bone marrow cells previously incubated with the hormone was found to be stimulatory in this system (Chang and Goldwasser, 1973). It is of interest also that neither cyclic AMP nor other cyclic nucleotides appear to be involved in the effect of Ep on RNA synthesis (Chang and Goldwasser, 1973).

BIOCHEMICAL CHANGES IN BONE MARROW ERYTHROBLASTS DURING DIFFERENTIATION AND CELL DEVELOPMENT

Following stimulation of erythropoietin-sensitive cells by the hormone progeny cells arise which are active in macromolecular synthesis until the cessation of cell division at the orthochromatic cell stage (Denton and Arnstein, 1973). Synthesis of both RNA and DNA appears to continue undiminished throughout the period of active cell division, as judged by the incorporation of labelled precursors into the macromolecules. On the other hand, there is a continuous decline in total protein synthesis per cell in parallel with a decrease in cytoplasmic RNA and protein content (Denton and Arnstein, 1973; Denton, 1973). The observed changes in protein synthesis and RNA content per cell are, however, mainly if not exclusively a consequence of the reduction in cell volume which takes place during this period of cell development, as discussed previously. If one considers macromolecular synthesis not on the basis of cell numbers but in terms of the cell volume of the progenitor proerythroblast, protein synthesis remains essentially unchanged and nucleic acid synthesis increases, as one would expect from the fact that the amount of genetic material is doubled at each cell division. Thus, the main result of the changes taking place in the dividing cell compartment is a considerable amplification in nucleic acid synthesis. Qualitatively, there are some alterations in the nature of cellular proteins, consisting principally of small increases in carbonic anhydrase and haemoglobin, but several other enzyme activities show little change during this period (Denton, Spencer and Arnstein, 1975; Hulea, Denton and Arnstein, 1975).

After the final cell division, which gives rise to the orthochromatic erythroblast, the synthesis of both DNA and RNA stops abruptly and condensation of the nucleus takes place, possibly as a result of the accumulation of haemoglobin (Yataghanas, Gahrton and Thorell, 1970). Although protein synthesis continues for a considerable time, there is a marked change in the pattern of enzyme activities present in these non-dividing cells. In particular, glucose-6-phosphate dehydrogenase, 6-phosphogluconate dehydrogenase, adenosine deaminase, nucleoside phosphorylase, ribonuclease, and, to a lesser extent, lactate dehydrogenase decreased, whereas adenylate kinase and subsequently catalase showed increased activity (Denton et al., 1975, Hulea et al., 1975). At this stage, there is also a marked increase in haemoglobin synthesis and in the mobilization of iron from the cell stroma (Denton et al., 1974).

The complex changes in the enzyme pattern after the final cell division cannot be due to regulatory mechanisms involving transcription, since there is no significant synthesis of RNA in the non-dividing erythroblast (Denton and Arnstein, 1973). In the case of changes in enzyme activities from an initially high value to a relatively low but stable level, it is tempting to speculate that this may be due to the preferential destruction of some enzymes but not others by proteases released as a result of the rupture of lysosomes, which appears to take place at this time (S.A. Hulea and H.R.V. Arnstein, 1975, unpublished observations). Similarly, lysosome breakage would be expected to liberate nucleases and it is possible to envisage that some messenger RNAs would then be degraded whereas others, including globin mRNA, might be resistant to inactivation. In any case, it is difficult to explain the observed decreases except on the basis of a unique event involving degradation of perhaps 90% of the relevant enzymes.

It is also necessary to account for the marked increase in globin synthesis by erythroblasts after the final cell division (Denton, 1973), as well as the increases in carbonic anhydrase and catalase activities, which may also arise from changes in the synthesis of these enzymes, although a process of activation of pre-existing enzymes cannot be excluded at present. One possibility is that competition between globin and other messengers is largely eliminated as a result of the degradation of unstable messengers, and the protein-synthesizing machinery of the cell thus becomes available entirely for globin synthesis. Short-lived mRNAs might include those molecular species coding for proteins involved in the cell cycle, for example histones and components of the mitotic spindle, which would, of course, no longer be required after cessation of DNA synthesis and cell division. The production of histone approaches 1 pg/cell/hour (Denton, 1973), which is similar to the rate of haemoglobin synthesis by reticulocytes. Thus, cessation of histone synthesis alone could lead to a doubling of haemoglobin synthesis.

SUMMARY AND CONCLUSION

The information available at present suggests that one may consider the development of bone marrow erythroid cells as taking place in three distinct phases. First, there exists an initial trigger mechanism consisting of the stimulation of the erythropoietin-responsive cell by the hormone. Secondly, stimulation of the ERC results in cell proliferation and hence amplification of the genetic material relative to cell size. During this phase, a limited amount of globin is synthesized, but only as one component of a wider spectrum of macromolecules. Following condensation of the nucleus, a profound alteration in the enzyme pattern of the erythroblast occurs. There appears to be a considerable breakdown of some proteins and possibly also of certain mRNAs, leaving the protein—synthesizing machinery of the cell entirely available for globin biosynthesis until eventually globin mRNA is also degraded and all protein synthesis stops.

ACKNOWLEDGEMENTS

It is a pleasure to acknowledge stimulating discussions with Dr. M.J. Denton during preparation of this paper. Financial support of the Medical Research Council and the Science Research Council for the experimental work carried out by Dr. Pat Clissold, Dr. M.J. Denton and Mr. S. Hulea in this department is also gratefully acknowledged.

REFERENCES

Barker, J.E., Last, J.A., Adams, S.L., Nienhuis, A.W. and Anderson, W.F. (1973). Proc. Nat. Acad. Sci. U.S.A. 70, 1739.
Bottomley, S.S. and Smithee, G.A. (1969). J.Lab.Clin.Med. 74, 445.
Chang, C.-S. and Goldwasser, E. (1973). Develop. Biol. 34, 246.
Clissold, P.M. (1974). Exp. Cell Res. 89, 389.
Denton, M.J. (1973). Ph.D. thesis. University of London.
Denton, M.J. and Arnstein, H.R.V. (1973). Brit. J. Haemat. 24, 7.
Denton, M.J., Delves, H.T. and Arnstein, H.R.V. (1974). Biochem. Biophys. Res. Commun. 61, 8.
Denton, M.J., Spencer, N. and Arnstein, H.R.V. (1975). Biochem. J. 146, 205.
Fowler, J.H., Till, J.E., McCulloch, E.A. and Siminovitch, L. (1967). Brit. J. Haemat. 13, 256.
Goldwasser, E. (1972). In "First International Conference on Haemotopoiesis. Regulation of Erythropoiesis", ed. by A.S. Gordon, M. Condorelli and C. Peschle, Il Ponti: Milan pp. 227-235.
Gross, M. and Goldwasser, E. (1969). Biochemistry, 8, 1745.
Gross, M. and Goldwasser, E. (1970). J. Biol. Chem. 245, 1632.
Gross, M. and Goldwasser, E. (1971). J. Biol. Chem. 246, 2480.
Hodgson, G. (1967). Proc. Soc. Exp. Biol. Med. 125, 1206.

Hodgson, G. and Eskuche, I. (1968). Ann. N.Y. Acad. Sci. 149, 230.
Hrinda, M.E. and Goldwasser, E. (1969). Biochim. Biophys. Acta, 195, 165.
Hulea, S., Denton, M.J. and Arnstein, H.R.V. (1975). FEBS Lett. 51, 346
McCool, D., Miller, R.J., Painter, R.H. and Bruce, W.R. (1970). Cell Tissue Kinetics, 3, 55.
Paul, J., Freshney, I., Conkie, D. and Burgos, H. (1972) In "First International Conference on Haematopoiesis. Regulation of Erythropoiesis", ed. by A.S. Gordon, M. Condorelli and C. Peschle, Il Ponti: Milan pp. 236-244.
Paul, J. and Hunter, J.A. (1969). J. Mol. Biol. 42, 31.
Porteous, D.D. and Lajtha, L.G. (1966). Brit. J. Haemat. 12, 177.
Till, J.E., Siminovitch, L. and McCulloch, E.A. (1967). Blood, 29, 102.
Yataghanas, X., Gahrton, G. and Thorell, B. (1970). Exp. Cell Res. 62, 254.

MATURATION DEPENDENCE OF ADENYLATE CYCLASE IN RED BLOOD CELLS

D. Maretzki, M. Setchenska[*] and A. G. Tsamaloukas

Institute of Biochemistry, Humboldt-Universität, Berlin, GDR

[*]Central Laboratory of Biophysics, Bulgarian Academy of Sciences, Sofia 1113, Bulgaria

Originally adenylate cyclase activity was found only in nucleated erythrocytes (1). Sheppard and Burghardt demonstrated its occurrence and its stimulation by catecholamines and NaF also in the cell membrane of mammalian erythrocytes (2). Gauger et.al. (3) concluded, that adenylate cyclase was confined mainly to the reticulocytes.

The aim of our work was to investigate which cell-type of heterogeneous reticulocyte populations, produced by anemia, had the highest activity of the adenylate cyclase.

Reticulocytosis was produced in rabbits by bleeding anemia over a period of 8 days. White cells were separated by means of cotton filled columns. The red blood cells were fractionated by density gradient centrifugation in dextrane (16 - 23.5%, pH 6,9) as described by Gross et. al. (4). We separated 3 fractions of red blood cells during the anemia as shown in Figures 1-4.

Fraction 1 contained nearly 90% reticulocytes (range of density: 1.028 - 1.053 g/cm^3)

Fraction 2 contained young erythrocytes, which are morphologically no longer identified as reticulocytes (range of density: 1.051 - 1.073 g/cm^3)

Fraction 3 contained erythrocytes (range of density: 1.071 - 1.101 g/cm^3)

The maturation of the cell fractions was characterized by determination of the RNA-content (see Fig.1) by the method of Coutelle (5), and of the GOT-activity (see Fig.2) The reticulocytes produced by bleeding anemia

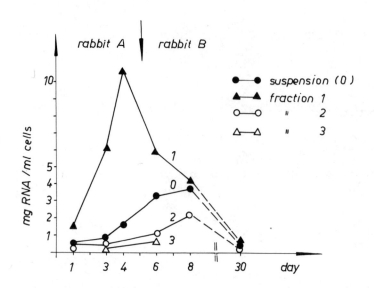

Fig.1. Changes of RNA-content during the course of rabbit anemia. The cells were fractionated in a density gradient (see text). Bleeding procedure: 50-60 ml blood per day and rabbit from the 1-st to 8-th day

could be differentiated in 3 populations as described by Rapoport et.al. (6)

1) On the first day normal sized reticulocytes with about 2-4 mg RNA/ml cells (Fig.1) and low lewels of several enzymes (for instance GOT, Fig.2) were found.

2) On the fourth day or even on the third a nearly pure population of cells appeared, which was characterized by large RNA-content (Fig.1) and was fairly rich in hemoglobin. These cells showed also a high GOT-activity (Fig.2). The cells of this population correspond to the line-2 cells as they were termed by Borsook (7).

From the 6-th day onwards appeared megaloreticulocytes with sizes 2 or 4 times as large as normal reticulocytes and with RNA-content of about 5 mg/ml cells (Fig.1). In contrast to the decline of RNA-content the GOT-activity remained high in megaloreticulocytes (Fig.2) which were released into the peripheral blood after skipped cell divisions.

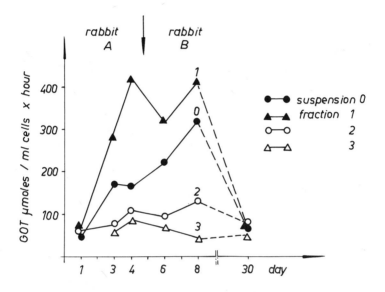

Fig.2. Time course of GOT activities in red blood cells during anemia.

We studied the adenylate cyclase activity in these distinct stages of red cells development to find out whether the cyclase was dependent on the maturation process. Cell membranes were prepared after hemolysis by frezing and thawing. The fluoride-stimulated adenylate cyclase activity was not influenced substantially if the membranes were prepared after osmotic shock or moderate sonications at 1-2 W/cm^2.

The adenylate cyclase activity stimulated by 10mM NaF was measured using the method of Krishna (8). The recovery was 10-20% with (^{3}H)-cAMP as internal standard after chromatography on DOWEX 50 WX 4.

Reticulocytes were counted in blood smears after staining with brilliant cresyl blue. Protein was determined according to Lowry et.al. (9).

In Figure 3 behaviour of adenylate cyclase activity during the anemia is demonstrated. The highest activity was found on the 3-rd and 4-th day when the line-2 cells were produced. The cyclase activity of megaloreticulocytes decreased on the 6-th and 8-th day.

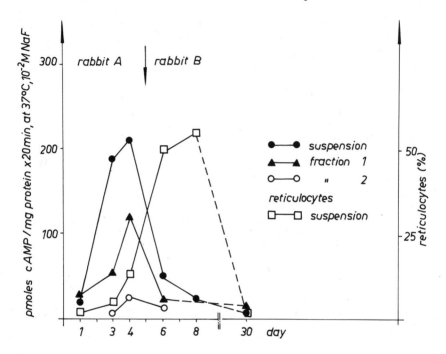

Fig.3. Changes of NaF-stimulated adenylate cyclase activities (left-hand ordinate) and of reticulocyte counts (right-hand ordinate) during anemia. Fraction 1 and 2 were obtained after density gradient centrifugation of the suspension (see text).

It may be assumed that these cells released after skipped cell division show less adenylate cyclase activity than the line-2 cells. Furthermore we could not find cyclase activity in the mature erythrocytes. It may also be seen from Fig.3 that the whole blood suspension has a higher cyclase activity than the separated fractions. We assume that the cyclase was inactivated due to numerous washings, which were necessary to discard the dextrane or that dextrane produced an inactivation of the enzyme. Therefore, the recovery in the reticulocyte fractions was poor. The cyclase activity was in correlation with the increase of the RNA-content in line-2 reticulocytes (Fig. 4). These cells had also high levels of GOT, cyto-

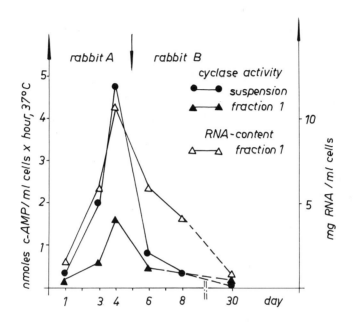

Fig.4. Corelation between adenylate cyclase activities (left-hand ordinate) and RNA content (Right-hand ordinate).

chromoxidase, 5-aminolevulinate-synthetase and inorganic pyrophosphatase (6).

Even if assumed that hormonal action approaches the stimulation obtained by fluoride one can estimate that the adenilate cyclase capacity in reticulocytes may produce levels of c-AMP in the range of µM/l cells. It has been reported that c-AMP in µM levels can enhance erythropoesis (10, 11).

Several authors have found that reticulocytes from rabbits (12) and also human erythrocytes (13) contain c-AMP stimulated protein kinases. Concerning the functional role of c-AMP in red blood cell, preliminary results indicate, that further reactions mediated by c-AMP ought to occur predominantly in reticulocytes and erythroblasts.

REFERENCES

1. Davoren, P.R. and Sutherland, E.W.:
   J. biol. Chem. 238, 3009 - 3015 (1963)

2. Sheppard, H. and Burghardt, C.:
   Biochem. Pharmacol. 18, 2576-78, (1969)

3. Gauger, D., Palm, D., Kaiser, G. and Quiring, K.:
   Life Sciences 13, 31-40 (1973)

4. Gross, J., Hartwig, A., Botscharowa, L.,
   Syllm-Rapoport, I. and Rosenthal, S.:
   Acta biol. med. germ. 29, 765-775 (1972)

5. Coutelle, Ch.: Analyt. Biochem. 32, 489-495 (1973)

6. Rapoport, S.M., Rosenthal, S., Schewe, T.,
   Schultze, M. and Müller, M. in :
   Cellular and Molecular Biology of Erythrocytes ed.
   by H.Yoshikawa and S.M. Rapoport,
   Univ. Tokyo Press, Tokyo, pp. 93-141 (1974)

7. Borsook, H., Lingrel, H., Scaro, J.L. and Milette, R. L.:
   Nature 196, 347 - 350 (1962)

8. Krishna, G. and Weiss, B.:
   Pharmac. Exp. Ther. 163, 379-385 (1968)

9. Lowry, O.H., Rosebrough, N.J., Farr, A. L. and Randau, R.J.:
   J. biol. Chem. 193, 265-275 (1951)

10. Schooley, J.C. and Mahlmann, L.J.:
    Proc. Soc. Exp. Biol. Med. 137, 1289-1292 (1971)

11. Norris, A., Gorshein, D., Besa, E.C., Leonard, R. A.
    and Gardener, F. H.,:
    Proc. Soc. Exp. Biol. Med. 145, 975-78 (1974)

12. Tao, M., Salas, M. L. and Lipmann, F.:
    Proc. Nat. Acad. Sci. U.S. 67, 408-414 (1970)

13. Rubin, C.S., Erlichman, J. and Rosen, O. M.:
    J. biol. Chem. 247, 6135-39, (1972)

CALCIUM AND CYCLIC AMP EFFECTS ON RABBIT EPIDIDYMAL SPERMATOZOA

Bayard T. Storey

Dept. of Obstetrics & Gynecology, School of Medicine

Univ. of Pennsylvania, Philadelphia, PA 19174, U.S.A.

The divalent cation Ca^{+2} plays a central role in many physiological processes through its control of many properties and functions of cell membranes (1,2). Morita and Chang (3) have reported that Ca^{+2} was required to maintain the motility of hamster spermatozoa incubated at 37°C, but that the motility of guinea pig, rat and rabbit spermatozoa was maintained in the absence of this ion. Effects of Ca^{+2} on rabbit spermatozoa might be expected, since rabbit oviduct fluid contains about 2 mM Ca^{+2}, as originally shown by Holmdahl and Mastroianni (4). Increased respiration by rabbit spermatozoa in rabbit oviduct fluid has been demonstrated (5,6), but this has been attributed to HCO_3^- in the fluid. The effect of Ca^{+2} in this regard has never been assessed. We have shown in this laboratory (7,8) a direct effect of Ca^{+2} on the mitochondria in hypotonically treated rabbit sperm cells in which mitochondria are intact and accessible to added reagents: respiration is increased as Ca^{+2} is taken up in an energy-linked process which competes rather poorly for the high energy intermediates of oxidative phosphorylation. The physiological significance of this mitochondrial Ca^{+2} uptake is as obscure for the sperm cell as it is for other mammalian cells in which the uptake is far more efficient (9-11).

One possible effect of exogenous Ca^{+2} on the intact sperm cell could be on the cyclic AMP of the cell via activation of adenyl cyclase. Morton et al. (12) have reported that Ca^{+2} activates motility in quiescent hamster epididymal sperm through increased cyclic AMP synthesis and that motility can be restored to poorly motile cells by 10 mM caffeine. The possible role of intracellular cyclic AMP in the initiation and maintenance of sperm motility has been recently discussed (13-15), as has the stimulatory effect of phosphodiesterase inhibitors on sperm metabolism and motility (16-18). It is evident from these reports that cyclic AMP can have profound effects on spermatozoa, and equally evident that the means by which these effects are produced are as yet unclear, including the possible role of Ca^{+2}. In this paper, we report some findings concerning the effects of Ca^{+2} and cyclic AMP on the metabolic state of intact epididymal rabbit spermatozoa.

MATERIALS AND METHODS

Spermatozoa were flushed from the excised epididymides of mature male White New Zealand rabbits and washed once with a medium composed of 75 mM NaCl, 75 mM KCl, 0.2 mM EDTA, and 20 mM Tris, pH 7.4, followed by centrifugation at 800g. The sedimented sperm cells were washed twice in a medium composed of 150 mM NaCl and 20 mM Tris, pH 7.45, and resuspended in this medium. The respiratory activity of both intact and hypotonically treated spermatozoa was determined on cell suspensions at 25° using the miniaturized oxygen electrode assembly described previously (19). Uptake of Ca^{+2} added to the suspending medium was monitored by means of the metallochromic indicator murexide (20) using the dual wavelength spectrophotometer (21) with the wavelength pair 507 and 540 nm. Changes in the redox state of cytochrome b were recorded with the same spectrophotometer with the wavelength pair 429 and 410 nm.

The uncoupler of oxidative phosphorylation, bis (hexafluoracetonyl) acetone, designated 1799, was kindly supplied by Dr. Peter Heytler of E. I. duPont de Nemours, Inc. The ionophore A23187 was the gift of Dr. Robert Hamill of Eli Lilly Company and the diesterase inhibitor SQ20009 of Dr. Sidney M. Hess of the Squibb Institute for Medical Research.

RESULTS

Spermatozoa flushed from the cauda epididymis of a mature rabbit in Tris-saline buffer containing 0.2 mM EDTA were motile; subsequent washing with Tris-saline buffer alone had no apparent effect on motility. The percentage of motile cells was variable, the range being 70% to 90%. Addition of pyruvate or glucose did not change the percent motility, nor did addition of Ca^{+2}.

If the vigor of motility in a sperm cell is increased by added Ca^{+2} or other agents, the greater activity of the flagellar ATPase should elicit increased respiratory activity as intracellular ADP rises and the mitochondria approach the fully active state of oxidative phosphorylation, State 3 (22). The lack of effect of Ca^{+2} on the respiratory rate of rabbit epididymal sperm isolated and resuspended in Tris-NaCl buffer is shown with glucose as substrate in Fig. 1. Addition of glucose gives a burst of oxygen consumption followed by return to a rate 1.6 fold greater than the endogenous rate (Fig. 1A), which is not affected by added Ca^{+2}. This respiration rate is sensitive to partial inhibition by oligomycin. The maximal respiratory rate is then observed on addition of the uncoupler of oxidative phosphorylation, 1799. In Fig. 1B, the sperm cells are treated with the antibiotic A23187 prior to Ca^{+2} addition. This compound has been shown by Reed and Lardy (23) to act as an ionophore for divalent cations, in particular, Mg^{+2} and Ca^{+2}: in its presence, Ca^{+2} can be rapidly transported across the plasma membrane of cells. Addition of Ca^{+2} to the A23187-treated spermatozoa stimulates the rate of respiration which is now not sensitive to inhibition by oligomycin. This effect is precisely that expected if Ca^{+2} passed through the cell membrane and was taken up by the mitochondria. Note that A23187 alone has no effect (Fig. 1B) and that Ca^{+2} alone has no effect (Fig. 1A); both must be present to give the stimulation of respiration. A similar result with bovine spermatozoa has been reported by Reed and Lardy (24). The effect of Ca^{+2} on the respiration rate of these cells utilizing pyruvate as substrate in the presence and absence of HCO_3^- is like that observed with glucose: oligomycin-insensitive stimulation of respiration by Ca^{+2} is observed only in the presence of the ionophore A23187.

Mitochondrial Ca^{+2} uptake in hypotonically treated rabbit spermatozoa competes poorly with oxidative phosphorylation for the mitochondrial high-energy intermediates generated by substrate oxidation and is best realized under highly non-physiological conditions: succinate as energy source, with oligomycin and rotenone present to block oxidative phosphorylation and energy-linked reverse electron transport from succinate to pyridine nucleotide, respectively (8). Stimulation of respiration is observed under conditions more comparable to those in intact cells, namely, KCl-Tris buffer containing phosphate, ATP, and Mg^{+2} with pyruvate plus malate as substrate. Under these conditions, a steady state of ADP concentration is established between ATP synthesis by oxidative phosphorylation and ATP breakdown by the flagellar ATPase which is activated by Mg^{+2} (19,24). The rate of Ca^{+2} uptake measured by the metallochromic indicator murexide, is shown in Fig. 2A. Uptake does occur, but is slower by a factor of about five compared to that observed with succinate plus oligomycin and rotenone. The oxygen consumption rate increases markedly over that attained in the steady state with ATP; but after about 3 minutes, there is a slowing of the rate (Fig. 2B). (Addition of 0.5 mM cyclic AMP instead of Ca^{+2} at the point indicated in Fig. 2B gives a slight inhibition of the respiration rate). The effect of Ca^{+2} on the mitochondrial energy state as reported by the redox state of cytochrome b is minimal under these conditions, as shown by the

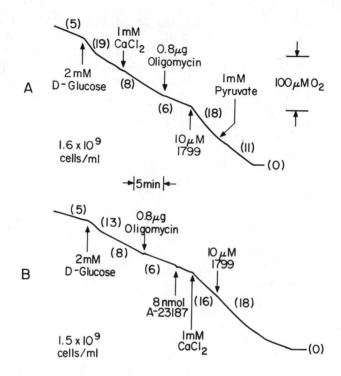

Fig. 1: Respiration rate of rabbit epididymal sperm with D-Glucose as substrate as determined with the oxygen electrode. The medium was 150 mM NaCl, 20 mM Tris, pH 7.45. Numbers in parenthesis give respiration rates as ng atoms O/min.-10^9 cells. A. Effect of Ca^{+2} addition followed by oligomycin. B. Effect of Ca^{+2} addition after treatment of cells with oligomycin and the ionophore A23187. Uncoupler 1799 was added at the end of the experiment to give the maximal rate. The point marked (0) indicates that the suspension has become anaerobic.

spectrophotometric record of Fig. 2C. A barely perceptible reduction of the cytochrome b component is observed, as compared to a marked reduction in the presence of oligomycin (trace not shown), and this is followed by a continuous reoxidation until anaerobiosis is reached. The slowdown in oxygen consumption after stimulation caused by Ca^{+2} addition in this concentration range has been observed in other mammalian mitochondria and shown to be due to complex formation by the intramitochondrial pyridine nucleotide and the Ca^{+2} taken up (25). This reaction would be expected to cause oxidation of cytochrome b, as is indeed observed. In

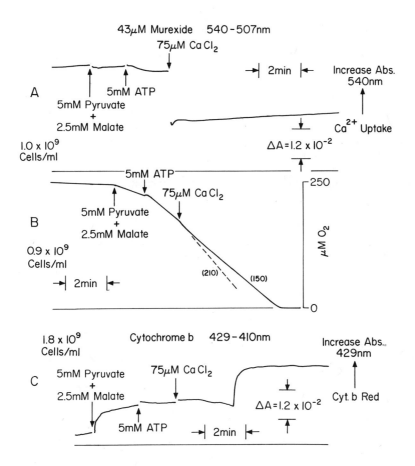

Fig. 2: Effect of added Ca^{+2} on hypotonically treated rabbit epididymal sperm utilizing pyruvate and malate as substrate in the presence of Pi, ATP, and Mg^{+2}. The medium was 113 mM KCl, 3 mM $MgCl_2$, 5 mM KPi, 20 mM Tris at pH 7.4. A. Uptake of Ca^{+2} from the suspending medium as monitored spectrophotometrically using the metallochromic indicator murexide. The rapid downward deflection of the trace is formation of the Ca^{+2} - murexide complex; the slower upward deflection represents dissociation of the complex as Ca^{+2} is taken up. Optical path is 0.5 cm. B. Respiration rate as determined with the oxygen electrode for the conditions in A. C. Changes in the redox state of cytochrome b of the hypotonically treated sperm cells under the conditions in A. An upward deflection of the trace corresponds to cytochrome reduction.

these mitochondria, as in mitochondria from other tissues, the reaction is seen with substrates requiring pyridine nucleotide, but not with succinate in the presence of rotenone.

The plasma membrane of intact rabbit epididymal spermatozoa is evidently impermeable to added Ca^{+2}: reactions mediated by calcium must involve binding sites on the outside of the plasma membrane. With pyruvate as substrate, a transient stimulation of respiration by Ca^{+2} addition is seen with spermatozoa treated with 2 mM EDTA (Fig. 3A), and this may reflect an active process of rebinding of Ca^{+2} removed by the high EDTA concentration. Subsequent addition of dibutyl cyclic AMP has no effect, but addition of A23187 causes the respiration rate to double after a lag. Addition of Ca^{+2} does not stimulate the endogenous rate (Fig. 3B), and the more rapid rate attained on subsequent addition of pyruvate is susceptible to inhibition with oligomycin. Addition of dibutyl cyclic AMP is without effect under both sets of conditions. The record of Fig. 3C is a control to that of Fig. 3A, showing that prior addition of dibutyl cyclic AMP does not interfere with the cycle of respiratory stimulation by Ca^{+2}. The sperm cells treated with 2 mM EDTA are motile as flushed from the epididymides, and dibutyl cyclic AMP has no perceptible effect on this motility.

In other experiments, 2 mM cyclic AMP was shown to be without effect on sperm respiration with either pyruvate or glucose as substrate, with or without added Ca^{+2}, even after preincubation of the cells with SQ 20009, a potent inhibitor of cyclic AMP phosphodiesterase (26). These reagents were also without perceptible effect on the motility of the cells. Contrary to expectation from the work of Morton et al. (12), however, incubation of freshly washed and resuspended sperm cells with 5 mM theophylline in Tris-saline medium decreased the percentage of motile cells to as low as 30%.

DISCUSSION

Mature epididymal sperm from the rabbit need only dilution in a simple aerated saline buffer to show full motility. External Ca^{+2} does not appear necessary to trigger motility, since the sperm can be flushed with a solution containing 2 mM EDTA which should complex all free Ca^{+2}, and full motility is still observed. Aeration and freedom from restraint due to crowding therefore appear to induce flagellar activity directly in mature rabbit spermatozoa.

Intact sperm cells are impermeable to Ca^{+2}. Alteration of the plasma membrane, whether by the ionophore A23187 or by hypotonic shock, is needed to give Ca^{+2} access to the cell's interior. This access is recognized experimentally by a stimulation of

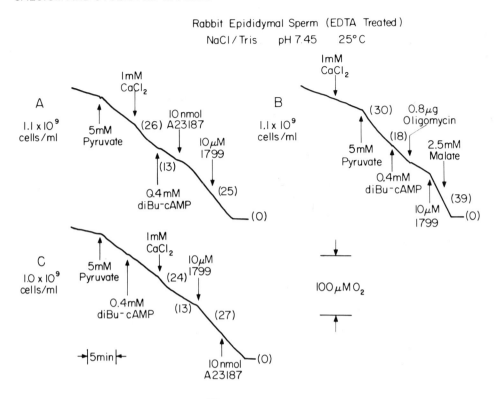

Fig. 3: Effect of Ca^{+2} and dibutyl cyclic AMP on the respiration rate of rabbit epididymal sperm utilizing pyruvate as substrate. These sperm were flushed from the epididymides with Tris-saline containing 2 mM EDTA. The medium and recording time scale was as in Fig. 1. A. Effect of Ca^{+2} followed by dibutyl cyclic AMP on respiration rate with pyruvate. B. Effect of Ca^{+2} pre-incubation on rate of respiration with pyruvate as substrate, followed by dibutyl cyclic AMP. C. Control experiment to A, in which dibutyl cyclic AMP is added prior to Ca^{+2} and uncoupler prior to the ionophore A23187.

respiration which is insensitive to inhibition by oligomycin. The respiration rate is a remarkably sensitive indicator of Ca^{+2} accessibility to the mitochondria, even when Ca^{+2} uptake is slow and perturbation of the mitochondrial energy state is minimal. An increase in rate of respiration of supposedly intact sperm cells observed in the presence of Ca^{+2} but absence of the ionophore A23187 indicates damaged plasma membranes which are permeable towards Ca^{+2}. The slowing of the oxygen uptake rate with pyruvate plus malate after initial stimulation by added Ca^{+2}, to hypotonically treated sperm cells (Fig. 3B), shows a deleterious effect of Ca^{+2}, namely, complex formation with mitochondrial pyridine nucleotide as shown by Vinogradov et al. (25). The impermeability of plasma membrane thus protects the cell by excluding the Ca^{+2} present in the oviduct fluid from the sperm cytosol, so that the intracellular Ca^{+2} concentration remains low enough not to effect mitochondrial function.

The lack of an immediate effect of added cyclic AMP on the respiration of rabbit spermatozoa is in accord with the observation of Hoskins (14) with bovine spermatozoa that fructolysis was not stimulated by direct addition of cyclic AMP to freshly washed cells, but that a 75 minute preincubation at 37° with the nucleotide did result in stimulation of fructolysis. The plasma membrane of both types of sperm cell seem to have restricted permeability to exogenous cyclic AMP. Rabbit and bovine spermatozoa thus appear to differ from hamster spermatozoa which respond rapidly to exogenous cyclic AMP as well as to exogenous Ca^{+2} (12). Such a difference is not unexpected in view of the earlier work of Morita and Chang (3). In rabbit and bovine spermatozoa, both cyclic AMP and Ca^{+2} may play a role in sperm capacitation rather than in motility, in which case neither would be permeable to the plasma membrane. It suffices to have Ca^{+2} bind to the outside of the plasma membrane surrounding the acrosome, thereby activating adenylate cyclase operative on the inside surface of this membrane. The resulting cyclic AMP generated in this region would then activate the protein kinase or kinases, known to be present in bovine sperm (27). These kinases would, in turn, promote phosphorylation of proteins in the inside of the plasma membrane and outside of the acrosomal membrane, thereby leading to capacitation and eventually to the acrosomal reaction.

ACKNOWLEDGEMENTS

The author is grateful to Mrs. Dorothy M. Rivers for skillful technical assistance. This work was supported by United States Public Health Service grant HD-06274 and National Science Foundation grant GB-23063.

REFERENCES

1. Bianchi, C.P. (1968). "Cell Calcium," Butterworths, London.

2. Manery, J.B. (1969). In: "Mineral Metabolism" (C.L. Comar and F. Bronner, eds.), Vol. 3, pp. 405-452, Academic Press, New York.

3. Morita, Z. and Chang, M.C. (1970). Biol. Reprod. 3, 169-179.

4. Holmdahl, T.H. and Mastroianni, L., Jr. (1965). Fertil. Steril. 16, 587-595.

5. Folly, C.W. and Williams, W.L. Proc. Soc. Exp. Biol. Med. 126, 634-637.

6. Brackett, B.G. (1968). Sixth Cong. Intern. Reprod. Anim. Insem. Artif., Paris 1, 43-46.

7. Storey, B.T. and Keyhani, E. (1973). FEBS Letters 37, 33-36.

8. Storey, B.T. and Keyhani, E. (1974). Fertil. Steril. 25, 976-984.

9. Chance, B. (1965). J. Biol. Chem. 240, 2729-2748.

10. Lehninger, A.L., Carafoli, E., and Rossi, C.S. (1967). Advan. Enzymol. 29, 259-320.

11. Lehninger, A.L. (1970). Biochem. J. 119, 129-138.

12. Morton, B., Harrigan-Lum, J., Albagli, L., and Jooss, T. (1974). Biochem. Biophys. Res. Comm. 56, 372-379.

13. Garbers, D.L., First, N.L., Sullivan, J.J., and Lardy, H.A. (1971). Biol. Reprod. 5, 336-339.

14. Hoskins, D.D. (1973). J. Biol. Chem. 248, 1135-1140.

15. Tash, J.S. and Mann, T. (1973). Proc. Roy. Soc. London B 184, 109-114.

16. Garbers, D.L., First, N.L., and Lardy, H.A. (1973). Biol. Reprod. 8, 589-598.

17. Morton, B.E. and Chang, T.S.K. (1973). J. Reprod. Fert. 35, 255-263.

18. Shoenfeld, C., Amelar, R.D., and Dubin, L. (1973). Fertil. Steril. 24, 772-775.

19. Keyhani, E. and Storey, B.T. (1973). Biochim. Biophys. Acta 305, 557-569.

20. Scarpa, A. (1972). Methods Enzymol. 24, 343-351.

21. Chance, B. (1972). Methods Enzymol. 24, 322-335.

22. Chance, B. and Williams, G.R. (1956). Advan. Enzymol. 17, 65-134.

23. Reed, P.W. and Lardy, H.A. (1972). J. Biol. Chem. 247, 6970-6977.

24. Tibbs, J. (1962). In: "Spermatozoan Motility" (D.W. Bishop, ed.), pp. 233-250, Publication No. 72, Amer. Assoc. Advance. Sci., Washington, D.C.

25. Vinogradov, A., Scarpa, A., and Chance, B. (1972). Arch. Biochem. Biophys. 152, 646-652.

26. Chasin, M., Harris, D.N., Phillips, M.B., and Hess, S.M. (1972). Biochem. Pharmacol. 21, 2443-2450.

27. Hoskins, D.D., Stephens, D.T., and Hall, M.L. (1974). J. Reprod. Fert. 37, 131-133.

ONTOGENINES AS CARRIERS OF INFORMATION DURING THE EARLY EMBRYONIC DEVELOPMENT

G.K. Roussev

Medical Research Institute Ovcha Coupel
Sofia, Bulgaria

All data support the idea that the embryonic development is a stricktly programmed process of carrying information. Francis Crick (1) proposes a model built on the rate of diffusion of embryoactive substances into the embryonic cells. Until now fully satisfactory explanations of the molecular mechanism of this programme are missing.

In 1953 (2) we began to investigate the activity of the extracts from embryos of Triturus cristatus collected in a definite stage.

We (3) observed that in some precisely fixed moments during the development the embryoactive substances are formed with a life span of about 5 to 30 minutes. We called these substances ontogenins. Using different methods (gel filtration, ion exchange chromatography, etc) the substances were purified and isolated from the embryos during a 100-hour development. Our investigations of their chemical nature are not completely finished. Most of the ontogenines are proteins, nucleoproteins, peptided or nucleopeptidates having the possibility to penetrate through the envelopes of the embryos. We also expect that on the surface of the embryos there are specific parts fixing the ontogenins. We (4) found four types of action of the ontogenins: A, B, C and D.

With the isotope study we observed that the low molecular weight ontogenins: type C or D have an unspecific effect on the cell cycle. The ontogenins bind to to the ribosomes, mitochondria and the yolk platelles. During this binding in some cases the ontogenins lose their activity or change their embryoactive properties.

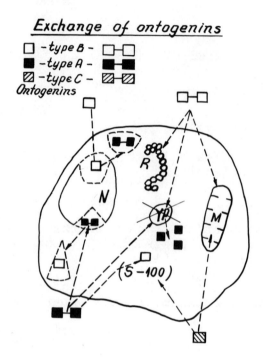

Figure 1

The binding of ontogenins to cell organelles is connected with morphogenetic changes: the stimulating ontogenins organize the single ribosomes in polysomes and accelerate the destruction of yolk platelles. The toxic ontogenins have an opposite action.

The ontogenines do not act only in the cells in which they are formed. The mutual influence is especially strong at the beginning of gastrulation. The different parts of embryos respond differently to the different ontogenins.

All these data lead us to the conclusion that the ontogenins are the basic carriers of information between the cell organelles during the embryonic development.

REFERENCES

1. Crick, Fr., Nature, 225, (1970), 420.

2. Roussev, G.K., Experimentia, 16, (1+60), 329.

3. Roussev, G.K., Agressologie, 12C, (1971), 13.

4. Roussev, G.K., Programme of Embryonic Development Directed by Ontogenins, Naouka i Izkoustvo, Sofia, 1974.

New Developments in Analytical Techniques and Apparatuses

SPECTROPHOTOMETRIC EQUIPMENT WITH FLASH-LIGHT EXCITATION

N. Koralov, T. Todorov, G. Kostov, A. Shosheva, V. Ivanova

Central Laboratory of Biophysics, Bulgarian Academy of Sciences, 1113 Sofia, Bulgaria

E. Fessenko, N. Orlov, V. Kulakov

Institute of Biophysics, Academy of Sciences of the USSR, Pushchino, USSR

The method of flash-light excitation is widely applied in studying the mechanisms of photochemical reactions, in investigating the properties of the free radicals, and in studying the processes of transport and degradation of the energy of electron excitation. The information about the processes which take place as a result of the flash-light excitation is obtained by recording the changes in the optical density, in the electric conductivity, or in some other parameter of the object studied. Most frequently used is the recording of the absorption spectra of the transient products obtained upon the illumination of the object with a short and powerful flash-light pulse. Furthermore, the duration of the flash-light pulse must not exceed the half-live of the transient products obtained, so that the latter could be recorded, while the energy must be sufficiently high to secure a concentration of these products sufficient for their recording. That is why the duration and the energy of the flash-pulse are the two basic parameters which have been used ever since the creation of the method (1).

The increase of the energy of the flash-pulses is obtained by increasing the electrical energy fed to the flash-lamp. Of course, one limiting factor is the capacity of the lamp to take high energies without being destroyed. On the other hand, it is expedient to increase the electric energy

$$E = \frac{1}{2} CV^2$$

by increasing the voltage, since any increase in capacity involves an increase in the parasitic parameters of the capacitor. At the initial moment of ignition of the lamp the entire voltage is borne by the inductivity of the electric circuit (since the active resistance of the lamp is much smaller - of the order of $10^{-1} \Omega$).

The process is described by the formula:

$$\frac{di}{dt} = \frac{V}{L_C + L_R + L'}$$

wherein: $L_C + L_R + L'$ is the inductivity of the entire circuit

L_C is the parasitic inductivity of the capacitor

L_R is the inductivity of the lamp

L' is the parasitic inductivity of the circuit.

The denominator does not increase upon increasing the voltage V, since L_R (which depends on the design of the lamp), L_C and L' do not depend on V.

The first flash-lamps made it possible to use high electric energies (hence the obtaining of a relatively high light energy, which is from 5 to 20 % of E), but the duration of the photo-pulse τ was rather long. By way of illustration we shall refer to Porter's lamp (2) with $E = 4$ kj, $\tau = 1$ ms, to the lamp devised by Lindquist (1) with $E = 10$ kj and $\tau = 150$ μs, etc.

For the purpose of creating a device for rapid pulse photolysis, a lamp with the following parameters was designed and built at the Institute of Biophysics of the USSR Academy of Sciences

$$E = 1,500 \text{ j}$$
$$\tau = 2 \text{ to } 5 \text{ μs}$$

The lamp has a cylindrical body made of brass. There is a quartz glass attached to the front part, and a partition insulator in the rear, with a wire passing through it for connecting the electrodes to the high voltage. There is a screen placed in front of the insulator for the purpose of protecting it against the shock wave accompanying the photo-pulse. The two high-voltage electrodes are situated vertically above one another on holders which make it possible to regulate the distance between them. The ignition electrode is situated between the high-voltage electrodes and is a molybdenum wire soldered in a glass tube. The design of the connections between the lamp and the capacitor is such as to allow the insertion of damping resistances in the high-voltage circuit (graphite rods in teflon insulation). Their introduction causes no practical increase in the inductivity of the discharge circuit (which is about 100 nH). The lamp can take several thousands flash-light-pulses without any need of cleaning the quartz glass.

Fig. 1 shows the dependance of the intensity of radiation on the time, which were obtained by the use of capacitor IM 20-8 (8 μF, 20 kV). The wavelength of recording is 450 nm, and the energy is 300 j. The lamp is filled with argon at a pressure of 3 atm. (Other gases were also used when the lamp was employed in equipment for flash-light excitation, such as nitrogen, oxigen, and air). Graph **(b)** illustrates the influence of the damping resistances.

A device for flash-light excitation was designed and build at the Central Laboratory of Biophysics, Bulgarian Academy of Sciences, on the basis of this lamp and of the capacitor IM 20-8. A method involving photomultiplier has been applied for the purpose of recording the information obtained from the object investigated. This method posesses a number of essential advantages over the photographic method, such as higher sensitivity, broader range of wavelengths, and more accurate quantitative data. The block diagram of the device is shown in Fig. 2.

The source of light used is a incandecent lamp with iodine cycle and capacity of 90 W (KIM-90), situated at a distance of 3 m from the photo-receiver on a solid rail. Installed on the same rail is the optical system consisting of quartz lenses, by means of which the photo metering light ray is focussed on the input slit of the monochromator type SPM-2. The flash-pulse is focussed on the object by means of a condenser made up of two quartz lenses, the pulse and the photometering lights being at an angle of 90°. The light passing through the object reaches the photo-multiplier type MQS-12 which has a spectral sensitivity from 200 to 800 nm. The electric signal carrying information about the changes in the object investigated, comes from the output of the photo-multiplier and reaches the input of the storage oscilloscope type C 8-1.

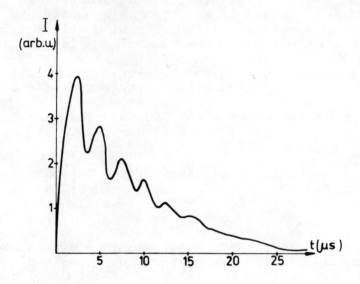

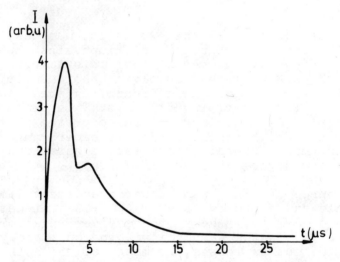

Figure 1

Time dependence of the light intensity
λ = 450 nm; energy 300j; pressure 3 atm.
gas - Ar. (a) - without damping resistances;
(b) with damping resistances in the high-voltage circuit.

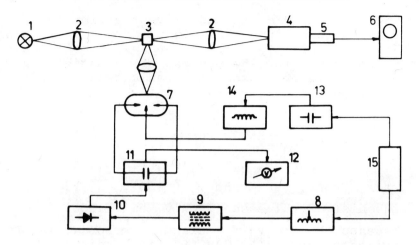

Figure 2

A block-diagram: 1- light source; 2 - quartz lenses; 3 - object investigated; 4 - monochromator; 5 - photo multiplier; 6 - storage oscilloscope; 7 - flash lamp; 8 - autotransformer; 9 - high-voltage capacitor; 12 - high-voltage rectifier; 11 - high-voltage capacitor; 12 - high-voltage voltmeter; 13 - low-voltage capacitor; 14 - step-up coil; 15 - control desk.

Instructions are issued from the control desk for feeding voltage to the high-voltage transformer, from whose secondary winding, through a high-voltage rectifier, charging voltage is sent for the high-voltage capacitor and for feeding a high-voltage (15 kV) starting pulse to the ignition electrode of the lamp. The device is designed for one-run operations, whereby the capacitor is discharged after the flash-pulse and can be charged to a new voltage also in a continuous mode of operation in which 5 to 6 sec after the flash-pulse the capacitor is automatically charged to the same voltage for obtaining a flash-pulse having the same energy.

The design adopted provides for positioning the flash-lamp on the high-voltage capacitor, for the purpose of reducing the parasitic influence of the connections.

In view of the power sources of high voltage involved in the device, provisions have been made for the requisite blockings and signals which ensure safe operation.

Using the device described, investigations were carried out of various rapid photo-induced processes in alkali halide crystals (5). Preliminary experiments have also been performed for the purpose of investigating certain photochemical processes in organic dyes and biological macromolecules.

REFERENCES

1. R.Norrish, G. Porter, Nature 164, (1949), 638
2. G.Porter, Trans. Farady Soc., (1950), 49, 2116
3. L. Lindquist, S. Klaesson, Arkiv kemi (1958), 12, 1
4. E. Fessenco, N. Orlov, V. Kulakov, Pribori i technika eksperimenta, No.5, (1972), 187 (in Russian)
5. T. Todorov, G. Dechev, M. Georgiev, Solid State Com., 11, (1972), 1731.

A RAPID-SCANNING SPECTROPHOTOMETER

P.Shishmanov, N.Koralov, G.Georgiev, J.G.Vassileva-Popova, S.Kovachev, N.Uzunov, T.Todorov, V. Neitchev
Central Laboratory of Biophysics, Bulgarian Academy of Sciences, 1113 Sofia, Bulgaria

A.Mogilevsky, V.Slavnyi
Institute of Biophysics,
Academy of Sciences of the USSR, Pushchino, USSR

M.Georgiev
Institute of Solide State Physics,
Bulgarian Academy of Sciences, Sofia, Bulgaria

Rapid-scanning spectrophotometry has scored great development in recent years as one of the efficient methods employed in investigating rapid chemical and physical processes.

Despite the good spectrophotometric accuracy of the single-ray spectrophotometric method, which consists of surveying the kinetics of the process investigated at one wavelength, it frequently leads to obtaining erroneous results in the cases of shifting the spectral maximums or when there are short-lived intermediate products during the process.

With rapid-scanning spectrophotometers it is possible to obtain successive absorption or emission spectra for a broad spectral range, i.e. to obtain fuller spectral information. This makes it possible to derive more kinetic information about the process, and thereby reducing the probability of measurement errors(1).

A rapid-scanning spectrphotometer in the visible and UV regions has been designed as a result of the cooperation in the field of rapid spectrophotometry between the Bulgarian Academy of Sciences and the Academy of Sciences of the USSR. The design of the spectrophotometer is based on the method of electronic scanning. The operation principle of the rapid-scanning spectrophotometer Model 170 can be seen on the block diagram in Fig.1. (2, 3)

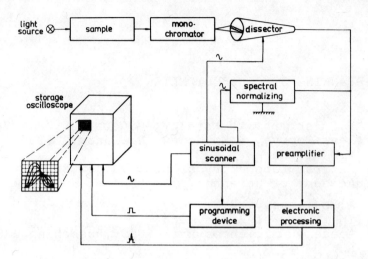

Figure 1

Block-diagram of the rapid-scanning spectrophotometer Model 170.

The following lamps were used as sources of light:
- Incandescent filament lamp Type KIM-9-75;
- Xenon lamp Type DKsŠ-150;
- Hydrogen lamp Type DDC-30;
- Mercury-helium lamp Type DRGC-12.

Each one of these lamps is used depending on the spectral range required for operation, while the last one - DRGC-12 - is used for preliminary calibration of the wavelengths scale.

The illuminating system provides for gradual and smooth abatement of the light flux of up to 100 times and for the instalation of filters.

The light flux, appropriately focussed, passes through the object of investigation and then reaches the input slit of the specially designed spectral apparatus, carrying full information about the absorption of the light by the object for all wavelengths of the spectral range investigated.

The operative spectral range of the apparatus is between 200 and 800 nm, a standard diffraction grid with 300 lines per mm being used for the purpose of resolving the light. The inverse linear dispersion is 15 nm/mm, while the resolution is not inferior to 1 nm. The spectral apparatus has no output slit and the decomposed light spectrum is focussed on the face of the phototransducer.

The choice of the phototransducer was preceded by thorough studies, the principal requirements involved being:
- Lack of inertia, with the aim of ensuring higher rapidity of operation of the spectrophotometer;
- The possibility of covering a wide range of durations of the processes investigated;
- Sensitivity;
- Linear amplification of the signal;
- Resolving capacity;
- Broad spectral range, etc.

The above requirements are largely contradictory in character, some of them being met to a larger extent by some of the phototransducers, at the expense of other requirements. Most satisfactory results were obtained for the electrostatic dissector Type LI-602M.

The optical image of the spectrum is transformed by the photocathode into an electric equivalent. The scanning of the electric information formed is realized by applying the sinusoidal voltage on the deflection plates of the dissector. The spectral information transduced into an electric signal is obtained at its output, whereby there exists a direct correlation between the time-lag from the beginning of the scanning of the successive spectrum and the wave-length, and also between the level of the electric signal and the light intensity for this wavelength.

The maximum spectral band which can be investigated by means of the rapid-scanning spectrophotometer Model 170 is 300 nm, and provisions have been made in the scanning device, by means of a stepwise decrease of the amplitude of the deflection voltage, to operate within spectral sections of 150, 75 and 37.5 nm. In view of the fact that an electrostatic dissector is employed, the speed of scanning can be varied within broad limits. In this case we have accepted the number of scannings to vary stepwise from 100 to 10,000 per second. The active part of scanning is of the order of 40% of the period T of scanning.

The choice of the number of scannings per second belongs to the experimentalist, ensuing from the preliminary orientation about the duration of the process investigated.

It we assume that a minimum of 100 spectrograms are necessary for the sufficiently accurate description of a process, whereby we take it for granted that no spectral changes occur in it during the scanning of the spectrum, i.e. that the time involved in scanning the spectrum is negligibly small compared with the duration of the process, then the most rapid process which can be investigated by the rapid-scanning spectrophotometer Model 170 is:

$$t_{pr.min.} = 100 \; T_{sc.min.} = 100.10^{-4} = 10^{-2} \; sec$$

Where:

$t_{pr.min.}$ is the minimum time for the duration of the process investigated;

$T_{sc.min.}$ is the minimum period of scanning of the spectrophotometer.

This time of 10 msec can be greatly reduced - to approximately 1 msec duration of the process investigated, in view of the fact of the onset of changes in the spectrum during its scanning, and it is possible, by means of a relatively simple mathematical apparatus, to restore from the data obtained the full spectral information about the process that has taken place.

The slowest processes for investigation are limited by the lowest speed of scanning and by the recording capacities of the device.

$$t_{pr.max.} = K \cdot T_{sc.max.} = 600.10^{-2} = 6 \; sec$$

Where:

$t_{pr.max.}$ is the duration of the slowest process which can be investigated with the spectrophotometer Model 170.

K is a coefficient which takes into account the performance of the recording device;

$T_{sc.max.}$ is the slowest period of scanning of the spectrophotometer.

By using relatively simple circuitries, this time can be prolonged for the purpose of investigating still slower processes.

The sensitivity of the photocathode of the dissector to the different wavelengths of the spectral range is not the same. That is why a specially formed voltage, synchronized with the frequency of the scanning voltage, is fed from the device for normalizing the spectrum, and it changes the load resistance at the output of the dissector. The circuitry provides for a change in sensitivity up to 10 times.

The normalized electric signal of the spectrum is amplified to an appropriate level and, at the same time, the possibility has been provided for additional electronic processing of the signal such as:

- Stepwise change of the time constant of the amplifier from 0.5 μsec to 1,000 μsec;
- Stepwise shift of the zero level of the signal.

The increase of the time constant of the amplifier leads to reducing the error from the noise, and this is possible when scanning is carried out at frequencies lower than the maximum one and also in those cases when the shape of the spectrum investigated makes it possible to operate with a resolution that is lower than the maximum for the working spectral range.

The stepwise shift of the zero level makes it possible, in the case of small changes of the spectrograms, to follow easily their succession on the screen of the recording device.

The photometric accuracy of the apparatus, upon measuring transmission coefficients close to 1, is not below 2 per cent, while its resolving capacity is not below 5 nm for 300 nm spectral range, and not below 2 nm for 150 nm or smaller spectral ranges.

The selection of the most interesting spectra for recording and the storage of part of the spectral information is done by the recording device which consists of a programming device and a storage oscillograph.

The programming device enables the research worker to select for recording the most suitable spectrogram from a reaction run. Any spectrograms from the 1st to the 60th can be recorded. A scanning frequency divider is provided, with degrees of division from 1:1 to 1:10, and this makes possible the overlapping of the periods of the processes investigated, increasing the choce of the spectrograms from 1 to 600.

A standard storage-type oscillograph C 8-1 is used for recording the spectrograms, with part of the sinusoidal scanning voltage for the dissector being used as a deflecting voltage. In this manner it is possible to attain a fairly high accuracy in determining the wavelengths. The spectrograms recorded can be reproduced continuously up to 30 minutes after the recording, without switching off the oscillograph. By way of exception they can be reproduced after seven days as well. The spectrograms can be photographed from the screen of the storage oscillograph by means of a standard camera and can be subjected to appropriate mathematical processing.

APPLICATIONS

One of the interesting applications of the rapid-scanning spectrophotometer Model 170 is the spectrophotometric study of chemical reactions taking place upon mixing two substances. The rapid-mixing device used by us operates on the principle of the "stopped flow". The reagents subject to mixing are put in 2 cm^3 syringes and are extruded under pressure through a double four-jet mixer. The volume ration of mixing is always 1:1. The solution obtained from the mixing passes into an optic cuvette. The windows of the cuvette are made of quartz glass. The length of the optic pathway can be varied from 2 to 20 mm. The efficiency of mixing is 95 per cent. The experimentally measured dead time of the rapid-mixing device is 4 msec. The necessary volume of solution is ~2 ml from each reagent for one experiment. The possibility for thermostat treatment has also been provided.

The rapid-scanning spectrophotometer and the device for rapid mixing were used in studying the reaction of complexing between ammonium molybdate and the propyl ester of gallic acid. The reaction was investigated in a medium of acetate buffer at initial concentrations of the two reagents of 0.025×10^{-3} M at $25°C$. Under these conditions the reaction lasts about 120 msec, with half-conversion time of 20 msec. The reaction was investigated

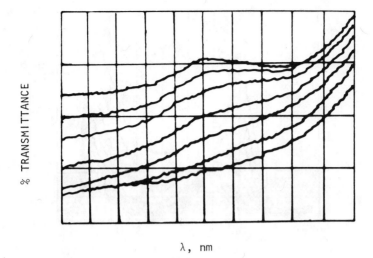

Figure 2

Reaction of complex formation between ammonium molybdate and propyl ester of gallic acid in a medium of acetate buffer, with initial concentrations for both reagents at 0.025×10^{-3} M at T = 25°C.

in a wavelength range of 380-530 nm, with a scanning speed of 1,000 spectra per second. Seven spectrograms were recorded by means of the programming device, and these spectrograms were situated at 4, 8, 12, 20, 32, 60 and 124 msec from the starting moment of the reaction. They were used for calculating the respective speed constants of first order for different wavelengths. The same value was obtained for $K_{exp.} = (150 \pm 15)$ sec^{-1}. These studies on the speed of complexing of the reaction of ammonium mollybdate - propyl ester of the gallic acid in the wavelength range of 380 to 530 nm showed that only one complex is formed with strictly determined composition.

The possibility is also demonstrated of recording intermediary products (Fig.3). The diagram of such a reaction is the following: A → B → C, i.e. substance A is converted into a final product C upon passing through the form B. In our instance the substances differ in their spectral properties, and this fact provides for their identification. The decrease of the transmission at 430 nm and the gradual shifting of the maximum towards the lon-

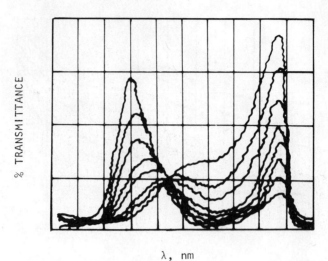

Figure 3

ger wavelengths are indicative for the formation of the B form. The final product of the reaction, the C form, has its maximum at 600 nm. A series of experiments were carried out in solvents of varying acidity and it was established that the spectral maximum of the B form shifted towards the longer wavelengths with the rise in acidity. This is paralleled by a change in the reaction rate as well.

Another interesting application of the rapid-scanning spectrophotometer Model 170 is the spectrophotometric investigation of processes initiated by means of flashlight excitation. We have been engaged in recent years in intense studies of various photochromic processes in alkali halides with the aim of clarifying their mechanisms and the possibilities of using them in the optical storage of information. One of these processes, the F - F' conversion in crystals of KCl, have been the object of very little study at temperatures close to the ordinary room temperature on account of the high thermal instability of the F' band. By means of the spectrophotometer described, combined with flash-light excitation, we obtained for the first time the absorption spectrum of the F' band at room temperature (4).

The block-diagram of the experimental setup used is shown on Fig. 4.

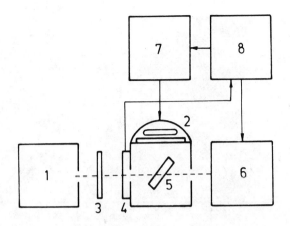

Figure 4

Block-diagram of the rapid-scanning spectrophotometer Model 170 with flash-light excitation.

To the standard equipment used with the rapid-scanning spectrophotometer Model 170 (blocks /1/ and /6/) were added a dark camera for the samples /5/, a flash-lamp with duration of the flash-pulse of 1 msec at energy of 40 wtsec /2/, a camera shutter with oppening time 1 msec /4/, and a synchronization block /8/ ensuring the starting of the programming device of the spectrophotometer, the opening of the camera shutter - i.e. the transmission of photometering light - and the switching on of the flash-lamp. The experimental setup has been created with the aim of ensuring the possibility of recording a trace in the dark, transmission spectrum of the sample prior to the flash-light excitation.

The samples, crystals of KCl about 0.1 cm thick, were placed at an angle of 45° to both the excitation and photometering lights. In accordance with the rate of thermal destruction of the F' band at room temperature, operation proceeded at a scanning speed of 1000 spectra per second. The spectra of the bands investigated were obtained in a spectral range of 630 - 780 nm.

Fig.5 shows a typical oscillogram on which the following has been recorded: line in the dark (lowest trace), spectrum of the light which has passed through a non-excited sample (the uppermost curve), and spectra selected

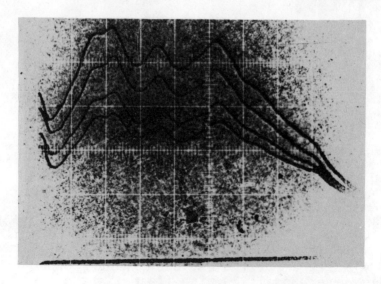

Figure 5

Oscillogram of the transmission spectra corresponding to a flash-induced F' band in KCl (see text).

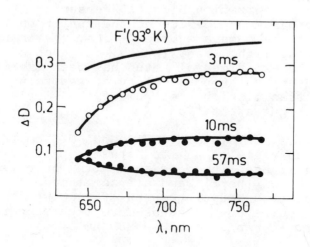

Figure 6

F' band absorption obtained from an oscillogram as in Fig.5 at different times after the flash-light excitation.

by ourselves at certain times after the flash-light excitation (the intermediate curves, the lowest of which corresponds to a spectrum obtained as the earliest one after the flash-light excitation). From these spectrograms, after convertion to optical density, the absorption spectra of the colour centres investigated were obtained (Fig.6), 3, 10, and 57 msec respectively after the flash-light excitation. The same Figure (the uppermost curve) shows also the spectrum of the F' band in that spectral range obtained at low temperature by a conventional spectrophotometer. We can see that the spectrum of the F' band preserves its form at room temperature.

The applications described can give an idea only about some of the possibilities of employing the rapid-scanning spectrophotometer Model 170.

REFERENCES

G.C.Pimental, Appl.Optics,7 (1968) No 11, 2155.

A.N.Mogilevsky, V.Slavnyi, E.Afanassiev, P.Shishmanov, S.Kovachev, N.Koralov and N.Uzunov, Nauchnye Pribory SEV, in press, (in Russian).

Pat. 621,397, 30 August 1973, Pat.Office USSR (in Russian)

T.Todorov, N.Koralov and M.Georgiev, Optics Comm., (1975) in press.

EQUIPMENT FOR DATA REGISTRATION AND PROCESSING

IN FAST SPECTROSCOPY

A. Stoimenov and M. Kotarov

Central Laboratory of Biophysics

Bulgarian Academy of Sciences, 1113 Sofia, Bulgaria

A special purpose equipment for experimental data registration and processing in fast spectroscopy is described. The high speed (up to 1,000,000 samples per second) the good accuracy (8 bits) and the flexible approach to the preliminary data processing are achieved by using IZOT-0310 minicomputer as a basis of the system.

INTRODUCTION

Recently great efforts are made to design new methods and equipment for fast processes investigation in physics, chemistry and biology. Fast spectroscopy is specialy suitable for biology with it's high accuracy, sensitivity and low desturbance of the object. For large spectral range a great amounts of high speed data are obtained. Equipment for registration and processing of such information is described in this paper.

HARDWARE

During the highspeed spectrophotometer action the optical image of the spectra investigated is projected on the photocathod of a TV pick-up tube, dissector type. This image is transformed in an electron image and scanned by means of sine-form voltage. After proper amplification the output signal is converted into digital form by a successive approximation analog-to-digital converter (ADC). The clock pulses are coming from tracing type

ADC, following the predetermined step-rising of the scanning voltage. In this way 8 lit binary codes equivalent to the instantaneous value of the light extinction for one of the 8 or 16 or,..., or 256 wavelenghts of the spectral range investigated.

Proper data registration is performed in the real-time by a direct memory access in the core of the minicomputer, packed in spectra files. After registration information is stored, procesed and visualized on the screen of a special graph-display unit.

SPECIALIZED OPERATING SYSTEM

The automatic action of hardware is provided by specialized operating system of programs in four modes of operation: registration, visualization, preliminary processing and data output. Briefly it includes:

- the input of all parameters (number of wavelengths, number of spectra, etc.);

- the automatic registration of experimental data in a predetermined order;

- the decelerated visualization of data recorded on the graph-display screen;

- the preliminary data processing (normalization of the spectra in reference to a pattern, scaling, etc.);

- the selected data file (spectrum) output on an intermediate holder (paper tape) for storage and further sophisticated processing

- the flexible dialogue between the operator and the system.

CONCLUSIONS

The equipment is now installed in our laboratory. Taking into account the experience accumulated during the operation in the field further improuvement of hardware and operating system will be made. A set of programmes for routine processing of spectral information and multicomponent system separation are prepared.

REFERENCES

1. IZOT-0310 Computer Handbook, Sofia, 1973 (in Bulgarian)

2. A.Stoimenov, M.Kotarov, Conference on Computers in Scientific Research, Novosibirsk, 10-12 June, 1974. (in Russian)

ECG TELEMETRIC DEVICE AND SYSTEM FOR POPULATION ATTENDANCE

G. Georgiev, A. Venkov, Vl. Prokopov, S. Kovachev

N. Hadjivanova, R. Mateeva, G. Kostov

Central Laboratory of Biophysics
Bulgarian Academy of Sciences, 1113 Sofia,
Bulgaria

The diseases of the cardiovascular system rank first among total morbidity and mortality of the population in our country, definitely tending towards further increase of their relative share in the last years. They are the most frequent lethality causes, responsible for about 50 per cent of all lethal cases per 100,000 population.

The problem of the treatment and prophylaxy of the cardiovascular diseases is associated with means for wide-speading and bringing specialized cardiological care closer to where needed. That is even more valid for the cases with acute coronary and rhythm disturbances, where the timely and qualified interference of the cardiologist is of vital importance for life preservation.

Among the complex measures aiming at a wide-spred and immediate cardiological aid for the population, the problem of the wide implementation and use of the electrocardiographic method in medical practice both for emergency diagnosis and prophylaxy requires a great number of cardiologists and well equipped cardiological consulting rooms, including the peripheral health services as well.

In order to meet these needs - a telemetric device for transmitting and recording a patient's ECG over telephone or radio (Telecard) was designed at the Central Laboratory of Biophysics, Bulgarian Academy of Sciences.

During the current year, experimental regional telemetric ECG systems were organized in all districts in the country.

Over 1800 telemetrically transmitted ECGs of emergency cases were recorded in the course of several months of operation, including 103 myocardial infarctions, 437 rhythm disturbances, 387 stenocardias and 873 other diseases. The favourable results and the experience gained through those systems, provided grounds for the Ministry of Public Health and the Central Biophysical Laboratory to approach jointly the formation of a complex telemetric ECG system for the whole country with a consultative centre and district, regional, interurban and interhospital telemetric systems.

The telemetric ECG systems, independently of their type, consist of two separate pocket-size units.

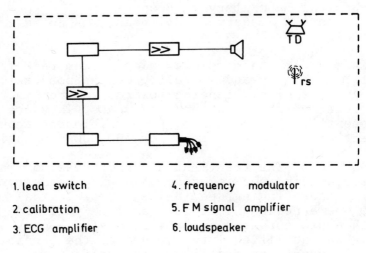

1. lead switch
2. calibration
3. ECG amplifier
4. frequency modulator
5. F M signal amplifier
6. loudspeaker

Figure 1
Block diagram of the modulator

I. Model "Telecard" - Modulator (Transmitter).

This unit takes up the patient's cardiac biopotentials through corresponding electrodes and a "patient cable" transmits them to an amplifier by means of an ECG position switch. The amplified biopotential pulses are then converted into frequency-modulated signals which after another amplification are fed to a loudspeaker coupled to the transmitter, and then to the microphone of a radio or telephone set.

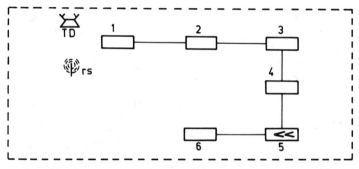

1. selective amplifier
2. Schmidt trigger
3. monostable vibrator
4. frequency detector
5. amplifier
6. ECG device

Figure 2

Block diagram of the demodulator

II. Model "Telecard" - Demodulator (Receiver).

This unit receives the frequency-modulated signals, amplifies them, attenuates all noises in the signal bandwidth, and transforms them into electrical pulses which are the fed to a recording mechanism of an one-channel electrocardiograph or some other type of recorder.

The receiver can operate a directly transmitted ECG signal or a typed record. It can also be used for magnetic tape recording of the signals for further processing or storage (in cases of mass prophylactic medical examinations).

The receiver may be connected to an oscilloscope for visual analysis of the signals.

Model "Telecard"-demodulator is mainly used in cases where direct telemetric ECG consultation is required between the patient's bedside and the receiving telemetric ECG center.

Power supply being provided by a 12V rechargeable dry battery makes the system much safer to use than conventional electrocardiographs.

Power supply may also be provided by a built-in storage battery in cases where electricity is not available.

MODE OF OPERATION

The medical staff requiring expert advice get in touch with the consulting ECG center through the telephone or radio, stating the anamnesis data and the doctor's findings (the patient being already connected to the instrument) and transmit consecutively signals for adjustment, testing and the corresponding ECG positions. After examining the ECG, the cardiologist from the center gives his expert advice regarding the diagnosis and the necessary treatment.

The telemetric ECG system "Telecard" offers the following benefits:

- quick qualified cardiac consultation outside specialized hospitals;

- limits the necessity for supplying peripheral medical units with ECG equipment;

- helps reduce the number of qualified cardiologists required in peripheral units;

- provides possibilities for a wide-range ECG screening for the purpose of prophylaxis and hospitalization of cardic patients;

- immediate ECG consultation in urgent cardiac cases;

- provides possibilities for developing inside the hospitals systems for telemetric ECG information, including monitoring of specific cases.

The equipment has a simple operation: an exellent sevice life of over 300 charge/discharge cycles. The battery is recharged without being taken out of the instrument. There is no AC interference and minimum reduction of noise without affecting the ECG waveform. Neither the patient nor the instrument should be earthed.

SPECIFICATIONS

Modulating type	FM/FM
Frequency deviation	7300 Hz
Signal band width	0,03 - 5 MB
Amplitude non-linearity	± 5%
Frequency range	0,1 - 100 Hz; 3 dB
Input resistance	10 Mohms
Time constant	1,2 sec
Power supply	rechargeable 12V dry battery
Lead selection	I, II, III, aVr, aVe, aVf, V_1, V_2, V_3, V_4, V_5, V_6.
Dimensions	230 x 100 x 62 mm
Weight - for each unit	1600 g

Accessories available.

REFERENCES

1. Alexander D.C., D. Worzmann, "Computer diagnosis of electrocardiograms. I. Equipment." - Comp. Biomed. Res., 1, 1968, 3, 348 - 365.
2. Buet A., "Bio-telemesures" - Electronique Medicale 49, 3-7, 1969
3. Demmerie A.M., S.I. Allen and K. Klempner, "Computor assisted heart surveillance using telephone" - Proc. 9th Int. Conf. Medical and Biological Engeneering, Aug. 1971
4. Ferris C.D., "A system for generating a stable dc baseline and suppressing noise when processing pulsed signals of physiological and other origin" - Med. Biol. Eng., 4, 1966, 4, 381-392.
5. Goldsmith A.N., "Biochemical engeneering" - IEEE Spectrum, 2, 1965, 12, 46.
6. Lou S. Davis "A system approach to medical information" - Methods of information in medicine, vol. 12, 1, 1973.
7. Michaelis, Dudech J., "Computer analyse des Electrokardiogramms II. Programmsysteme zur automatischen EKG Auswertung"- Klin. Wschr. 1971 49/13, 729-738.
8. Pordy L. et al., "Computor analysis of the electrocardiogram: a joint project"- I. Mt. Sinai Hosp., 24, 1967, 69.
9. Watts R.N., "Some designe considerations for narrow-band medical telemetry over the switched-message network" - IEEE trans. Commun. tech., vol. COM - 19, pp. 246-225, VI-1971.

MODERN INSTRUMENTATION IN PROTEIN CHEMISTRY

W. Zainzinger

Beckman Instruments Austria, Stefan-Esders-Platz 4, A-1191, Vienna, Austria

Instrument makers are claiming that our recent progress in biochemical research is predominantly due to the fact that scientists have at their disposal a highly sophisticated, reliable and specialized instrumental and computational equipment which puts them into the position of collecting and handling a tremendously increased amount of data. In addition, new instrumentation has promoted new experimental techniques leading efficiently to more reliable results.

Beckman Instuments International whose headquarters are located in California, USA has devoted almost four decades of research and development work to provide reliable and proper instruments to researchers, scientists and technicians of routine laboratories. It is the definite wish of each instrument maker to maintain close contact with the instrument users in order to match the laboratory demands and needs with the design features of the devices. In doing so and by keeping in close touch and good cooperation with leading research laboratories in the United States and in Western Europe, Beckman Instruments is able to offer the world's longest line of compatible apparatus for biochemical work.

Almost every biochemical research program starts with a living cell which is to be preserved through the experiment as long as possible. The cell and all its subunits under investigation are to carry out their functions and biological activities in order to enable the experimentator to obtain unaltered information from the in-vivo behaviour of the particles in question. Even the separation of whole cell as there are erythrocytes suspended in blood serum or bacteria and microbes in some liquid suspension can result in an irreversible change of the cell's properties by changing the medium as far as viscosity, pH, temperature and ion strength are concerned just to establish optimum separation parameters. The newly introduced elutriation technique constituting a very gentle counterflow centrifugation was developed to separate cells from their original medium avoiding any mechanical stress as friction and shocks against tubing walls. This technique uses a specially designed centrifugation rotor holding two conical containers with their axes parallel to the centrifugational force generated by the rotation of the rotor in a cooled centrifuge. During the run the sample suspension is pumped through both of the cones in a way that flow counteracts the centrifugational field. Thus the particles in question will find a zone where the influences of the fluid motion and the applied centrifugal field are in equilibrium and these cells are then kept in their position as long as the experiment parameters are maintained. This separation principle, which was introduced by Beckman Instruments in 1972, has proved to become an experimental tool of extraordinary significance.

Once the cells are collected or the fraction of tissue to be examined is extracted, a cell homogenate is produced. Whereas in many cases environmental parameters, i.e. intracellular conditions can be maintained during homogenisation the following extraction of certain cellular subgroups represents a critical task. Almost every chemical treatment often irreversibly changes biological activities of nuclei, mitochondriae, ribosomes, microsomes and other cell organs. With the commercial availability of preparative ultracentrifuges right after the Second World War from the Beckman SPINCO group a new era of separating and concentrating biological materials has begun. A purely phisical method was introduced avoiding all undesiredl chemical interactions. The separation medium could be matched with the in-vivo conditions to a very large extent. The classical techniques of sedimentation in fixed angle rotors and the fractionating density gradient techniques for separation of various components in one single swinging bucket rotor run have been complemented by large scale zonal and continuous-flow ultracentrifugation. Volumes as large as 5 liters per hours can be pumped through a running continuous-flow rotor and many commercial applications as the industrial purification of leukemia viruses and the preparation of vaccine concentrates have made this technique a technically and also economically fully accepted scientific aid. Beckman large volume zonal rotors are generating more than 250 000g, fixed angle rotors even more than half a million of gravitational force then spinning 75 000 rpm. Beside that, several Beckman laboratories all over the world have established application data sheets optimising run parameters for all commonly known separation problems.

By the aid of these modern highly efficient preparative ultracentrifuges experiments finally will yield concentrated and homogeneous portions of nucleic acids or proteins of different cell fractions. To evaluate the changes of these compounds of the properties due to experimental parameters, a further step of measurements has to determine the phisical and chemical behaviour of these compounds. Electrophoretic mobility on paper, cellulose acetate, acrylamide gels and agarose gel membranes can be checked and used for related separation problems on a universal "Microzone" electrophoresis system with computerized data handling. Sedimentation coefficients, Svedberg constants, molecular weights and degrees of polydispersity will be determined on the world's first and still modern Model E Analytical Ultracentrifuge. Three different optical systems (Toepler's Schlieren Optics, Interference Fringe Optics and UV Absorption Optics) together with a variety of optical cells can solve almost every problem involving sedimentation studies at high gravitational forces. New dedicated optical systems were developed which can easily be adapted to preparative ultracentrifuges and which offer the applicational possibilities of an analytical ultracentrifuge to biochemists and scientists owning a preparative ultracentrifuge at considerably cut down expenses.

The chemical composition of proteins and nucleic acid may be the next question which has to be answered. Modern Microcolumn Liquid Chromatographs as the Model 121 M give complete answer to 20 μl of injected protein hydrolyzate sample within less than 90 minutes even if one of the amino acids might be present in as little as 0,1 nanomals only. Sample injector mechanics, fully automatic electronic control of operation, and computer-calculated output of mol-content or percentage values are additional features of this kind of time-saving restless assistants at biochemical institutes.

But liquid chromatography has proved in Beckman application laboratories not only to be the most efficient and specific detection and separation methods for amino acids, likewise nucleotides, organic bases as well as oligosaccharides were successfully injected onto the columns of Beckman Liquid Chromatographs.

In a good many cases of genetic and immunological studies the primary stucture of various proteins is of overwhelming interest. Tedious manual degradation work has summitted in Dr. Edman's rotating-cup method. Beckman's Spinco group pushed forward Dr. Edman's development and soon released the first "Sequencer". Today, seven years after Edman and Begg's publication some 150 "Sequencers" all over the world are automatically degrading long protein and short peptide chains day and night. At an average rate of 24 degradation steps during a 24-hour working day these instruments are continuously producing extremely valuable data particularly on globular and fibrous proteins, mainly hormones and the heavy and light chains of immunoglobulins.

In the field of immunology and related disciplines it turned out that the synthesis of polypeptides can yield considerable information about binding sites nechanisms and enzyme activities in general. The manual construction of polypeptides represents a time-consuming sometimes rather critical manipulation sequence in a biochemical laboratory. Dr. Merrifield's solid-phase method has proved to allow automation to a certain extent. So again, Beckman's Spinco group has designed a device some 30 units of which in the United States and in Western Europe are producing polypeptide chains of 20 amino acids and longer in biologically relevant quantities. Average coupling rate is 3-4 coupling reactions per 24-hour working day in a fully automatic mode.

Besides these dedicated equipment which is specialized for application in biochemical research, also necessary environmental instrumentation as pH-meters and spectrophotometers has attained a level of perfection and reliability that saves valuable research time and costly labourable manipulation. A modern UV-Vis spectrophotometer as Beckman's ACTA-M series is capable of continuous multi-wavelength monitoring of up to four different enzyme kinetic reactions. Fully automatic control of temperature gradients allows the study of denaturation behaviour of thermo-sensitive materials with a minimum of the operator's attention. Digital output facilities will very simply couple the instrument into computer or data handling systems.

As an example to the statement that the availability of reasonably priced and technically reliable measuring equipment brought about new detection methods we will in a short present an introduction into Radioimmunoassy. This new but rapidly expanding technique combines the specifity of an immunological reaction with the sensitivity of radioactivity detection. Beckman has undertaken extensive efforts to put into the hands of scientists and doctors a series of Liquid Scintillation Counters of Gamma-Counters, sets of chemicals and ready-to-use test kits as well as all the know-how for detecting hormones, vitamines in the pg-range in human and animal body. Up to now working-sheets and kits are available for the detection of triiodotyrosine, tetraiodotyrosine, Thyroid Stimulating Hormone, Thyrocalcitonine, Renin, Angiotensin, Insulin, Luteinizing Hormone, Follicel Stimulating Hormone, Human Placental Lactose, Human Growth Hormone, Gastrin, Digixin, Digitoxin, Vitamin B_{12}, Carcinoembryonic Antigen, different Gamma Globulines and others. This technique is based on the competitive binding activity of an unknown amount of antigen in the sample and a known amount of radioactive labelled antigen of the same kind as the corresponding antibodies. The results of this technique have been so outstanding that classical methods of quantitative immunology will very soon be almost completely replaced.

Physical and Chemical Transfer in Reproductive Processes

HYPOTHALAMIC - PITUITARY FUNCTION AND AGEING

M.R.P. Hall

University of Southampton

Southampton, England

Although there is nothing new about ageing it is only in comparatively recent times that large numbers of people have survived to reach old age. Havighurst, at a meeting on Leisure and the 3rd Age held in Dubrovnik in May 1972 estimated that two-thirds of those who have reached the age of 65 in the last 2,000 years are alive today. This gives some idea of the current explosion of old age. In the United Kingdom about 15% of the population are over the age of 65 years. Most European countries have similar figures while other developed countries such as the United States, Australia and New Zealand have aged populations which approach 10% of the total population. Even in the underdeveloped countries the population is beginning to show signs of ageing and many of these have more than 5% over the age of 65 years.

In 1891 Weissman wrote as follows: "I consider that death is not a primary necessity but that it has been secondarily acquired as an adaption; I believe that life is endowed with fixed duration not because it is contrary to its nature to be unlimited, but because the unlimited existence of individuals would be a luxury without any corresponding advantage. The above mentioned hypothesis upon the origin and necessity of death leads me to believe that the organism did not finally cease to renew the worn out cell material because the nature of the cells did not permit them to multiply indefinitely but because the power of multiplying indefinitely was lost when it ceased to be of use. I consider that this view if not exactly proved, can at any rate be rendered extremely probable".

This non-renewal of cells by the organism leads to increased disease and disability. The incidence of all diseases, particularly those like cancer and vascular disorders which are likely to produce considerable disability increases, consequently those over the age of 65 years tend to put an excessive strain on the Health Services. Statistics for the National Health Service in the United Kingdom showed that in 1972 48% of all the beds available each day in hospitals, excluding mental hospitals, were occupied by the over 65's. This was an increase of 4% on the previous year and if such an increase continues, by 1990 90% of all the beds in the country will be occupied by the over 65's.

Consequently there is the need for better understanding of the ageing process. We need to understand the basic principles relating to physiology and biochemistry. Ageing is a complex subject, there is no simple answer. We need to identify the various ageing processes and see how they are controlled and if they are controlled by clocks or a clock as some people suggest, we may be able to modify the process of ageing and postpone 'vigour loss' by tampering with these clocks.

Gerontology may be defined as the applied study of the processes which cause the human body to deteriorate with time. Geriatric medicine or medical gerontology is the branch of general medicine concerned with the clinical, preventive, rehabilitative and social aspects of health and illness in the elderly. Personally I believe that the major function of medical gerontology is to preserve the health of the elderly individual at its maximum. I believe that we must maintain the individual's vigour so that he can enjoy life right until the very end. If we look at life span curves over the years (Fig. 1) we can see that the curve is becoming more rectangular and our objective is to make it even more so. Whether or not we can shift the curve further to the right remains to be seen and this will depend on whether it is possible by genetic engineering to alter the basic programming of the organism. At the present there would seem to be tissue limiting factors. Though organ reserve in the very old still exists, it would seem that from the study of lung and kidney that life is not possible beyond a maximum of 150 years.

WHY HORMONE STUDIES?

The question must be asked why should the study of hormone levels in old people be necessary. Does it really matter what the hormone levels are, when the organism is so near to death. Well of course, deficiency or excess may be treated and the terminal years of life be made much more comfortable. For example myxoedema may present as loss of vigour and mental slowing, both symptoms which are common in old age. Nevertheless, if due to myxoedema, these are reversible and improvement in performance can be seen after so short a time as six weeks.

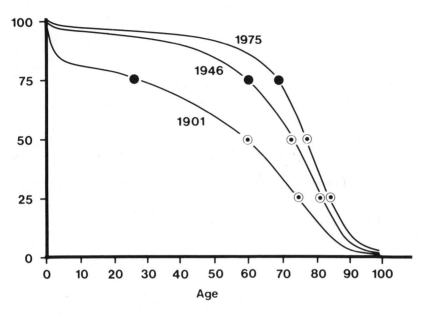

Fig. 1.   SURVIVAL CURVES FOR 1901, 1946 and 1975

Diagnostic levels of hormone dysfunction may vary with age, consequently it is necessary to know whether or not we need age specific ranges of hormone levels.   For instance if we take the thyroid again as an example, the T3 resin uptake and T4 tests which thereby give rise to the free thyroxin index, which is probably the most accurate biochemical assessment of thyroid function, are unaltered.   However I 131 uptake is diminished (normally) and consequently patients with an apparently normal range may in fact be thyrotoxic and will have high T.4 levels. Serum T3 also drops with age and consequently a patient with a T3 serum level which would have been normal in a younger person will indicate thyrotoxicosis in an older person.   The T.S.H. response to T.R.H. also diminishes with age.   Consequently there is need when looking at various parameters of thyroid function to take age into account when making a diagnosis.   In conjunction with my colleague Professor Hall in Newcastle we have been looking at fasting T.S.H. levels in over 85 year olds.   These were all hospitalised patients in a long stay unit whose disease processes had reached a steady state and were otherwise in good health.   We found that 39% had levels higher than 5 µ units per ml and 20% had levels higher than 6 µ units per ml (the upper limit of normal for this laboratory is 5 µ units per ml).   The highest level of 30 µ

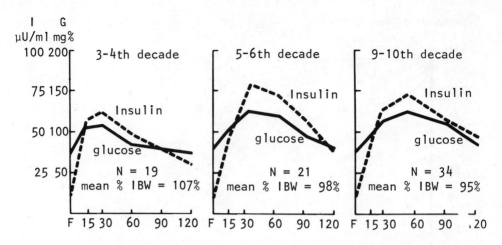

Fig. 2 PLASMA GLUCOSE AND SERUM INSULIN RESPONSE IN NORMAL SUBJECTS AT VARYING AGES

units was in a patient who had previously had a carcinoma of the thyroid, treated by surgery and radiotherapy 20 years earlier. Only this patient and two others had evidence of thyroid disease (i.e. a total 3%).

If we look at another endocrine gland, i.e. the beta cell of the islets of Langerhans we find that carbohydrate tolerance in response to a 50 gm glucose challenge, i.e. the standard glucose tolerance test, is altered in old age. Fig. 2 shows that while young people will peak at 30 mins. old people peak at 60 mins. This suggests that insulin is secreted more slowly or is less effective in controlling glucose levels. Further, Smith and Hall (1973) have shown that if glucose tolerance is studied in normal old people over the age of 85 that there are three types of response. Fig. 3. When the insulin levels are studied in these groups it is found that they are higher than in the normal group. Consequently in spite of increased insulin secretion, blood sugar levels were higher, Fig. 4. This suggests that the insulin is ineffective in controlling blood sugar levels and this could be due either to the physiologically inactive insulin or to an insulin antagonist, or perhaps due to a defective insulin receptor action in lymphocytes as suggested in the round table discussion by Bletcher earlier in this colloquium.

Consequently the more one looks at hormone action in the aged the more queries and abnormalities one finds. This seems to be a subject of considerable interest and with considerable research potential.

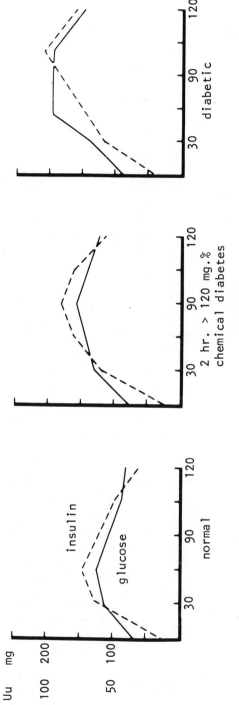

Fig. 3   50 gm. ORAL G.T.T. IN ELDERLY SUBJECTS

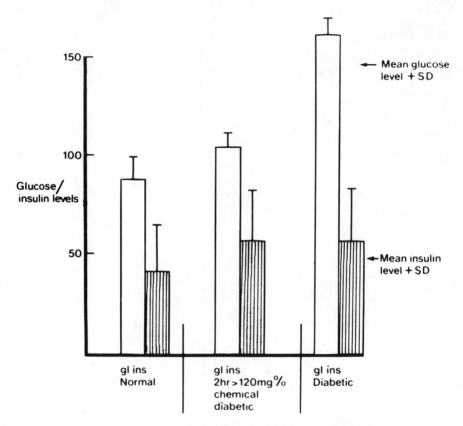

Fig. 4 MEAN GLUCOSE AND INSULIN LEVELS DURING 50 gm ORAL G.T.T. IN ELDERLY SUBJECTS

FUNCTION OF THE HYPOTHALAMIC-PITUITARY-ADRENAL AXIS IN AGEING

There are at present many gaps in our knowledge of the function of this axis in old age. The subject has been well reviewed by David Gusseck (1972) and Everitt (1973), the latter proposes the hypothesis that the site of the ageing 'clock' is the hypo-thalamic-pituitary system. He suggests that environmental factors produce a stress which causes a secretion of 'ageing hormone' which stimulate target organs and tissues and this accelerates the rate of ageing. It also suggests that the hypothalamus is less sensitive to feed back. However, much of the evidence presented in this paper is conjectural and some is conflicting.

Semantics are important in relation to ageing. It is important to know exactly what we mean. Many papers are referable to growth or development, others relate to mature animals such as two year old rats. Few studies have in fact been done on really old people, or for that matter really old animals, since the maintenance of ageing colonies is both difficult and expensive. A good deal of work is currently being focused at the cellular level. Professor Arnstein's paper given earlier in this colloquium raises the possibility that erythropoiesis could be a suitable model of ageing at the cellular level which might be a fair one to study. Certainly Smith's model (1926) in which he described rapid ageing changes in hypophysectomised rats is not a true model of ageing changes since replacement with hormones would render target tissues normal. Further, Everitt (1973) suggests that hypophysectomy may actually protect the organism against ageing.

EFFECT OF ADRENO-CORTICAL HORMONES ON AGEING

Adreno-cortical hormones have both good and bad effects on the ageing process. For instance mammalian life span may be prolonged in some species. Bellamy (1968) showed that prednisolone phosphate will prolong life in a strain of short lived mice. Prednisolone also has other life-prolonging effects on cells. Post mitotic life span of cells seems fixed after the closing of a developmental "gate" but the interval of the position of the gate may be modifiable. Further, adrenal enlargement has been reported following feeding B.H.T. (di-t-butyl hydroxytoluene) and it has been suggested that this and other anti-oxidants prolong life span though they may not alter specific age. However, animals fed B.H.T. also eat less so the life prolonging effect may be that of subnutrition.

The immuno-suppressant action of hydro-cortisone may also be beneficial in various ways. For instance it releases the epigenetic blocking of cells thereby altering tissue maturation and subsequent ageing (Gelfont and Graham Smith 1972). It has also been suggested that prednisone may act as a lysosome stabiliser, and ageing may be mediated through lysosomal or membrane damage (Allison 1966). There is perhaps a possible parallel here with Arnstein's earlier paper in this colloquium, when he showed that when haemoglobin concentration rose the level of many cell enzymes dropped. This was probably associated with lysosomal disappearance and it is possible that haemoglobin causes lysosomal damage as it accumulates in the cell. The question can be posed, is this similar to the accumulation of lipofuscin granules which accumulate in ageing cells and does prednisone in some way interfere with this process. Alternatively is the lysosomal damage related to a misspecified self-propogating portion which produces a "slow virus" effect. If so then lysosomal stabilization might be a result of the immunosuppressant action of prednisone.

Autoimmune phenomena are of course, common in old animals and humans. The parallel between autoimmunity and ageing has been drawn by many people. Cortico steroids are well recognised to have a beneficial effect on the course of many diseases which are autoimmune in origin and to prevent or postpone death which would occur if these conditions were allowed to run their natural course. We have studied the incidence of thyroglobulin antibodies in a 100 old people over the age of 85. We have found a titre greater than 1 in 25 in 1/3 of the women and in 1/8 of the men. This work is in the process of being repeated and other antibodies and the immune profile of these patients studied in greater detail. The exact relation between hormones and the immune response is however, in some doubt since it has been reported (Fabris et al 1972) that a progeria like state in hypopituitary mice can be abolished by the establishment of normal lymphocytic function. This is a very interesting observation in view of some of the remarks that have been made at the round table discussions during this colloquium.

On the other hand we know that corticosteroids have harmful effects in that they cause salt and water retention and reduce the collagen content of tissues such as bone and skin. It has been known for years that subjects with Cushings syndrome have a low skin collagen content while those with Acromegaly have a higher than normal skin collagen content. Hall (1973) has recently shown that the skin of subjects receiving treatment with Prednisone contains considerably less collagen than skin from normal subjects of a similar age. Solubility of collagen is altered being increased in subjects receiving corticosteroids. The amount of collagen lost however, while increasing with age is not dose related.

ABILITY OF THE AGED TO RESPOND AND ADJUST TO STRESS WITH AGE

Much work suggests that the aged lose the ability to respond and adjust to stress with age. As we have seen earlier in this paper the T.S.H. response to T.R.H. diminishes with age. In spite of this some doubt still exists as to whether or not the pituitary adrenal axis is in fact affected by age. Animal work tends to support the hypothesis that function is reduced with age. The functional reserve capacities of young and old male and female Long Evans rats has been studied by measuring the increase in plasma cortico steroid concentration following ether vapour stress and ACTH stimulation (Hess and Riegle 1970). Old male and female rats showed significantly lower plasma cortico-steroid levels than the young animals when maximally stressed with either ether vapours or doses of exogenous ACTH. Adrenocortical function in dogs also tends to decrease slightly in response to ACTH with age (Breznock and McQueen 1970) similar findings have been shown in older cattle (Riegle and Nellow, 1967) and goats (Riegle et al, 1968).

Similarly levels of circulating corticosterone have been found in ageing mice (Grad and Khalid 1968).

Human research is quite rightly complicated by ethical problems. Studies involve tests which directly provoke hormonal secretion, followed by measuring plasma levels by radio immune assay techniques. Pituitary function is assessed by response to acute stress and adrenal function by ACTH injection. Stimuli provoked must be strong enough to produce adequate response but not be dangerous to the patient. The insulin tolerance test is almost certainly the safest and best test of pituitary/adrenal function. It is however, probably wise to modify the test if hypopituitary function is suspected. It is probably wise therefore, to estimate fasting plasma cortisol levels prior to performing the test. The intravenous metyrapone test has also been described as being safe and effective. Blichert-Toft and Jensen (1970) using this test did not find any difference in plasma cortisol and plasma compound S levels between young and old. Laron and his colleagues (1967) found impaired human growth hormone response to insulin hypoglycaemia in four out of the 19 subjects they studied. However in three of these the hypoglycaemic stimulus was not satisfactory. The cortisol response to insulin and ACTH as well as the effect of dexamethasone has been studied by Friedman, Green and Sharland (1969). While they found no evidence for a decrease of the functional integrity of the glands in old age they did find an excessive response to exogenous ACTH gel and they also found that the midnight plasma cortisol levels were much higher than in younger subjects. Cartlidge and his colleagues (1970) investigated the function of the pituitary-adrenal axis in 8 elderly subjects aged 80-95. Plasma cortisol, plasma free fatty acids (FFA) and serum human growth hormone (HGH) were measured following insulin induced hypoglycaemia. The results were compared to those which had been obtained in younger subjects. The serum HGH response to insulin was normal and as one would expect the FFA recovery index was also normal. The maximal plasma cortisol level achieved after insulin was also normal though the maximum increment of cortisol was in fact reduced in four subjects.

Human growth hormone output has also been studied by Laron and his colleagues (1970) in association with exercise. While human growth hormone was excreted many subjects appeared to have a reduced reserve. Significance however of this cannot be judged accurately since similar findings can be found in younger patients.

Other studies (West and his colleagues 1961 and Romanof and her colleagues 1961) suggest that cortisol is removed from the circulation at a progressively slower rate with advancing age. Consequently stimulation of the adrenal cortex might be expected to produce higher levels of circulating cortisol than one gets in the young. This was not found and consequently they concluded that the adrenal cortex was less responsive in old age. The reason for the decreased rate of cortisol utilisation is however, not clearly understood. Certainly decreased renal function may exaggerate this, but probably the decreased consumption of cortisol is related to the decreased muscle bulk which occurs with age.

While many of the studies in old humans and old animals suggest that in fact there is diminished function in the pituitary adrenal axis in old age, others have shown normal response. The problem however, with nearly all these studies is that they have been performed on aged hospital populations and the subjects are all suffering from chronic disease. Cook and his colleagues in 1964 showed that chronic subnutrition is often associated with a considerable degree of pituitary depletion. Recent studies of our own in long stay hospital patients have shown that in spite of what appears to be an adequate diet considerable vitamin deficiency exists. It is possible therefore, that the subjects studied do not reflect what is truly happening in the normal aged subject. Further Carroll (1969) has shown that patients with depressive illness also suffer from hypothalamic-pituitary insensitivity and it is quite possible many of the patients studied may have been suffering from major or minor degrees of depression in association with their illness or chronic disability.

SUMMARY

On balance, therefore, while function in the hypothalamic-pituitary-adrenal axis is pretty good in old age there are some indications to suggest that some diminution or alteration of function may exist. The results are inconclusive. Further it is important to realise that these are point of time studies and only reflect age differences. True age changes can only be determined if prospective cohort studies are undertaken. A lot more work needs to be done in this field. Perhaps the biggest gain to the patient so far is the knowledge that cortisol half life is prolonged so that steroid therapy should be given with caution and only small doses used.

There would seem to be no evidence at present that a clock exists in the hypothalamic-pituitary region and problems such as peripheral utilisation of hormones, protein binding, and receptor action have yet to be studied. This colloquium has provided a forum at which these views can be expressed and discussed with others in different disciplines and I am sure that it will act as a great stimulus to further work which may perhaps throw some light on some of the problems which have been raised.

REFERENCES

Havighurst R. "Life Style and Leisure Patterns" pp 35-48, Proc. 3rd Int. Course of Social Gerontology, 1972, Leisure and the 3rd Age, Int. Center of Social Gerontology.

Weissman A. Essays upon Heredity and Kindred Biological Problems. Oxford University Press, London & New York 1891

Smith M.J. and Hall M.R.P. Carbohydrate Tolerance in the Very Aged, Diabetologia 9, 387-390, 1973

Gusseck D.J. In "Endocrine Mechanisms and Ageing" pp 105-66 Advances in Gerontological Research, ed. B.L. Strehler, Academic Press, New York and London 1972

Everitt A.V. The hypothalamic-pituitary control of Aging and Age-related Pathology, Exp. Geront. 8, 265-277, 1973

Smith P.E. Anat. Rec. 32, 221, 1926

Bellamy D. 'Long term action of prednisolone phosphate on a strain of short lived mice' Exp. Gerontol. 4, 327-334, 1968

Gelfont S. and Smith J.G. Ageing: Noncycling cells - an explanation, Science, 178, 357-361, 1972

Allison A.C. 'The role of Lysosomes in Pathology' Proc. Roy. Soc. Med. 59, 867-868, 1966

Fabris N., Lymphocytes, Hormones and Ageing, Nature, 240, 557-559, 1972

Hall, David A., Reed F.B., Nuki G., and Vince J.D. Effects of age and corticosteroid therapy on skin collagen in human subjects, Communication to the British Society for Research on Ageing, October 1973.

Hess, G.D. and Riegle G.D. Adrenocortical Responsiveness to Stress and ACTH in Aging Rats, J. Geront., 25, 354-358, 1970

Breznock, E.M. and McQueen R.D. Adrenocortical Function during Aging in the Dog, Amer. J. Vet. Res. 31, 1269-1272, 1970.

Riegle G.D. and Nellor J.E., Changes in adrenocortical function during aging in Cattle, J. Geront. 22, 83-87, 1967.

Riegle G.D., Przekop F. and Nellor J.E., Changes in Adrenocortical Responsiveness to ACTH infusion in Aging Goats, J. Geront. 23, 522-528, 1969

Grad B. and Khalid R. Circulating Corticosterone levels in Young and Old, Male and Female C57Bl/6J Mice. J. Geront. 23, 522-528, 1969

Jensen H.K. and Blichert-Toft M. Pituitary Adrenal Function in Old Age evaluated by the intravenous metyrapone test, Acta. Endocrin, 64, 431-438, 1970.

Laron Z., Doron M. and Amikan B. Serum Growth Hormone in Old Age. J. Israel. Med. Assoc. 73, 375-378, 1967.

Friedman, M., Green M.F. and Sharland D.E. Assessment of Hypothalamic-Pituitary-Adrenal Function in the Geriatric Age Group, J. Geront. 24, 292-297, 1969

Cartlidge N.E.F., Black M.M., Hall M.R.P. and Hall R. Pituitary Function in the Elderly, Geront. clin. 12, 65-70, 1970.

Laron Z., Doron M. and Amikan B. Plasma Growth Hormone in men and women over 70 years of age. In Medicine and Sport IV p.126-131 Petah Tigra - Karger (Basel) 1970.

West C.D., Brown H., Simons E.L., Carter D.B., Kumagai L.F. and Englert E. Jr., Adrenocortical Function and Cortisol Metabolism in Old Age, J. Clin. Endocrinol & Metab. 21, 1197-1207, 1961

Romanoff, Louise P., Morris, Carol W., Welch Patricia, Rodriguez Rosa M. and Pincus G. The metabolism of Cortisol-4-C^{14} in excretion of tetrahydrocortisol allotetrahydrocortisol, tetrahydrocortisone and cortolone (20α and 20β) J. Clin. Endocrinol & Metab. 21, 1413-1425, 1961.

Carroll B.J., Pearson, Margaret J. and Martin, F.I.R. Evaluation of three acute tests of Hypothalamic-Pituitary-Adrenal Function, Metabolism, 18, 476-483, 1969

ADENOHYPOPHYSIS AND THE MALE GONADAL AXIS FOLLOWING PARTIAL BODY γ - IRRADIATION

G.S. Gupta

Department of Biophysics, Panjab University

Chandigarh 160014, India

INTRODUCTION

Of all cellular elements of vertebrate testis, spermatogonia are by far the most sensitive to ionising radiation. Sertoli and interstitial cells on the other hand are relatively radio-resistant (Rugh, 1960; Ellis, 1970). So much so that the interstitial tissue gets hypertrophied (McEnery and Nelson, 1953; Lindsey et al, 1969; Gupta and Bawa, 1974a, 1975a) and hyperplasia of interstitium, at times is converted into neoplasia (Lindsey et al, 1969). The possible explanation for the origin of the neoplasia is related to the endocrine imbalance and to the altered testis pituitary axis following localised irradiation (Lindsey et al,1969; Ellis, 1970; Gupta and Bawa, 1974b; 1974c). However, evidences of the altered cytomorphology of the pituitary after irradiation of testis are lacking (Lindsey et al, 1970). The present report describes the molecular changes associated with hyperplastic rat testis interstitium subjected to localized γ-irradiation and their relations with altered testis-pituitary axis.

MATERIALS AND METHODS

(a) Radiation procedure and treatment of tissues: Normal white albino rats (160-220 g) were divided in different groups with 6-8 rats in each group and anaesthetised with Nembutal and the scrota irradiated (70 - 80 R/min) at a distance of 65 cm from the source. Sham irradiated animals acted as controls (Gupta and Bawa, 1974a, 1975a). The feeding of animals was stopped 6-8 h prior to sacrifice and they were killed by decapitation according to a schedule. Testes in a group of rats were pooled and frozen. A portion of randomized parenchymal tissue was homogenized in Potter-Elvehjem homogenizer to yield 10% aqueous homogenates. Supernatant fraction obtained at 3500 g was used for the assay of glucose-6-phosphate dehydrogenase.

(b) Assay of Glucose-6-phosphate dehydrogenase (D-glucose-6-phosphate: NADP oxidoreductase EC 1.1.1.49): The activity of G6PDH was determined according to the method of De Moss (1955) using 0.1 ml of 10 per cent testis supernatant at 37°C. 3 ml reaction mixture consisted of 0.2 ml sodium salt of glucose-6-phosphate. (13.3 μM), 1.5 ml tris-buffer (0.1 M) at pH 7.6, 0.1 ml NADP(0.0027M) and 0.1 ml $MgCl_2$ (0.1 M). Absorbancies were measured for a period of 5 min. Unit of enzyme is defined as the amount of change in optical density per min per mg protein at 37°C.

(c) P^{32} incorporation studies: Rats were separated in different groups with 6 rats in each group and irradiated as described above. Before sacrifice of animals, each rat was injected interaperitoneally 40 μCi/100 g body weight carrier free radioactive phosphorous (P^{32} from BARC, Trombay) in 0.5 ml saline and the activity was stabilized for 4 hr. Tissue fractionation of P^{32} for alcohol soluble and RNA fractions was carried out as described elsewhere (Gupta and Bawa, 1974c; 1975b).

RESULTS

It is evident that the activity of glucose-6-phosphate dehydrogenase in testis is enhanced on 3-5 days after irradiation and continues to rise upto a period of 11 months. After day 7, the activity of the enzyme was 3-6 folds at a lethal dose (p 0.01) and 1.5-6 folds at a sublethal dose (p < 0.01) /Fig.1/. However, the enhancement of the enzymatic activity per testis was more conspicuous at 720 R than that observed at 2000 R (Gupta, 1973). At a subcellular level, the increase in the activity was 10 times that of the normal activity in the cytoplasm which was devoid of the nuclei and the mitochondria (Gupta, 1973). It was possible to demonstrate that the increase of the enzyme was mostly localized in the dissected interstitial tissue (Gupta,1973).

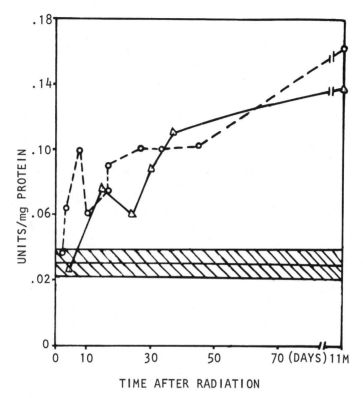

Fig. 1: Glucose-6-phosphate dehydrogenase activity/mg protein in testis on different days and 11 months (M) after partial body γ-irradiation at 720 (△—△) and 2000 R (o...o). Each observation is the average of 6-8 rats. Hatching shows the standard deviation in normal population of rats.

With the increase in the activity of glucose-6-phosphate dehydrogenase, there was a parallel rise in the P^{32} activity in alcohol soluble and RNA fractions subsequent to day 16 (Figs. 2 & 3).

It was observed that the increase in the labelling of P^{32} in alcohol soluble and RNA fractions was more evident at a lethal dose than that at a sublethal dose, 74 and 30 days after irradiation respectively (Fig.2&3). The initial decline of RNA labelling on days 8-16 was presumably due to the loss of germinal cells from the seminiferous tubules (Gupta and Bawa, 1974a).

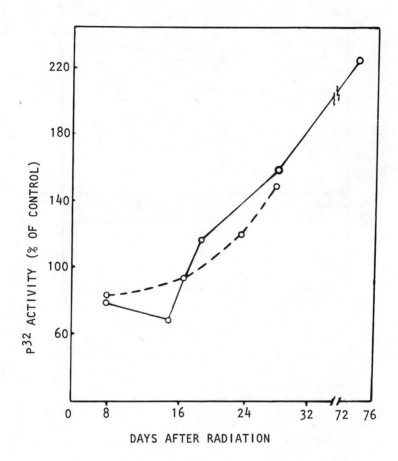

Fig.2 - P^{32} activity in alcohol soluble fraction (relative counts/g tissue).

Fig.2 & 3: Activity of P^{32} in alcohol soluble and RNA fractions of rat testis on different days and 11 months (M) after partial body γ-irradiation at 720 (o...o) and 2000 (o——o)R. Each observation in the average of 6 rats. Values are expressed as percent of controls.

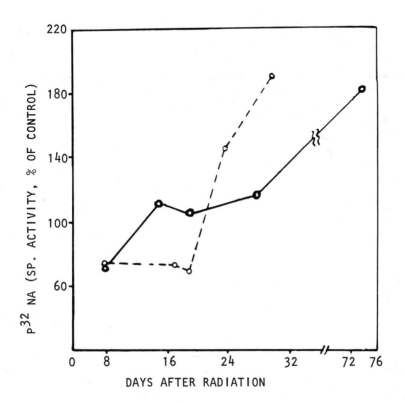

Fig.3- P^{32} activity in RNA fraction (relative counts/mg RNA).

DISCUSSION

The increase in the activity of glucose-6-phosphate dehydrogenase following irradiation of testis can be related to the (i) shrunkenness of testicular tubules (Ellis, 1970) and (ii) loss of feed back control between the testis and the pituitary (Gupta and Bawa, 1974b), or it may be due to the imbalance of local factors caused by irradiation (Mark and Banks, 1960; Hitzeman, 1962). An indirect induction of glucose-6-phosphate dehydrogenase due to the increased pouring in of pituitary gonadotrophins (Heller et al, 1965) after localized irradiation of testis seems to be a distinct possibility. Since it is known that the glucose-6-phosphate dehydrogenase of testis is influenced by gonadotrophins (Bollman and Gosler, 1969; Elkington and Blackshaw, 1970), it can be suggested that the rise in RNA and phospholipid labelling by P^{32} subsequent to a minimum value, is associated with the de novo biosynthesis of the enzyme in the nongerminal cells which abound in histological preparations (Gupta and Bawa, 1974a, 1975a). Concerning the local factors which may influence the activity of glucose-6-phosphate dehydrogenase, it can be stated that glucose-6-phosphate dehydrogenase and other NADP-linked dehydrogenase are probably under the control of local mechanisms in which small molecules like pregnenolone, dehydroepiandrostenedione (Marks and Banks, 1960) and glucose-6-phosphate (Steiner and Williams, 1959) seem to play a significant role. The loss of androgens leading to the accumulation of cholesterol and loss of precursors of steroid hormones following localized irradiation is a subject of investigations of Ellis and Coworkers (Ellis, 1970). The possibility of accumulation of glucose-6-phosphate due to the inhibition of glucose-6-phosphatase has been suggested by us (Gupta and Bawa, 1970; Gupta and Bawa, 1975c). It is possible that the loss of precursors hormones, viz., pregnenolone and dehydroepiandrostenedione which are the inhibitors of the enzyme in vitro (Marks and Banks, 1960) and the accumulation of glucose-6-phosphate, an activator of the enzyme (Steiner and Williams, 1959), are some of the factors which are partially responsible for the enhanced activity of glucose-6-phosphate dehydrogenase in rat testis following localized irradiation. A possibility of local control of the enzyme has been suggested earlier (Bawa and Gupta, 1974).

The enhanced activity of the glucose-6-phosphate dehydrogenase is related to the hyperplasia of rat interstitial cells which predominate in the synthesis of net work of membranes as shown by P^{32} activity in alcohol soluble fraction. In respect to phosphate shunt enzymes, the hyperplasia of testis interstitium resembles the malingnant tissues of different origins (Chayen et al,1962; Kit, 1956; Ayre & Goldberg, 1966; Cohen, 1964; McLean and Brown, 1966; Krishnamurthy et al, 1968). The hyperplastic rat interstitial tissue is associated with enhanced levels of glycogen (Gupta and Bawa, 1974a) and enhanced endogenous respiration (Steinberger and Wagner, 1961). It is suggested tnat the glycogen acts as a reserve substrate for the activity of shunt pathway in hyperplastic tissue (Gupta and Bawa, 1974a).

ACKNOWLEDGEMENTS

The author wishes to express his thanks to Professor S.R.Bawa for his keen interest in the work and to Dr P.N.Chhuttani and Dr B.D.Gupta for providing radiation facilities at PGI.

REFERENCES

1. Ayre, H.A. and Goldberg, D.M., Brit.J.Cancer 20, 743-750 (1966).

2. Bawa, S.R. and Gupta, G.S., J.Steroid Biochem. 5, 347 (1974).

3. Bollman, V.R. and Goster, H.G., Acta Histochem. 33, 7-12 (1969).

4. Chayen, J., Bitensky,L., Aves, E.K., Jones, G.R., Silox, A.A. and Cunningham, G.J., Nature 195, 714-715 (1962).

5. Cohen, R.B., Cancer 17, 1067-1075 (1964).

6. DeMoss, R.D., In: Methods in Enzymology, Vol.1 (S.P. Colowick and N.O. Kaplan, eds.) pp 328-334, Academic Press, New York,(1955).

7. Elkington, J.S.W. and Blackshaw, A.W., J.Reprod.Fert. 23, 1-20 (1970).

8. Ellis, L.C., In: The testis, Vol.III (A.D. Johnson, W.R. Gomes and N.L. VanDemark, eds.) pp 333-376 (1970), Academic Press, New York.

9. Gupta, G.S., Ph.D. Thesis, Panjab University, Chandigarh, India,(1973).

10. Gupta, G.S. and Bawa, S.R., J.Reprod.Fert. 27, 451-454 (1970).

11. Gupta, G.S. and Bawa, S.R., J.Reprod.Fert. 41, 185-188 (1974a).

12. Gupta, G.S. and Bawa, S.R., In: Proc. 5th Asian and Oceania Congress of Endocrinology, Chandigarh, pp 26 (1974b).

13. Gupta, G.S. and Bawa, S.R., Strahlentherapie 148, 420-424 (1974c).

14. Gupta, G.S. and Bawa, S.R., J.Reprod.Fert. 42, 29-34 (1975a).

15. Gupta, G.S. and Bawa, S.R., Strahlentherapie, In Press (1975b).

16. Gupta, G.S. and Bawa, S.R., Radiat.Res. (Communicated, 1975c).

17. Heller, C.G., Wootton, P., Rowley, J., Lalli, M.F. and Brusca, D.R., Proc. 4th Pan.Am Congr.Endocrinol., Mexico City, Intern.Congr. Ser.No.112, 408, Excerpta Med. Found, Amsterdam (1965).

18. Hitzeman, J.W., Anat. Rec. 143, 351-359 (1962).

19. Kit, S., Cancer Res. 16,70 (1956).

20. Krishnamurthy, A.S., Russfield, A.B. and DeLisle,M.P., Endocrinology 82, 989-994 (1968).

21. Lindsey, S., Nichols, C.W.Jr., Sheline, G.E. and Chaikaff, T.L., Radiat. Res. 40, 366-378 (1969).

22. Lindsey, S., Nichols, C.W.Jr. and Sheline, G.E.,Proc. Soc.Expt.Biol.Med. 134, 523-526 (1970).

23. Marks, P. and Banks, J., Proc.Natl.Acad.Sci. (US),46, 447-452 (1960).

24. McEnery, W.B. and Nelson, W.O., Endocrynology 52, 104-111 (1953).

25. McLean, P.M. and Brown, J., Biochem.J. 98, 874-882 (1966).

26. Rugh, R., In: Mechanism in Radiobiology, Vol.2, p.2-94 (M.Errera and A.Forssberg, Eds.) Academic Press, New York, (1960).

27. Steinberger, E. and Wagner, C., Endocrinology 69, 305-311 (1961).

28. Steiner, D.F. and Williams, R.H., J.Biol.Chem. 234, 1342-1346 (1959).

THE TESTICULAR FUNCTION EFFICIENCY DETERMINED BY THE SPERMATOGENIC ACTIVITY TEST (SAT) IN DIABETIC AND ALCOHOLIC PATIENTS

P.Kolarov, D.Panayotov, M.Protich, D.Strashimirov, S.Milanov

Institute of Endocrinology, Gerontology and Geriatrics, Medical Academy, Sofia, Bulgaria

As far back as 1962, Czerniak /1/ proposed a new test to rate the spermatogenic activity of the gonads by means of radioactive phosphorus (^{32}P). The predilectional accumulation of high-energy phosphate compounds in the tissues which have a pronounced metabolite activity underlies it. As these compounds, being carriers of considerable bioenergies, participate in the synthesis of nucleic acids, their accumulation is in keeping with the mitoses in a given tissue. This accounts for their accumulation in organs and tissues with a clear-cut active cell proliferation, such as testicles, bone marrow, and malignant tumors. The advantages of the SAT are connected with the opportunity of a separate evaluation of the functional capacity of each gonad as well as of establishing the intensity level of spermatogenesis even in patients from whom it is impossible to obtain sperm ejaculation.

MATERIAL AND METHODS

The test was performed in 106 men, from 20 to 49 years of age, namely in 40 cases of diabetes mellitus, 56 of chronic alcoholism, and in 10 healthy individuals (control group). The frequent sexual disorders in the diabetic and the alcoholic determined our choice. In order to establish whether there is a correlation between the sexual disorders and those of the generative function as rated by the SAT we subdivided the diabetic and the alcoholic under test into two groups depending on whether they had, or had not troubles in their sexual dynamics (changes in their libido, erection, orgasm and ejaculation).

Except for antidiabetic drugs (if necessary), no other medication was administered to the patients under test prior to and during its performance. Diabetics in a state of decompensation of their carbohydrate metabolism, or of keto-acidosis, were eliminated in view of removing the influence of the sharply impaired metabolic disorder on the sexual function.

The SAT, in a modification of our own, was performed by ingestion of a tracer ^{32}P-dose of 5 μC/10 kg of body weight, on an empty stomach in the morning. It has been stated (Czerniak, /1/; Panayotov et al., /2/) that the aforementioned dose is quite harmless. The accumulation of the isotope was read separately for each gonad by counting the impulses emanating from it. An M-counter VA 310 served the purpose being put in contact with the scrotum streched manually by the experimentator.

Readings were made after 24 hours, on the sixth and eighth day respectively. After extraction of the background activity, the net one obtained at the first measurement was taken as the initial activity, while the arithmetical mean value of the readings on the sixth and eighth day respectively was rated as a simple average value of the seventh day. The biologic half-life of ^{32}P in the gonad, by twenty-four hours, was rated too by means of the curve of radiation activity decrease thus obtained.

The results of our observations were analysed according to the methods of variation statistics, using the T test of Student and Fisher.

RESULTS AND DISCUSSION

The three SAT index values checked by us afforded a significant difference bilaterally, both between the diabetic and control individuals, on the one hand, and between the alcoholic and the control group, on the other

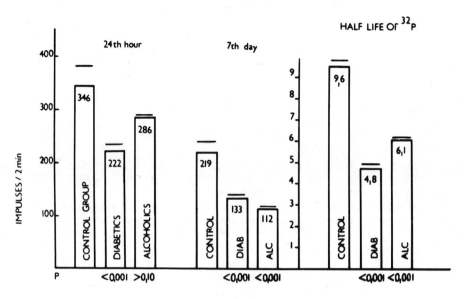

Fig. 1

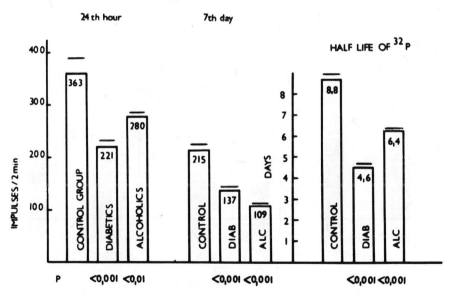

Fig. 2

hand, except only for the insignificant difference of isotope accumulation on the seventh day in the right gonad between the alcoholic and the control group. (Figs. 1 and 2).

The differences observed between the diabetic and the alcoholic (either with or without sexual complaints) were assessed after the first two parameters (radiophosphorus accumulation after 24 hours and on the seventh day respectively) and found insignificant statistically, moreover discrepant in some of the groups. Furthermore, the differences between the groups with and without sexual disorders when rated by the half-life index value of ^{32}P in the gonads, were indicative of a bilaterally significant impairment of spermatogenesis in the diabetic group with disorders of their sexual dynamics(Fig.3). The same group of alcoholic patients, contrary to all expectations, showed a significant prolongation of the radioisotope half-life bilaterally (Fig.4).

Findings resulting from the application of the SAT in our groups of diabetics and alcoholics have afforded convincing proof of the alteration of spermatogenesis in both these conditions. Additional investigations are ne-

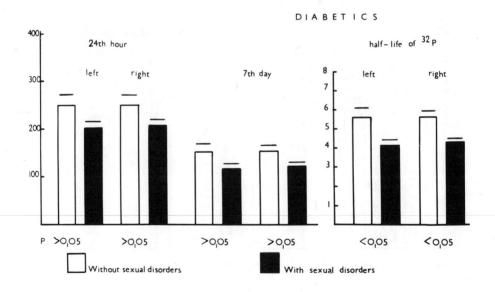

Fig. 3

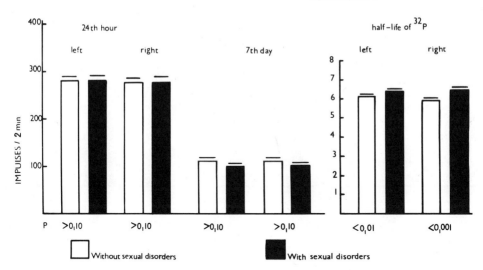

Fig. 4

cessary to get a more precise notion of the intimate mechanism of these injuries, however. It is fair comment to guess a direct toxic influence in the germinative epiterium, or an involvement of hypothalamus and pituitary; last but not least trophic blood vessel lesions, nerve lesions, etc.

The differences between the groups with and without sexual disorders among the diabetic and the alcoholic are rather interesting. After the first two parameters, they are insignificant (as already mentioned), often even discrepant. This fact does not afford sufficient reasons for drawing conclusions. After the third index value (which is obviously of greatest informational importance), statistically significant differences resulted between these two groups. Yet, while the diabetic with sexual desorders displayed a significantly shortened half-life of the radioisotope in their gonads indicative of detrimental effects on their spermatogenesis, this same group of alcoholic test persons showed a significant prolongation of the half-life. Hence, we venture to draw the conclusion that in the diabetic the disorders of sexual dinamics are caused by gonadal injuries or, at least, take a parallel course with them. In chronic alcoholism, both these processes are obviously out of keeping pace with each other. Disorders of libido, erection, ejaculation, and orgasm in alcoholic patients appear under the influence of the toxic effect of alcohol, yet not in association with the lesion of the germinative epithelium.

REFERENCES

1. Czerniak, P.: Am.J.Roentgenol., 1962, 88, 327.

2. Panayotov, D., P.Kolarov, G.Dashev, S.Milanov and M.Protich: Polish Endocrinology, 1969, 20, 144.

NEW PROTEIN- GLYCOLIPOPROTEIN COMPLEX FORMATION IN DILUTED SEMEN

Georgi Kichev

Institute of Biology and Pathology of Reproduction
73 Lenin Ave., Sofia 1113, Bulgaria

The aim of these investigations was to examine the physicochemical interactions between seminal plasma glycoproteins and glycolipoproteins of lactose- yolk glycerine diluents, used for semen deep freezing (1, 2, 3, 4).

METHODS

Samples, each of 0,2- 6 μL of diluent supernatants, diluted and undiluted seminal plasma were separated using microultramicroelectrophoresis on Cellogel at 200 V for 20 min. The protein bands were stained with Amido Schwarz 10B or Poanco S. Glycoproteins were identified by treatment with periodic acid-Schiff reagent, Lipoproteins- by staining with Sudan III or Fettrot 7B and phospholipids- by vapours of iodine (2).

RESULTS AND DISCUSSIONS

The electrophoretic investigations revealed no correlation between the protein distribution (glyco- and lipoproteins) and the degree of dilution after semen dilution. New complexes of protein- glycoproteins were formed rapidly for 30-60 sec due to unknown physico-chemical interactions. The participation of the yolk in the biocomplex formation was confirmed by immunoelectrophoresis on cellogel.

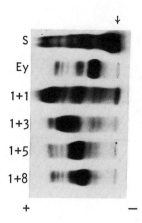

Figure 1

The influence of the dilution degree on the bicomplexes of bull semen (S), diluent-Ey, dilutions: 1+1, 1+3, 1+5, 1+8.

Fig. 1 shows the newly formed glycolipoprotein complexes in bull semen diluted as follows: 1+1, 1+3, 1+5, 1+8. The band of lipoviteline (one of the diluent components) and the band of the major seminal plasma fraction are of a low protein concentration after dilution. Biocomplexes are formed between yolk glycolipoproteins and anion glycoprotein fractions of the seminal plasma. Cation seminal plasma fractions i.e. phospholipids do not take part in the complex formation. These biocomplexes are not stable. They depend on the degree of dilution and yolk concentration in semen diluent but they do not undergo any changes during equilibration, deep freezing and thawing.

The physico-chemical characteristics of the biocomplexes, formed in diluted bull semen, is the same for the biocomplexes in ram and stallion semen.

Those results give the possibility for the control of the degree of semen dilution of the three species mentioned above by ultramicroelectrophoresis.

It is also evident that the complex formation is a species specific process. The electrophoregrams of 1+3 diluted semen obtained from different species are presented in Fig. 2. Four to five biocomplexes are formed in bull semen (A), one biocomplex- in ram semen (B) and none- in boar semen (C) which indicates that boar semen is a cation. Boar seminal plasma proteins and the dilu-

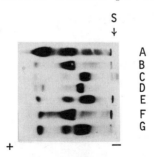

Figure 2

Species specifity in complex formation. A - bull, B - ram, C - boar, D - turkey, E - diluent, F - stallion, G - cock.

ent are moving in different directions from the start of the electrophoregrams (Fig.2 - C). It is possible that the unsuccessful deep freesing of boar semen is due to the lack of physico-chemical cohesion between glycolipoproteins of the seminal plasma and the diluent.

Fig.2 shows alsc the results of the electrophoretic separation of turkey semen (D); Nagase- Niva diluent (E); stallion semen (F) and cock semen (G). The cock seminal plasma proteins overlap these of the diluent.

The characteristic features of the biocomplexes might be used for easy control on the species origin of diluted semen from bull, ram, stallion, turkey, boar and cock. (Fig.1).

The investigations on the character of the complex-formation in bull, ram, and stallion semen revealed that one glycolipoprotein biocomplex was formed between the yolk, on one hand and the semen of the three species studied, on the other hand (S+Ey) (Fig. 3, Fig. 4, Fig. 5). The distruction of this biocomplex, when lactose was added, occured in bull semen only (S+Ey+L). Many investigators indicate the great protective effect of carbohydrates at low temperatures for bull semen. The mechanism of this effect has not been clarified (8). It is evident that the lactose, acting as a strong complex dispersing reagent, changes significantly the dispersion of the coloidal system of yolk diluent and seminal plasma. The formation of a greater number and different types of glycolipoprotein biocomplexes facilitates the physico-chemi-

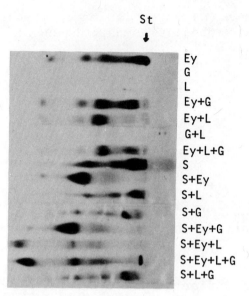

Figure 3

Complex formation in bull semen. Ey-yolk, G - glycerine, L - Lactose, S - semen. Possible combinations between them at 1+3 dilution and 20 per cent yolk concentration

cal cohesion between them and spermatozoa membrane (3, 6, 7). The addition of carbohydrates to the diluents facilitates the coacting of bull spermatozoa with biocomplexes. This process does not occur in the case of ram and stallion semen. (Fig. 4, Fig. 5).

Glycerine is a component of the diluent that plays a special role in the complex formation in stallion semen- Its addition to the combination of semen and yolk (Fig.5 - S+Ey and S+Ey+G) or semen, yolk and lactose leads to a concentration of the biocomplexes at one band on the electrophoregrams. This band is crescent and includes above 80 per cent of the glycolipoproteins of the medium (Fig. 5 - S+Ey+G).

The same form will be found in bull and ram semen when the yolk concentration increases above 20 per cent and the rate of the dilution- above 1+5 (Fig. 6).

Such configuration of the biocomplexes are in connection with the bad preservation and low survival of bull, ram and stallion spermatozoa after deep freezing.

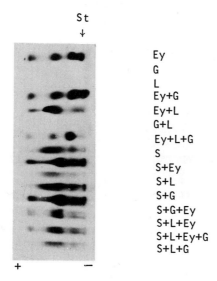

Figure 4

Complex formation in ram semen. Ey - yolk, G - glycerine, L - lactose, S - semen. Possible combinations between them at 2+3 dilution and 20 per cent yolk concentration

In conclusion, our experiments indicate undoubtfully that the spermatozoa deep freezing is a physico-chemical phenomenon.

The main processes of the physico-chemical cohesion or the coating of spermatozoa membrane with glycolipoprotein complexes occur during semen equilibration, a period of time necessary for the deep freezing of semen. It is probable that glycerine, filling the spaces left free from the angles formed after interaction, increases the protective effect by lowering of the freezing point.

In order to support this new physico-chemical theory, several points should be indicated:

1. Using the more disperse medium of ready biocomplex for ram semen dilution we achieved better results in deep freezing by 15 - 20 per cent higher than found before.

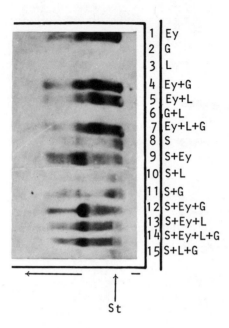

Figure 5

Complex formation in stallion semen.
Ey - yolk, G - glycerine, L - lactose,
S - semen. Possible combinations between
them at 1+3 dilution and 20 per cent
yolk concentration

2. An 80 per cent correlation was found between bull semen deep freezing and the characteristics of the newly formed biocomplexes.

3. The disturbances in sperm production are preceded and accompanied by disturbances in the biocomplex formation (infections epidymitis in ram).

4. The physico-chemical theory described above explains many questions connected with technologies of sperm dilution and the choice and contents of the diluents used for deep freezing of semen from species, whose semen up till now is not successfully preserved at low temperatures (9).

These findings give ground to conclude that the search for new components and proper concentrations for the semen diluents as well as investigations on the optimal dispersion between molecules may throw light on the

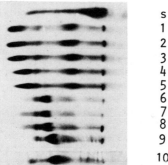

Figure 6

Bull semen (S): 1+1 dilution (1-5), 1+5 dilution (6-9), 10 per cent yolk concentration (1,6), 15 per cent yolk concentration (2,7), 20 per cent concentration (3,8), 25 per cent concentration (4,9), 30 per cent cocentration (5,10).

mechanisms of deep freezing of semen, tissue and cells.

An interesting question arises from the observations described above about the possibility of spermatozoa creating at the time of equilibration physico-chemically nonpermeable glycolipoprotein insulating coats that keep the spermatozoa in frozen state.

REFERENCES

1. Kichev, G.Vet. med. nauki, XI (1974), No. 10, 59 (in Bulgarian).

2. Kichev, G. and Danov, T. Compt. rend. Acad. Bulg. Sci. 26 (1972), No. 10, 1421.

3. Kolb, G. in: Biophysical Aspects of Organism Response in Tuberculosis, (1974) Belorus, Minsk (in Russian).

4. Yordanov, Zh. in Hen Egg Cytobiology, (1969) Izd. BAN, Sofia (in Bulgarian).

5. Kichev, G. and Danov, T., Compt. rend. Acad. Bulg.Sci., 2   26, (1973) No. 11, 1557.

6. Kichev, G., Kalev, G., Ivanov, I., Petkov, Z. and

Mihov, M., Compt. rend. Acad. Bulg. Sci., 27, (1974) No. 11, 1577.

7. Afonski, S. in: Biocomplexes and their Importance, (1965), Kolos, Moscow (in Russian).

8. Müller, E., VII Intern. Kongress fur Tier. Fortpfl., München, (1972), B. II. (in German)

9. Nischikava, Y., Iritani, A., Shimonia, S., VII Intern. Kongr. für Tier. Fortpfl. (1972), München (in German).

CHANGES IN EEG AND IN THE ACTIVITY OF THE WHOLE CERVICAL VAGUS UPON APPLICATION OF SEX HORMONES

Anna Varbanova, Pavlina Doneshka and Julia G. Vassileva-Popova[*]

Institute of Physiology and [*]Central Laboratory of Biophysics, Bulgarian Academy of Sciences, 1113 Sofia, Bulgaria

In the course of experimental work carried out with cats we have observed for years on end that the periods of estrus in animals take place with increased amplitude of the tonic vagal impulses. This fact suggested to us the idea to study the changes in the bioelectrical activity of the cervical vagus and in different cortical and subcortical structures at various periods of the reproductive cycle (pregnancy, parturition, suckling) as well as in coital and precoital states.

There is evidence the relevant literature indicating changes in the bioelectrical brain activity during pregnancy and parturition but only in human subjects. Hypersynchronous alpha-rhythm of a high amplitude was found at the end of pregnancy and during the intervals among contractions (Levinson 1959; Lebedeva 1963). Depression of the alpha-rhythm and appearance of beta-rhythm of increased amplitude and higher frequency have been observed upon contraction of the uterus and upon movement of the foetus (Jakovlev et al. 1952; Syrovatko and Jahontov 1953; Levinson 1959; Lebedeva 1962).

Experiments were performed with animals to study the changes in the EEG during coitus and suckling. The coitus as well as the vaginal stimulation led to a hypersynchronous EEG activity which occured 7 to 13 min after the stimulus was applied and continued for several minutes (Sawyer a. Kawakami 1959; Kawakami a. Saito 1965; Varbanova, Doneshka a. Vassileva 1973). Hypersynchronous high-amplitude oscillations have been observed during suckling as well (Ljubimov 1968).

There is only report which treats of the changes in the nerve impulses in different periods of the sexualhormonal cycle (Krijanovska 1964). The author has studied the afferent activity of n. hypogastricus in cats and finds an increased ampitude of the impulses during estrus and 2 to 6 days after the administration of folliculin.

We performed 83 experiments on cats under chronic conditions (5 female cats and 2 male). In all animals electrodes were implanted in the cervical vagus and in different cortical and subcortical brain formations (gyr.orbit., gyr.sygmoid., nVPL, Hvm, for.ret.mes.).

We undertook systematic investigations of the bioelectrical vagal activity in the above mentioned formations at the end of pregnancy (12 to 18 days prior to parturition), on the days of parturition itself, and long after that (40 to 60 days). In all animals the recordings from the end of pregnancy, from the days of parturition and 6 to 8 days after that showed similar characteristics and differed essentially from the recordings made late after parturition. Thus, predominant in the electroencephalogram of the period of parturition was a high-amplitude synchronous beta-activity of considerably higher frequency compared with the period late after parturition. These changes were seen in all recorded brain formations and they were most clearly manifested in the cortical ones. A statistically significant ($p < 0.001$) redistribution was established in the frequency spectrum of the electroencephalograms. It was evident that on the day prior to parturition the percentage of beta-waves of higher frequency was considerably larger. The simultaneously recorded bioelectrical activity of the cervical vagus showed substantial increase in the amplitude of the tonic vagal impulses (the amplitude was below 20 μv in the background records and reached 50 μv in the period of parturition itself). No significant changes were seen in the frequency of the vagal impulses from the two periods.

Parallel with the described amplitude and frequency alterations the period of parturition was characterized also by the appearance of a large number of paroxysmal patterns in the EEG (spikes and slow oscillations).

At the last days of pregnancy and particularly on the days of parturition itself there appeared distinct changes in the vagal and EEG activities related to the uterine contractions. These changes mainly affected the amplitude of the rhythmic vagal impulses: their amplitude increased severalfold during contraction, and in some cases it reached 120 μv. The changes in the EEG accompanying the uterine contraction depended highly on the initial state of the brain activity. In some cases there were patterns of desynchronization in the EEG whereas (Fig.1) in others slow oscillations and spikes were prominent.

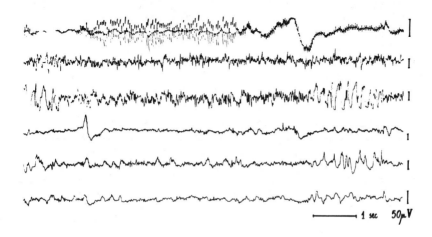

Fig. 1: Vagal and brain activities during uterin contractions. From top to bottom: n.vagus; orbital gyrus; sigmoid gyrus; n.VPL; mesencephalic reticular formation; lateral hypothalasmus.

We have established similar changes in the EEG and in the activity of the servical vagus in coital and precoital states as well. A statistically significant redistribution was found in the frequency spectrum of the encephalograms prior to and after coitus. At the same time the vagal activity after coitus displayed a considerable increase in the amplitude of the rhythmic vagal impulses with a frequency of 26 to 34 cps (Fig.2).

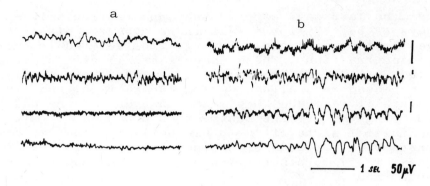

Fig. 2: Vagal and brain activities before (a) and after (b) coitus. (cat ♂) From top to bottom: n.vagus; sigmoid gyrus; n.VPL; orbital gyrus

The conclusions emerging from these data that in different periods of the reproductive cycle there occur similar changes in the electroencephalographic activity which affect both the frequency and amplitude of the waves and are accompanied by tonic vagal activity of much higher amplitude. In our earlier experiments we have observed simultaneously occurring changes of similar pattern in the vagal and EEG activities by continuous interoceptive stimuli, abundant feeding, electrical stimulation of the vagus, vibration, etc. (Varbanova 1967, Doneshka 1971). All this suggests that also in the phases of the reproductive cycle the effect described is a matter of non-specific reaction of the different parts of the nervous system in relation to to the hormonal factor. This non-specific reaction is likely to reflect the altered level of homeostasis of the organism.

REFERENCES

Doneshka P.   C.R.Acad.bulg.sci. 24, 1971.

Jakovlev J., Lisovskaja G., Shminke G.J.Obstetr.a. Gynaecol. 5, 37, 1952.

Kawakami M. a. Saito H.   Jap.J.Physiol. 17, 466, 1965.

Krijanovskaja E.   Physiol.J.USSR 50, 106, 1964.

Lebedeva L.   Physiol.J. USSR   48, 290, 1962.

Lebedeva L.   Physiol.J. USSR   49, 24, 1963.

Levinson L. (after Lebedeva L.   Physiol.J.USSR   48, 290, 1962).

Ljubimov E.   Physiol.J.USSR 54, 1199,   1968.

Sawyer Ch.H. a. Kawakami M.   Endocrinol. 65, 622, 1959.

Syrovatko F. a. Jahontov V.   J.Obstetr. a. Gynaecol. 1, 9,   1953.

Varbanova A. "Interoceptive Signalization", Sofia,1967.

Varbanova A., Doneshka P. a Vassileva-Popova J.   Bull. Inst. Physiol. Sofia, 15,13,   1973.

EPINEPHRINE RECEPTION BY SPERM CELLS

Penka Tsocheva-Mitseva, Kostadinka Veleva and Julia Vassileva-Popova

Central Laboratory of Biophysics, Bulgarian Academy of Sciences, 1113 Sofia, Bulgaria

Since the time of the first experiments of Oliver and Shäfer (1), the investigation on the epinephrine (adrenaline, adrenin) mechanism of action mark a new achievment in the field of physiology, biochemistry and recently also in biophysics.

Epinephrine (adrenine, adrenaline)

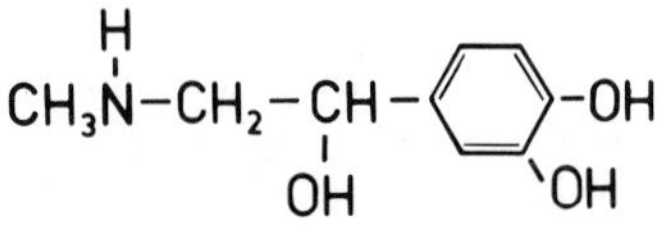

$$CH_3\overset{H}{\underset{}{N}}-CH_2-\underset{OH}{CH}-\underset{}{\bigcirc}\begin{matrix}-OH\\ \diagdown OH\end{matrix}$$

Tyrosine derivative
Regulatory hormone

Fig. 1

As one of the non-protein products in the biosynthesis of aromatic acids, (Fig.1) the catecholamine hormones are characteized by a high chemical and biological activity. Probably for the same reason, epinephrine also preserves an effect on the adenohypophysis (2).

The epinephrine effect is particularly clearly expressed under physiological stress which is manifested as a recombination of the blood pressure. In this extraordinary physiological state, the level of the epinephrine released comes up to 15 mg/min.

One of the mechanism of action of the catecholamine hormone is via a presumable cyclizing enzyme system in the cell membrane and the production of 3´,5´-cyclic AMP from ATP (Fig.2). The next stages of the epinephrine action is to enhance the activity of the kinase system and to maintain the phosphorylase in active form followed by the conveyer propagtion of this activation in the cell. In this connection the enzyme mediation is contributive (3).

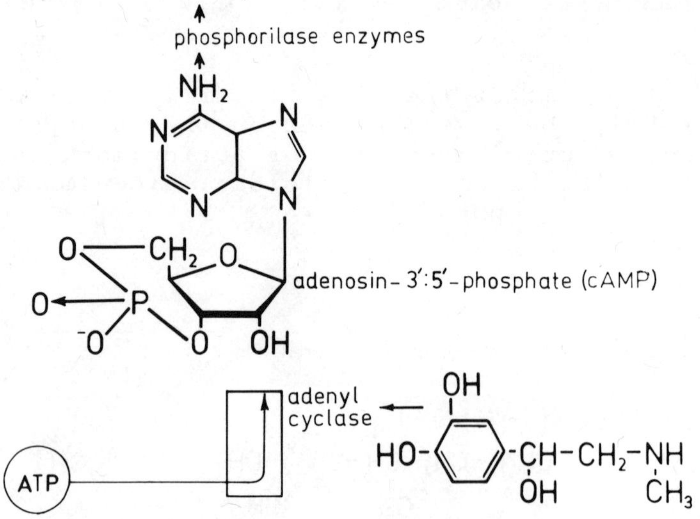

Fig. 2

It is interesting to note that epinephrine as a non-protein hormone mediates its action via similar pathways (membrane adenyl cyclase-3´,5´-cyclic AMP) which is accepted as characteristic for the protein hormone action. On the basis of the 3´,5´-cyclic AMP as a second messenger for interacellular communication the epinephrine appears as a cellular communication of the first order: the transfer of non-genetic stages of information.

Since the existence of the epinephrine receptors is accepted (for non-adrenaline predominantly-α, for isoprenoline - β and for epinephrine α and β), it is of interest to find an answer to the question whether an epinephrine reception is present in cells different from the muscle cells (4) and to understand some other epinephrine functional tasks (Fig.3).

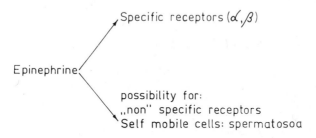

Fig. 3

The question we ask is whether or not there is an epinephrine reception in cells with self motility containing contractile proteins. As a most suitable model we used sperm cells. There is a support (5-7) of the presence of contractile protein in the spermatozoa. For its similarity to the myozine the contractile protein in sperm filaments was called spermosine. Of interest for our study is the fact established (8-19) that sperm metabolism is similar to that in muscles. The methods used for studying the spermatozoa are principally the same which were applied (20) for the investigation of glycerol extracted muscle system.

According to the findings regarding the common features in contractile proteins and the metabolism between sperm cells and muscle cells it is assumed that the spermatozoa should have possible acceptors for the epinephrine action. In this respect the purpose of the present study is to investigate whether or not there exists an epinephrine reception in non-target (in the classical sense) elements.

EXPERIMENTAL DATA

The investigation has been carried out <u>in vitro</u> with 10^{-5} M/ml epinephrine added to undiluted semen. A preliminary laboratory estimation of the ejaculate was made. Only semen classified as normal was used in the experiments.

The incubation period lasted 52 hours and more at a temperature of 18°C and the behaviour of the sperm cells was observed. Parallel experiments were also carried out with semen diluted with a diluant used in the practice of artificial insemination. Since the experimental procedure and the results were in the same direction, they will not be discussed separately here.

The experimental data were obtained by the classical methods of:

1. Microscopic evaluation of the sperm cells self motility;

2. Redox potential test based on the application of methilene blue;

3. Histochemical and electron microscopic investigation of the phospholipid membrane of the sperm cells. The methods are decribed in our previuos work on sperm cells characteristics **and** photobiology (21, 22).

RESULTS AND DISCUSSION

<u>Sperm cells self motility.</u> The actively moving spermatozoa in the experimental samples with epinephrine were by 12-14 % more than in the control samples with an accuracy of $p < 0.001$ (according to Student) (Fig.4).

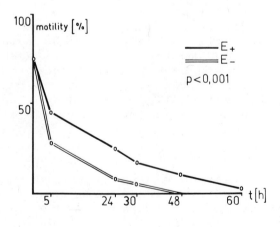

Fig. 4

The experimental samples showed also a longer than in the control absolute survival for more than 52 hours. The change in the self motility of the sperm cells was expressed not only quantitatively, but also qualitatively. In the control samples for a period of about 2 to 5 hours helicoidally moving cells were observed, characterized by a partial paralysis of the fibrillar apparatus and an incoplete function of the contractile proteins. In the samples preserved in contact with epinephrine, a more prolonged and normal function of the contractile protein was observed demonstrated as a full capacity of the sperm cells motility.

A possible explanation of the differences found is that the epinephrineproduced a stimulation by means of additional signals directed to the contractile proteins.

Another characteristics of the sperm cells preserved in epinephrine contact is that at a relatively high temperature (18°C) the metabolism did not fall down rapidly as it happened in the control samples. It seems that the epinephrine produced an effect of conservation and a more economical energetic balance as well.

<u>Investigation of the redox potential system.</u> An increase in the respiratory function of the sperm cells connected with the enzyme activity of the Krebs cycle was observed (Fig.2). The decreasing pH of the medium was the result of lactic acid accumulation (Fig.5).

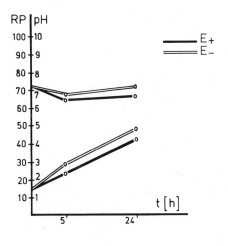

Fig. 5

For the epinephrine action in the muscle an accumulation of lactic acid was also observed (16). It looks that the epinephrine action was accompanied by an increase of the lactic acid not only in muscle cells. It is quite possible that this kind of end-effect is due to the reverse Pasteur effect. There is a probability for a feedback between the enzymes from the respiratory and carbohydrate chains which are influenced by epinephrine. The data indicate that epinephrine activates the enzymes from the carbohydrate metabolism and to a relatively small degree those of the respiratory chain. At this stage we do not have enough evidence to explain the data for the activation of enzymes from the respiratory chain demonstrated by the redox potential method.

These results lead us to assume that the predominance of the carbohydrate metabolism in facultative sperm cells is in connection with the prolongation of sperm cells survival observed in the samples with epinephrine.

<u>Cytochemical and electron microscope investigation of the sperm cells membrane</u>. During the prolonged incubation the double cellular membrane is preserved complete while the cell membrane in control samples shows a destruction. This was found by the fluorescent microscopic estimation, as well as by the electron microscope study. The method is described in our previous work (21).

As a result of a present investigation a pragmatic conclusion should be suggested: that the application of epinephrine as a constituent of the semen diluent should produce a positive effect with regard to achieving longer preservation of the sperm cells motility. On the basis of a correlation between the degree of spermatozoa self motility and the fertilization capacity good results may also be expected.

However, the following comments may be given:

1. Epinephrine reception should exist also in non-classical targets for the epinephrine action. In our case in free moving cells (spermatozoa) containing contractile proteins..

2. It is not clear whether the epinephrine effect on spermatozoa should be sought in the similarity between proteins in the contractile system of semen cells and the muskle tissue, especially in view of the possibility of specialized receptors in the muscle tissue apparatus from contractile proteins (4).

In conclusion the concept about the possible side-role of epinephrine in the reproductive process might be suggested. For some wild animals being often in stress situations an increase of the reproductive capacity is observed. It is a question whether the epinephrine as a side-factor contributes to the complex of reproductive stimulation for the prolongation of the species' life-time.

REFERENCES

1. Oliver,G. and Shafer,A.J., Physiol., 18, 230 (1895).

2. Russell,J.A., in: Physiology and Biophysics, Eds. Ruch T.C. and Patton H.D. W.B.Sounders Co., Philadelphia-London (1960).

3. Krebs and Fisher, Biochim. Biophys. Acta, 20, 150, (1956).

4. Ashley,R., J.Theor.Biol., 36, 339, (1972).

5. Engelhardt,V.A., Advances in Enzymology, 6, 147, (1946).

6. Engelhardt,V.A. and Burnasheva,S.A., Biokhimia, 22, 554, (1957),(in Russian).

7. Burnasheva, S.A., Biokhimia, 23, 558,(1958),(in Russian).

8. Nelson,L., Biochim.Biophys.Acta, 14, 312, (1954).

9. Nelson,L., Biol.Bull.Woods Hole, 107, 301, (1954).

10. Nelson,L., Biochim.Biophys.Acta, 27, 634, (1958).

11. Nelson,N., Biol.Bull.Woods Hole, 117, 327, (1959).

12. Mohri, H., J.Fac.Sci.Univ.Tokyo, 8, 307, (1958).

13. Utida,S., Maruyama,K. and Nanao,S., Jap.J.Zool., 12, 11, (1956).

14. Tibbs,J., Biochim.Biophys.Acta, 28, 638, (1958).

15. Tibbs,J., Nature, Lond.,193, 686, (1962).

16. Mann,T., R.C.I-st Sup. Sanit., 21, 16, (1958).

17. Mann,T., Bull.Soc.Chim.Biol., 40, 745, (1958).

18. Mann,T., Proc.Soc.Study of Fertility, 9, 3, (1958).

19. Masaki,J. and Hartree, E.F., Biochem.J., 84, 347, (1962).

20. Szent-Gvöryi,A., Biol.Bull.Woods Hole, 96, 140, (1949).

21. Vassileva-Popova,J., Photobiology of Semen, Dissertation, Sofia, 1968 (in Bulgarian)

22. Vassileva-Popova,J. and Tsocheva,P., Proc.Symp. on Sheepbreeding, Russe, 1968 (in Bulgarian).

Physical and Chemical Information Transfer in cAMP and Enzyme Network

CONTROL OF SIZE AND SHAPE IN THE RODENT THYMUS

D. Bellamy

Department of Zoology
University College
Cardiff

INTRODUCTION

Since the pioneer work of Selye (1) on the initiation of thymic involution by the administration of adrenocortical extracts, many studies have been made to clarify the role of adrenocorticosteroids on the life-cycle of thymic lymphocytes. (2-9) It was soon established that cortisol-type steroids caused the death of lymphocytes in suspension and that their relative toxicity in vitro matched their thymolytic potency in vivo. (10-15) Although suspensions of lymphocytes are still used for biochemical work on the action of corticosteroids, with the tacit assumption that all lymphocytes are identical, (16) the cytotoxic action on suspended cells appears to involve interactions between two distinct lymphocyte populations. (17) Further the involution of intact thymus after single injections of cortisol occurs with a complicated shift in proportions of the various cell-types, with the likelihood that some aspects of the total cellular response are of an indirect nature. (18) Despite much work on the acute response to cortisol, little has been carried out on changes in the thymus following the chronic administration of corticosteroids. Information on the effects of long-term treatment with corticosteroids is important in relation to the possible selection of a population of lymphocytes resistant to steroids.

In order to maintain a simple system and yet retain some properties of the intact organ, thymic explants (19-21) may be utilized. The ease with which explants may be prepared, and the relative simplicity of the gross functional organization of the thymus, would appear to make thymic explants particularly suited to histological studies of the differential action of corticosteroids on lymphocytes.

A great deal of work has also been published on thymus grafting in experimental animals, mainly in relation to the immunological and leukaemogenic roles of the gland. (22-28) The implanted thymus may also be taken as a useful model for investigating problems of differentiation and ageing of organs composed mainly of mitotic cells. (18,29-31,37) This work indicated the usefulness of the implantation technique for gerontological studies and suggested that growth and involution of the thymus are largely independent of systemic factors. The research did not, however, examine the influence of host age and sex, nor did it clarify the mechanisms underlying thymus autonomy. The present work deals with an extension of the previous investigations carried out notably by Selye and Metcalf and opens up new questions concerning the role and relative importance of intrinsic factors in the control of growth and form.

RESULTS AND DISCUSSION

The histological response of thymus to repeated injections of cortisol was complex and proceeded in at least four phases. During the first two days, there was a selective loss of cortical lymphocytes. Between the 2nd and 6th days, some medullary lymphocytes were lost and there was a rise in the collagen concentration, to stabilize the weight of the thymus at about 12% of its original weight. During the next two days, the density of medullary lymphocytes increased, with a corresponding drop in collagen concentration. Between days 8 and 12, some of the medullary lymphocytes were replaced by collagen.

This series of events is interpreted in terms of differential sensitivity of cortical and medullary lymphocytes to cortisol with the operation of a mechanism for compensatory proliferation of resistant medullary lymphocytes.

Thymic cortical explants were characterized by a high rate of cellular death of lymphocytes, with no significant change in their mitotic index. Cellular death in explants was further increased by the addition of cortisol; and from the dose-response relationship, there were at least three types of lymphocyte which differed with regard to cortisol sensitivity.

Two weeks after grafting, whole thymus implants had the characteristic appearance of a normal young thymus. They increased in weight until 6 weeks post-operation and then underwent involution at maximum weights that were below those attained by the host thymus.

Maximum weights attained by grafts in male hosts were significantly higher than those attained in female hosts. Initial rates of graft involution were higher in males than in females. Weights attained by grafts in young hosts of either sex reached maxima that were significantly higher than those attained by grafts of the same age in old hosts. The number of thymuses grafted separately into a single host did not influence the weight of the host's own thymus or the mass of individual grafts. However, when multiple grafts were linked together on transplantation, they behaved as an individual graft with respect to general morphology, growth rate and maximum size.

Evidence obtained from the regeneration of different-sized transplants suggested that the size and shape attained by the regenerated graft was dependent on the size and shape of the donor tissue.

When donor rat thymuses were transplanted in Millipore chambers the lymphocyte population did not reappear and after seven days only reticular-epithelial cells remained, retaining their normal appearance. However, when these thymic remnants were removed from the chambers and transplanted into secondary hosts, the thymus regenerated normally, suggesting that the lymphocytes in the regenerated gland were derived from the host. Thymic remnants after cortisol treatment of donors also formed distinct organs after grafting despite the fact that they contained few donor lymphocytes.

From the differential effects of cortisol on host and transplanted thymus and the different growth characteristics of transplants it appears that transplants differ in their growth/involution control system from the host thymus.

From the dose-response relationship _in vivo_, it is likely that thymic explants contained at least two kinds of lymphocyte which differed greatly with respect to their sensitivity to the cytotoxic action of cortisol; the more sensitive lymphocytes were found mainly in the cortex. The dose-response curve for the _in vitro_ action indicated that the latter lymphocytes may be further subdivided into two groups, with respect to their sensitivity. The more resistant lymphocytes comprised about 20% of the total population of thymic cells, and appeared to be mainly confined to the medulla. Besides the initial loss of cortical lymphocytes _in vivo_, there was an associated mobilization of collagen. These findings fit previous results obtained following short-term treatment _in vivo_ (17) and the addition of cortisol to suspension of thymic cells. (16)

The concentration of circulating corticosteroids of the cortisol type ranges from about 0.01 to 0.1 ug/ml in unstressed rodents. This may be increased after stress by a factor of 10. The effects obtained for cortisol-treated explants were obtained at doses within this physiological range.

At the moment, the major physiological difference relating to a differential sensitivity to cortisol cannot be linked with any morphological diversity, but the evidence presented from the treatment in vivo suggests that there may not be an absolute distinction between the two lymphocyte populations of cortex and medulla in terms of their response to cortisol. From the reduced density of medullary lymphocytes during the early stages of chronic treatment in vivo, it appears that some lymphocytes in this zone are killed by cortisol.

The subsequent rise in the density of lymphocytes in the medulla remnant was associated with a drop in collagen concentration with no change in thymus size, which suggests that at this stage of the experiment, the medullary lymphocytes showed compensatory proliferation despite continued cortisol treatment, with a linked mechanism for collagen mobilization. Does this represent the preferential division of a population of lymphocytes which is resistant to cortisol? And if so, what are the implications of this with regard to the regeneration of cortex which is known to occur when cortisol treatment ceases?

Previous work on long-term studies of both intact and adrenalectomized animals has pointed to the possible existence of compensatory mechanisms to activate lymphocyte proliferation, which counteract the cytotoxic action or corticosteroids. (7,32-34) However, the relationship between lymphocyte depletion and proliferation is far from being understood and much more work is needed in this area, particularly in relation to the possibility of clonal selection of resistant lymphocytes, where the findings would have an important bearing on many areas of cell biology. (35)

Bearing in mind the high rate of lymphocyte proliferation in the thymus and that cell division appears to be largely balanced by cellular death, (36) the present findings offer a mechanism for promoting rapid and specific changes in the make-up of lymphocyte population, according to the type and duration of environmental stresses.

With respect to growth autonomy of grafts, the results are in agreement with those of Metcalf (30) on mice. Additionally, the findings indicate that transplantation modifies the normal pattern of growth and involution. Sex of host in young animals also affected this pattern, suggesting that sex hormones have differential effects in both processes. Ageing of the host influenced regeneration and involution, but this is unlikely to be due to changes in sex hormones as previous work showed that sex hormones influence both mitosis and death of lymphocytes in rats only up to 30 weeks of age. It may be that prior ageing of host lymphocytes which eventually re-populated the transplant is responsible for the difference of growth between grafts, in that their proliferation capacity is impaired due to cellular ageing. (18) Current views on the process of thymus graft regeneration are that lymphoid elements of the grafts consist mainly of host cells (37) whilst the medullary tissue, essential for the functional integrity of the graft remains donor type (38) and regulates lymphopoiesis within the thymus by the production of humoral factor, (39) thus implying a considerable element of host autonomy and organizing capacity in the thymus graft. On the other hand, the effects of host sex and age suggest systemic factors are also important.

It is difficult to resolve the paradox of intrinsic proliferative autonomy against the obvious strict limits on gland size. A key experiment in this respect is that of the linked implants. The simplest explanation for the fact that multiple, separate implants all grow to the same size, independently of the developmental state of the host thymus, whereas multiple linked implants grow as a single small implant, is that the size of the vascular bed at the time of regeneration is the factor limiting subsequent regeneration. It is difficult to measure the magnitude of the potential vascular bed, but a good approximation would be the minimum size of the transplant prior to growth. For all thymic grafts, whether single or multiple, the various minima were all proportional to the initial size, but not related to the maximum size after regeneration. This indicates that vascular bed differences were largely overcome by some other factor governing the growth rate and maximum permissible size of the regenerating graft. With regard to multiple linked grafts, it is also important that both the total initial and minimum size were less than those of single grafts from older animals that were successfully established to reach the same maximum regenerated size.

A second theory to account for a multiple linked graft behaving as a single graft is that thymic mass in normal thymus (separate implants and linked implants) is maintained by an inhibitory mechanism which, in some way, sees each implant as being a separate entity; it regulates each one to a constant size which depends on age, sex and developmental age of the thymus. Bearing in mind the independent organizing ability of the thymic reticulum, the control system may rest on a balance between intrinsic stimulation and peripheral inhibition.

The inhibitory mechanism could be of a subtle chalone type but, if so, it must rest on there being different regulator substances being produced in each regenerated population. The necessary differences between thymic populations may be based on chance differences between donor cells that survive the first few days of degeneration or, on the hypothesis of host cell control, based on the characteristics of the first few host lymphocytes that enter the implant at random. Thus, the differences between multiple separate transplants may be due to random clonal selection, with growth control operating through the detection of differences between resulting cell populations. The situation observed with multiple-linked implants may fit this picture, in that the various reticular cell populations compete for resources and that when coupled in the same 'organ environment' only one reticular framework is able to survive and give rise to a regenerated thymus.

The differences between multiple implants may be similar to those involved in the establishment of clones during an immune response. Thus, it is possible that a micro-environment is generated within the donor thymus which controls the balance between cell division and cell death but which is independent, to a great extent, of the host's age and own internal environment, and other lymphocyte populations. There is some supporting evidence for this idea from comparative studies on thymus anatomy. For example, avian thymic tissue is composed of a series of discrete masses in the neck, and each apparently identical organ has to be able to compete for a share of the body's resources with its neighbour. However, definitive evidence for this theory awaits the further chemical characterization of thymic clones.

Although the results may be taken as further evidence for the host origin of cortical lymphocytes in regenerated implants, the importance of the reticular-epithelial cells in regeneration is emphasized by the ability of the thymic remnant, which survives cortisol treatment *in vivo*, to give rise to a new thymus on transplantation. However, the fact that cortical lymphocytes of the regenerated tissue in the latter experiments showed a different dose-response relationship to cortisol compared with the donor is evidence for repopulation of the graft by cells with properties different from those of host thymic lymphocytes. There is also some evidence from the present study that there are population differences between lymphocytes in multiple transplants.

The most interesting new findings relate to the properties of the system that exerts differential control over cellular proliferation in order to produce a regenerated thymus of the same relative dimensions as that which was implanted. At its minimum size, shortly after being placed in the host the medullary tissue of the transplant is all that survives and this appears to lack any kind of ordered structure. Out of this ill-defined mass of reticular-epithelial cells arises a new thymus of a characteristic shape.

It appears that implants do not grow in phase with the host thymus, and do not follow the same pattern of development particularly with regard to the maximum size attained before involution indicating that the donor intrinsic control system is greatly modified by implantation. Nevertheless, the morphology of the established implant seems to be directly related to the shape of the piece of donor thymus that is implanted. The importance of shape of donor tissue is demonstrated by the fact that whole thymus glands of different implanted weights grow to the same extent whereas half a gland only reached 50% of the weight of the intact glands. This, taken together with the obvious importance of the reticular-epithelial cells indicates that the specifications for the relative dimensions and mass of the various parts of the gland emanate from thymic medulla and this precise spatial organization survives transfer. These instructions are probably realised through differential regional control of the rates of cell division within the population of cortical lymphocytes. It is for future studies to determine how the requisite pattern of cellular reproduction is specified from within the medulla.

References

1. Selye,H.,Endocrinology,21,169, 1937.
2. Simpson,M.E.,Li,C.H.,Reinhardt, W.O. & Evans,H.M.,Soc.exp.Biol. Med.,54,135,1943.
3. Sayers,G.,White,A. & Long,C.N. H.,J.biol.Chem.,149,425,1943.
4. Dougherty,T.F. & White,A., Soc. exp.Biol.Med.,53,132,1943.
5. Dougherty,T.F. & White,A., Endocrinology,35,1.1944.
6. Dougherty,T.F. & White,A.,Amer. J.Anat.,77,81,1945.
7. Yoffey,J.M. & Baxter,J.S., J. Anat.Lond.,80,132,1946.
8. Baker,B.L.,Ingle,D.J.&Li.C.H., Amer.J.Anat.,88,313,1951.
9. Ringertz,N.,Fagraeus,A.& Berglund,K., Acta path.microbiol.scand.,30(Suppl.93)44,1952
10. Heilman,D.,Proc.Mayo Clinic,20 310,1945.
11. Hechter,O.& Stone,D.,Fed.Proc. 7,52,1948.
12. Schrek,R.,Endocrinology,45,317, 1949.
13. Feldman,J.D.,Endocrinology,46, 552,1950.
14. Feldman,J.D.,Endocrinology,51, 258,1952.
15. Trowell,O.A.,J.Physiol.,119, 274,1953.
16. Munck,A.,Mosher,K.M.& Wira, C. R.,in:Hormones in Development, ed.Hamburg,M.& Barrington,E.J. W.,p.191,N.Y.,Appleton-Century Crofts,1971.
17. Bellamy,D.,J.Endocrinology,30, v,1964.
18. Bellamy,D.& Alkufaishi,H.Age and Ageing,1,88,1972.
19. Pinkle,D.,Soc.Exp.Biol.Med., 116,54,1964.
20. Tomatis,L.& Wang,L.,Soc.exp. Biol.Med.,118,1037,1965.
21. Tweel,J.G.Vanden.,The thymus in vitro.Academic Thesis, Univ. of Utrecht,The Netherlands,1971
22. August,C.S.,Rosen,F.S.,Filler, R.M.,Janeway,C.A.,Markowski,B. & Kay,H.E.M.,Lancet,2,1210, 1968.
23. Cleveland,W.W.,Fogel,B.J.Brown W.T.& Kay,H.E.M.,Lancet,2,1211, 1968.
24. Dempster,W.J.,Lancet,1,1294, 1970.
25. Green,I.,J.Exp.Med.,119,581,1964
26. Schlesinger,M.& Hurvitz,D.,J. Exp.Med.,127,1127,1968.
27. Stutman,O.,Yunis,E.J.& Good,R. A.,J.Immunol.,103,92,1969.

28. Stutman,O.,Yunis,E.J.,Teague, P.O.& Good,R.A.,Transplantation,6,514,1968.
29. Bellamy,D.& Alkufaishi,H., Differentiation 1,425,1974.
30. Metcalf,D.,Aust.J.Exper.Biol. & Med.Sci.,41,437,1963.
31. Pepper,F.,J.Endocrin.,22,349, 1961.
32. Weir,D.R.& Heinle,R.W.,Soc.exp. Biol.Med.,75,655,1950.
33. Dougherty,T.F.& Kunagai,L.F., Endocrinology,48,691,1953.
34. Wintrobe,M.M.,Cartwright,G.E., Palmer,J.G.,Kuhns,W.J.& Samuel, L.T.,Arch.Intern.Med.,88,310, 1951.
35. Bellamy,D.,Gerontologia,19, 162,1973.
36. Metcalf,D., in: The thymus in Immunobiology,ed.Gord,R.A.& Gabrielsen,A.E.,New York: Harper Row,1964.
37. Biggar,W.D.,Stutman,O.& Good, R.A.,J.Exp.Med.,135,793,1972.
38. Dukor,P.,Miller,J.F.A.P.,House, W.& Allman,V.,Transplantation, 3,639,1965.
39. Metcalf,D.& Ishidate,M.,Aust. J.Exper.Biol.& Med.Sci.,40, 57,1962.

THE DNA POLYMERASE AND THE cAMP CONTENT OF HUMAN TONSILLAR LYMPHOCYTES

Mária Staub, Anna Faragó and F. Antoni

1st Institute of Biochemistry, Semmelweis
University Medical School, Budapest, Hungary

Up to now the role of the tonsils in the immune-system has been a matter of discussion.

Cooper et al.(1) have shown different types of antibodies on the surface of tonsillar cells, Ogra (2) suggested the local production of IgA antibodies by tonsillar lymphocytes.

In our Institute the synthesis of the anti-bacterial enzyme lysozyme has been investigated by Puskás and Antoni (3). Faragó et al.(4) isolated the cAMP dependent and independent histon-kinases from human tonsillar lymphocytes. All of these experimental results suggest active cellular functions. As it is known all of the functions of immune-competent cells are accompanied by cellular proliferations i.e. by DNA synthesis.

For the study of in vitro DNA synthesis lymphocytes offer several advantages: first of all the cell can be stimulated in vitro by different mitogens, and second, the exceptionally low deoxyribonuclease activity in these cells facilitates experiments with partially purified enzymes, too.

In our Institute we have investigated the DNA-synthesis of human tonsillar lymphocytes at the level of ^{3}H-Thymidine incorporation and at the DNA-polymerase activity.

The tonsillar lymphocytes were isolated by the method of Piffkó et.al.(5) from freshly removed tonsils of children of 3-6 years. From one pair of tonsils 10^9 lymphoid cells can be obtained, which is a relatively high quantity, necessary for biochemical preparations. The cells consist almost of 95-98 per cent of lymphoid cells, which are morphologically heterogen. According to Zucker-Franklin (6) and to our own findings 70-80 per cent of them are small - or middle-sized lymphocytes and 25-30 per cent are lymphoblasts.

The incorporation of ^{3}H-Thymidine has been measured to get some information about the DNA synthesising capacity of the cells. (Fig. 1.)

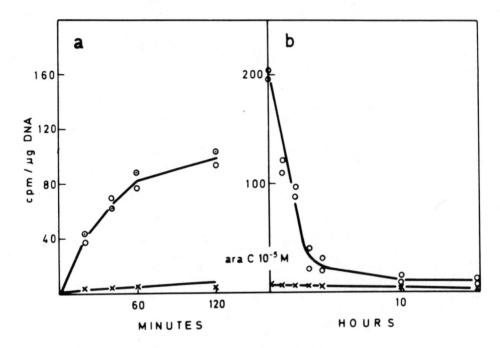

Fig.1: The incorporation of ^{3}H-Thymidine in the presence and in the absence of araC.
A. 10^7 freshly prepared cells were incubated in the presence of 1.5 µCi ^{3}H-Thymidine (Sp. a. 25 Ci/mmole) in 1.0ml Hanks' solution. The incorporated activity was measured in the acid insoluble material in a toluene based cocktail.
B. 10^6 cells/ml were incubated in Earls' medium in the presence of 10% human AB serum. At the indicated times the cells were pulsed for 20 minutes with 1.5 µci of ^{3}H-Thymidine.

In the experiment demonstrated, 10^7 viable lymphocytes were incubated in Hanks's solution, in the presence of 1.5μCi of ^{3}H-Thymidine for the indicated time. The Thymidine concentration was 1.5μM. The incorporated Thymidine was measured in acid insoluble material and expressed as cpm per μg DNA. As it can be seen, the tonsillar lymphocytes incorporate Thymidine, while normal human lymphocytes have only low capacity for synthesising DNA. In the presence of 10^{-5} M arabinosylcytosine the incorporation was stopped in the first minutes demonstrating, that the measured incorporation was specific for eukaryotic cells. In the right-hand side of Fig.1 an experiment is demonstrated where cells were kept in Eagle's medium for 24 h in the presence of human AB serum. At the indicated time the cells were pulsed for 20 minutes with ^{3}H-Thymidine. As we can see the DNA synthesis dropped sharply though the survival of the cells was about 50 per cent, and they could be stimulated by Phytohaemagglutinin or by Concanavalin A.

In order to get more direct information about the DNA synthesis of our cells the DNA-polymerase activity was also measured. After freezing and thawing, the cells were homogenised in a buffer containing EDTA, mercapthoethanol and 20 per cent glycerol. The extract was centrifuged for an hour at 100 000 x g and the activity was measured.

The requirements for the cell-free DNA synthesis are shown on Table I.

The complete reaction mixture contained all the four deoxynucleotide-triphosphates. As template calf-thymus DNA was used activated with pancreatic DNase according to the method of Aposhian and Kornberg (7). There is a substantial incorporation in the presence of a single deoxynucleotide--triphosphate too, but the incorporation is stimulated by addition of other dNTP-s.

If one compares the DNA-polymerase activity (measured under the same conditions) the activity of the human tonsillar lymphocytes is about the same as that of normal spleen cells (8). The most important enzymological properties of the tonsillar DNA-polymerase have been also measured in our laboratory.

The enzyme has a pH optimum in the region of 7.5- 8.1. The reaction rate is linear in time up to 60 minutes, and it is also linear with the enzyme concentration. The enzyme has an absolute requirement for divalent cations. The

Table I

REQUIREMENTS FOR DNA SYNTHESIS

REACTION CONDITIONS	^{3}H-TMP INCORPORATED (pmol)	% CONTROL
COMPLETE	8.3	100
— DTT	7.1	86
— Mg^{2+}	0.5	6
— K^+	2.61	30
— TEMPLATE	0.26	3
— ENZYME	—	—
— d ATP	2.61	30
— d ATP, dCTP, dGTP	1.6	20

Mg^{2+} ion concentration required for the maximal activity is shown in Fig. 2. The maximal activity has been reached in the presece of 10 mM Mg^{2+} ions or in the presence of 2 mM of Mn^{2+} ions.

In prokaryotic cells three different DNA-polymerases have been isolated, and the studies with mutants have proved their real existance. In different eukaryotic tissues the multiplicity of DNA-polymerases is a matter of the current investigations. Multiple activities of a high molecular weight DNA-polymerase could be detected in mouse myeloma by Matsukage et al. (9), in rat liver and in rat spleen by Holmes (10).

We have also purified the DNA-polymerase of tonsillar lymphocytes by several methods. The most fruitful procedure yielded an 500-1000 fold purified enzyme compared to the crude extract.

The main steps of the purification were: the extraction by 0.8 M KCl, centrifugation at 100 000 g, and then a chromatography on a DEAE-cellulose column to remove the nucleic acids. The enzyme was eluted by 0.3 M KCl from the

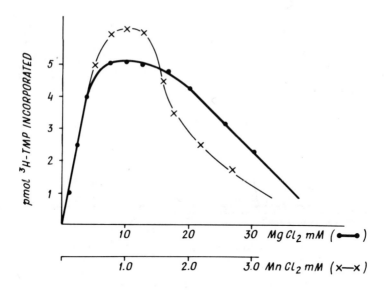

Fig.2. The optimal concentration of divalent ions for the DNA-polymerase reaction.
The reaction mixture contained 50 mM TRIS-HCl buffer pH 7.8; 0.12 mM, dATP, dCTP, dGTP and 0.004 mM. ^{3}H-dTTP (700 cpm/pmole), 4.0 mM KCl; 1.2 mM DTT; 50 μg calf thymus DNA and 5-20 μg of enzyme in a final volume of 200 μl. The mixture was incubated at 37ºC for 20 minutes and stopped by the addition of 5% TCA-0.1 M PP$_i$. The incorporation into acid insoluble fraction was measured.

column and after dialysis put on a phospho-cellulose column. The chromatography on P-cellulose will be demonstrated in Fig.3. The enzyme was eluted by a linear gradient of 0.1 - 0.6 M of KCl. Two DNA-polymerases could be separated by this method, the first active peak was eluted by 0.23 M and the second by 0.45 M of KCl. Incorporation of ^{3}H-Thymidine thriphosphate could be measured only in the presence of activated DNA. No incorporation has been detected either in the presence of native yeast RNA or of synthetic poly (A). The elution profile of the two DNA--polymerases is very similar to that of polymerases iso-

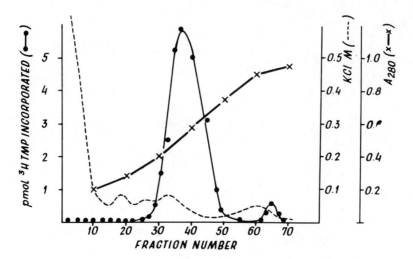

Fig.3. Chromatography on P-cellulose column.

The dialysed, nucleic acid free enzyme (after DEAE-cellulose chromatography) was put on a P-cellulose column (1.0 x 9.0 cm) and eluted with a 0.05 M pH 7.2 TRIS-HCL buffer containing 1 mM EDTA, 1 mM MEA and 20% of glycerol. The enzyme was eluted by a linear gradient of 0.1 - 0.6 M of KCl. The volume of the fractions was 2.0 ml.

lated from PHA stimulated peripherical lymphocytes by Smith et al. (11).

In the case of tonsillar cells the second peak could only be detected when cells were extracted with a high salt concentration and the enzyme appeared as the polymerase located in nuclei and had a low molecular weight.

Recently it has been shown that not only the ^{3}H-Thymidine incorporation varied according to the cell phases, but the DNA-polymerase activity and the cAMP content too (12, 13, 14). In response to the cellular prolypheration the amount of the cAMP increases (13) in the early S phase. In the case of non-transformed fibroblasts when DNA synthesis reaches the maximal level the cAMP content is also maximal (13). After the S phase of the cells the cAMP content drops sharply, while the activity of DNA-polymerase decreases in a slower way.

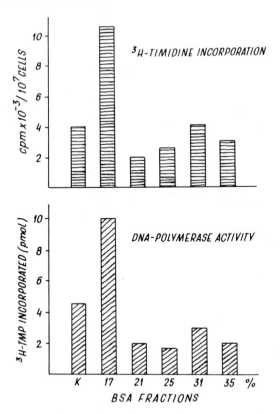

Fig.4. The ^{3}H-Thymidine incorporation and the DNA-
-polymerase content of the cell fractions
after separation on an albumine gradient.

Cells were collected from the inter phases
of the gradient and washed 3 times in 10 ml
of Hanks' solution. The ^{3}H-Thymidine incorporation and the DNA-polymerase was measured as in Fig.1. and Fig.2.

To get any information about the state of our cells
referring to the cell phases we separated the tonsillar
lymphocytes on a discontinuous albumine gradient. Bovine
serum aibumine was used dissolved in Hanks's solution and
a gradient was prepared consisting of five fractions (35%,
31%, 25%, 21%, 17%). 5 x 10^8 cells were suspended in 8%
albumine and put on the top of the gradient. Then the cells
were centrifuged at 600 g for 15 minutes and collected from
the interphases and washed. The incorporation of ^{3}H-Thymi-
dine, the DNA-polymerase activity and the cAMP content of
10^7 cells were measured.

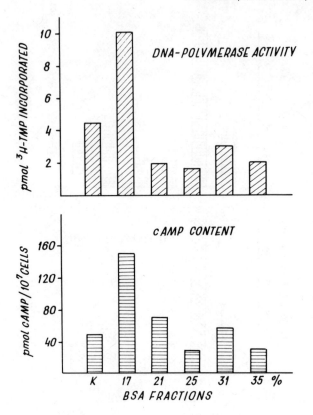

Fig.5. The DNA polymerase activity and the cAMP content of the cell-fractions.
The cAMP content was measured by the method of Gilman (15), the receptor protein was isolated from tonsillar lymphocytes (16).

The DNA synthesis was found to be the highest in cells sedimented in the less dense fractions of the gradient (Fig.4.). Cells sedimented in 17% of albumine had the highest capacity to incorporate ^{3}H-Thymidine, and had about a five times higher DNA-polymerase activity than the unseparated cells.

Fig.5. shows the cAMP content of the fraction in connection with the DNA-polymerase activity. The cAMP content of the cells was measured by the method of Gilman (15). The cAMP receptor used in our experiments was isolated from tonsillar lymphocytes and the K_D of the receptor-cAMP complex was 8×10^{-9} M.

As it can be seen in Fig.5. cells with the highest capacity for synthesising DNA contai the highest amount of cAMP too.

Our data regarding the DNA synthesis and the cAMP content of the tonsillar lymphocytes point to their active role in immunological processes. The isolation of a cell fraction containing five times more DNA-polymerase activity and the highest cAMP content is in good agreement with recent findings (13, 14). According to these data the increase of the cAMP concentration in cells and the increase of the DNA-polymerase activity may be a consequence of cell stimulation, perhaps by an antigen. According to our assumption, cells sedimented in the 17% albumine fraction are in the S phase of the cell-cycle.

REFERENCES

1. Cooper, A.G., Brown, M.C., Derby, A.H., Wortin, H.H.: Clin. Exp. Immunol. 2973, 13, 487.

2. Ogra, P.L.: J. Med. 1971, 284, 59.

3. Puskás, M., Antoni, F., Staub, M., Farkas, gy. and Piffkó, P.: Pract. Oto-rhino-laryng. 1972. 34, 160.

4. Faragó, A., Antoni, F., Takáts, A., Fábián, F.: Biochem. Biophys. Acta 1973, 297, 517.

5. Piffkó, P., Köteles, G.J., and Antoni, F.: Pract. Oto-rhino-laryng. 1970, 32, 350.

6. Zucker-Franklin, D., and Berney, S.: J. Exptl. Med. 1972, 135, 533.

7. Aposhian, H.V. and Kornberg, A.: J. Biol. Chem. 1962. 237, 519.

8. Tyrsted, G., Munch-Petersen, B. and Cloos, L.: Exptl. Cell Res. 1973, 77, 415.

9. Matsukage, A., Bohn, E.W. and Wilson, S.G.: Proc. Nat. Acad. Sci. 1974, 71, 578.

10. Holmes, A.M. Heslewood, J.P. and Johnston, J.R.: Eur. J. Biochem. 1974, 43, 487.

11. Smith, R.G., Gallo, R.C.: Proc. Natl. Acad. Sci. U.S.A. 1972, 69, 2879.

12. Spadari, S. and Weissbach, A.: J. Mol. Biol. 1974, 86, 11.

13. Seifert, W.E., Rudland, P.S.: Nature, 1974. 248, 138.

14. De Rubertis, F.R., Zenser, T.V., Adler, W.H. and Hudson, T.: J. Immunol. 1974, 113, 151.

15. Gilman: Proc. Natl. Acad. Sci. U.S.A. 1970. 67, 355.

16. Faragó, A., Antoni, F., Takáts, A.: in press

INSULIN AND ADRENALIN EFFECT ON LUNG TRIGLYCERIDE LIPASES

Nadejda B. Hadjivanova and G.A. Georgiev

Central Laboratory of Biophysics, Bulgarian Academy of Sciences, 1113 Sofia, Bulgaria

Lipid metabolism in lung tissue has been the object of various investigations. Many details are known today about this metabolism and its enzymatic systems (1).

It is common knowledge that the ability of lung tissue to oxidize fatty acids is much greater than that of other tissues (2). Lungs, just like liver, can synthetize all lipids they need both for energy reserves formation and for building up subcellular elements (3).

The lipolytic function of the lungs is also well known (4), but less is known about its regulation, with hormones having a basic role.

Recently, there have been investigations on the hormonal regulation of lipid metabolism in liver and adipose tissue.

Two kinds of lipase, an alkaline - associated with the microsomal fraction, and an acid one - associated with the lysosomal fraction, are known for liver and adipose tissue (5) which realize the hydrolytic degradation of the triglycerides.

The cyclic 3'5'-adenosine monophosphate (cAMP) is supposed to play a key role in hormonal regulation of lipid mobilization, meditating in fact the hormonal action (6).

The object of the present study is the in vivo and

in vitro effect of insulin and adrenalin on lung triglyceride lipases: the alkaline with pH optimum 7.4 and acid one with pH optimum 4.6.

MATERIAL AND METHODS

The investigations were carried out on Wistar male rats, 10 for each group, fasted for 15 hours. For the experiments in vivo the animals were subdivided into three groups as follows: first group - control animals; they were injected intravenously 0.5 ml 0.9 per cent NaCl solution; second and third groups - the rats were administered intravenously 0.5 ml saline containing 2IU insulin or 10 ɣ adrenalin, resp. The animals were killed and their lungs removed after one hour (for the second group) and three hours (for the third) after the injection.

For the experiments in vitro the animals were subdivided into two groups: control and experimental. The rats from both groups were not pretreated. The hormones were added to the incubation medium (0.4IU insulin and 0.1 ml 0.10 mM solution of adrenalin for 2 ml incubation medium). The enzyme preparation used was a crude homogenate and in some experiments a 105,000 g supernatant. The activity of lung tissue triglyceride lipases was determined by the method Müller & Alaupovic (7). The enzymatic activity was expressed in µmoles of fatty acids released per gramm of tissue per hour. The effect of cyclic N^6-2'-O-dibutyril adenosine monophosphate (cDBAMP) and caffein was studied by adding to the incubation medium 1 mM solution of cDBAMP and 0.8 mM solution of caffein, respectively.

RESULTS AND DISCUSSION

The results reveal that both lung triglyceride lipases, alkaline and acid, are affected by insulin and adrenalin.

The marked inhibitory effect of insulin on lung triglyceride lipases in vivo and in vitro could be seen on Fig. 1, persisting also after the first hour. The alkaline lipase (pH 7.4) in vivo showed an inhibition of 69 per cent, and in vitro about 100 per cent; the acid triglyceride lipase (pH 4.6) reached almost the same values -
- 76 per cent in vivo and 90 per cent in vitro.

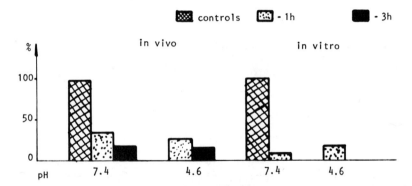

Effect of insulin on the activity of lung triglyceride lipases

Figure 1

This effect of insulin in lung tissue is supposed to to be due to the reduced intracellular concentration of cAMP, being caused either by the direct inhibition of adenylate cyclase or by the stimulation of a specific phosphodiesterase. The slow endogenous cAMP formation and its comparatively rapid degradation by the normally high active cyclic 3'5'-phosphodiesterase (cPDE), could be a probable explanation of the persisting inhibitory effect of insulin after the first hour (8).

Adrenalin in our experiments, as seen on Table 1(a), Fig.2a raises the activity of the inhibited by insulin lung triglyceride lipases with about 30 per cent for the alkaline and about 20 per cent for the acid lipase, resp. This comes to support the opinion that cAMP is the hormone mediator for insulin as well.

On the other hand, the data also show, that the addition of cDBAMP to the incubation medium does not restore the activity of both alkaline and acid triglyceride lipases, neither in vivo nor in vitro. This gives an idea that insulin may act also by way of some other mechanism.

Studying the effect of cDBAMP and insulin on alkaline triglyceride lipase activity in 105,000 g supernatant of lung tissue homogenates (Table 1b with Fig.2b), we found a considerably elevated activity with cDBAMP: 46 per cent

Effect of adrenalin and cDBAMP on the activity of lung triglyceride lipases inhibited by insulin (µmoles/gr/h)

pH	in vivo		in vitro	
	7.4	4.6	7.4	4.6
K	88,80±7,14	66,62 ± 7,60	58,80 ± 17,05	59,26 ± 10,84
I	27,14±9,78	31,66 ± 6,73	6,73 ± 4,56	11,94 ± 5,80
I+A	76,11 ± 25	43,14 ± 6,53		
I+cDBAMP	10,17 ± 5,34	15,47 ± 5,35	7,97 ± 3,72	11,25 ± 5,10

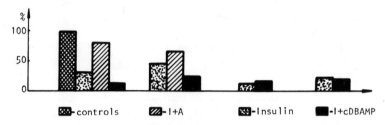

Figure 2a

above the control values. This again supports the opinion that the lung cAMP mediates the activation of the triglyceride lipase. We also found the already established for lung homogenates inhibition with insulin: here about 45 per cent. The fact that the adenylate cyclase has been sedimented with the membranes by this centrifugation, indicates that it is not this enzyme's inhibition that leads to diminished concentrations of intracellular cAMP and to inhibited triglyceride lipase, respectively. The cyclic 3'5' - PDE with its predominantly cytoplasmic location remains in the supernatant and is markedly activated in the presence of insulin.

Cyclic DBAMP does not restore the lung triglyceride lipase activity, after its inhibition by insulin in 105,000 g supernatant too.

It is accepted that insulin has an effect on the cPDE activity mainly by augmenting the enzyme synthesis and also by activating the enzyme itself (9). Probably in our case too, the raised activity and the stimulated synthesis of cPDE help the rapid degradation of the exogenous cyclic nucleotide.

Effect of cN^6-2-0-dibutyril AMP on the activity of lung triglyceride lipase in 105,000 g supernatant in controls and after insulin (μmoles/gr/h FFA)

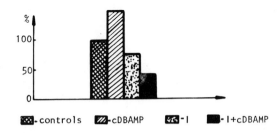

	in vitro
pH	7,4
K	109,12 ± 18,12
cAMP	186,67 ± 15,16
Ins	50,08 ± 23,77
Ins + cAMP	33,76 ± 9,20

Figure 2b

The in vivo inhibited triglyceride lipase (pH 7.4) in lung homogenates and in 105,000 g supernatant by insulin (Table 2 with Fig.3) shows a marked restoration of its activity in the presence of caffein; in lung homogenates the activity raises almost two and a half times, compared with the controls.

Caffein stimulates indirectly the activation of lung triglyceride lipase, inhibiting strongly the cPDE and in this way elevating the intracellular concentration of cAMP. When first treated with caffein, the lung tissue homogenates show a raised triglyceride lipase activity, compared with the controls, which does not change in the presence of insulin.

This competitive antagonism of caffein and insulin, observed by other authors as well (10), shows one and the same object of action, namely the cPDE.

The effect of adrenalin on the triglyceride lipase activity in lung tissue homogenates in vivo and in vitro

Effect of caffein on the activity of lung triglyceride lipase (pH 7.4) in vivo, after insulin in 105,000 g supernatant, and in vitro (μmoles/gr/h FFA)

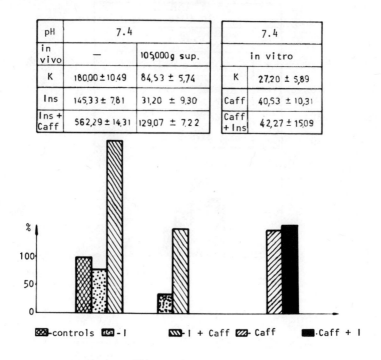

Figure 3

was also studied. The results (Table 3 with Fig.4) show a moderatly stimulated activity of both triglyceride lipases (pH 7.4 and 4.6), being more pronounced in vivo. So, in lung tissue too, we observe the well known lipolytic action of adrenalin for adipose tissue and liver (11), due to its direct adenylate cyclase activation.

In conclusion we can say, that the effect of insulin and adrenalin in lung tissue, most probably, is mediated by the intracellular concentration of cAMP. The prevailing activity of the cPDE could be accepted to have a more important role for insulin action. Insulin stimulation of cPDE activity has already been suggested as a mechanism responsible for the decreased cAMP levels, but often the results have been quite difficult to interpret and even contraversial. However, the existence of another mechanism of insulin action cannot be excluded. The problem needs further elucidation.

Effect of adrenalin on the activity of lung triglyceride lipases

pH	in vivo		in vitro	
	7.4	4.6	7.4	4.6
K	67,40 ± 3,12	66,50 ± 2,82	49,39 ± 3,24	92,69 ± 2,82
Adr.	91,98 ± 6,17	106,50 ± 15,29	80,34 ± 6,81	111,10 ± 9,86

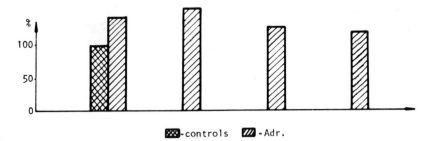

- controls - Adr.

Figure 4

Some light is hoped to be thrown on the role of these hormones upon lipid metabolism in lung. Their action is affected, as shown above, by means of the intracellular triglyceride lipases, which hold a key position in lipid transformations and are the actual regulators of lung tissue triglycerides.

REFERENCES

1. B.Wolfe, B.Anhalt, J.Beck, C.Rubinstein, Can. J.Biochem. 48, 1970, 170.

2. G.Mameya, Acta Med. Nagasaki. 11, 1967, 257.

3. J.Felts, Med. Thorac., 22, 1965, 89.

4. H.Yomada, Acta Med. Nagasaki. 8, 1964, 94.

5. S.Mahadevan, A. Tappel, J.Biol.Chem. 243, 1968, 2849.

6. E. Sutherland, Pharmacol. Rev. 12, 1960, 265.

7. L. Müller, P.Alaupovic, FEBS Letters, 10, 1970, 117.

8. E.Lotten, J. Sneyd, Biochem. J. 120, 1970, 187.

9. G. Senft, G.Schultz, K.Munske, M.Hoffmann, Diabetologia, 4, 1968, 322.

10. K.Hepp, L.Menahan, O.Wieland, R.Williams, Biochim. Biophys. Acta, 184, 1969, 554.

11. E.Sutherland, I.Oye, R.Butcher, Recent Progr. Hormone Res. 21, 1965, 625.

MAGNETIC PUMPING OF BLOOD IN THE VASCULAR SYSTEM

V.K. Sud and R.K. Mishra

Department of Biophysics, All-India Institute of Medical Sciences; New Delhi-110016, India

INTRODUCTION

The subject of Magnetohydrodynamics (MHD) is the study of the motion of electrically conducting fluids in the presence of electro-magnetic fields. When a conductor moves in a magnetic field, the latter induces an electric field in the conductor and thereby electric currents are generated. The interaction of the applied magnetic field and the induced electric currents produce a mechanical force, called the 'Lorentz force', which alters the motion of the conductor. Thus if an axially moving magnetic field is applied to a tube of a flowing conducting liquid, such as blood, in such a way that the velocity of the moving field is greater than that of the liquid, the motion of the liquid will be accelerated. On the other hand, if the velocity of the moving magnetic field is smaller than that of the liquid, the motion of the liquid will be retarded. In physiology, this subject of MHD application was first introduced by Kolin (1936) who used it for measuring the rate of blood flow. The possibility of regulating the movement of blood was first mooted by Korchevskii et al.(1965). This idea of magnetic pumping of the blood is of tremendous potential in its biological applications. To cite only a few instances, one might mention the pathological states like thrombosis, shock and collapse, haemorrhages, vascular insufficiency, and decreased urinary filtration due to decreased renal blood flow. However, the published literature still lacks any exact analysis of the effect of a moving magnetic field on the flow of blood in human arteries. In the present attempt, a human artery which is elastic, non-uniform in diameter and branched, is replaced for the sake of mathematical convenience, by a straight rigid tube of uniform tubular cross section. Blood is of course realistically considered a viscous, incompressible and electrically conducting liquid. Also since only the pumping effect is intended to be investigated, the blood flow is considered steady, and not pulsating.

BASIC EQUATIONS, ANALYSIS AND SOLUTION

We consider the flow of a viscous, incompressible electrically conducting liquid in a non-conducting rigid tube of uniform cross-section. The polyphase windings are arranged around the tube so that at $r = r_o$, where r_o is the radius of the tube, a magnetic field moving in the axial (z-direction, see Fig.1) is produced and is given mathematically as:

$$H_z(r_o, \theta, z, k) = H_o \cos(\omega t - kz)$$

where r, θ, z are in cylindical polar coordinates, $\omega = 2\pi f$ is the circular frequency of the current, f is the frequency in cycles/sec of the alternating current producing the magnetic field, k is the propagation constant in radians/cm., Ho is the amplitude of the applied magnetic field in oersteds.

The equations of moting for steady MHD flow (see Cowling /1957/).

Momentum Equation:

$$(\underline{V} \cdot \nabla)\underline{V} = -\nabla p + M_f \nabla^2 \underline{V} + \underline{J} \times \underline{B} \qquad (1)$$

Continuity Equations:

$$\nabla \cdot \underline{V} = 0 \qquad (2)$$

Maxwell's Equations: (neglecting the displacement currents)

$$\nabla \times \underline{E} = -M_o \frac{\partial \underline{H}}{\partial t} \qquad (3)$$

$$\nabla \cdot \underline{H} = 0 \qquad (4)$$

$$\nabla \times \underline{H} = \underline{J} \qquad (5)$$

Ohm's Law

$$\underline{J} = \sigma (\underline{E} + \underline{V} \times \underline{B}) \tag{6}$$

where

$$\underline{B} = \mu \underline{H} \tag{7}$$

The vectors and scalars used above are defined as:
$\underline{E}$ is the electric field strength, $\underline{B}$ is the magnetic induction, $\underline{H}$ is the magnetic field, $\underline{J}$ is the current density, $\underline{V}$ is the velocity, p is the pressure, σ is the electical conductivity, μ_0 is the magnetic permeability of the liquid and t is the time and μ_f is the dynamic viscosity of the fluid.

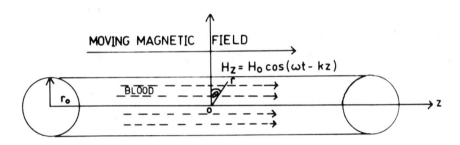

Fig. 1: Model of blood flow.

It may be noted that the symmetry of the motion around the z-axis is assumed at each instant and hence the result of operation $\frac{\partial}{\partial \theta}$ on any quantity will be zero. The flow is also assumed to be laminar and at each instant as the motion develops. From the Eq. (2) it follows that

$$\frac{\partial V_z}{\partial z} = 0$$

showing that the velocity is independent of z. V_r, V_θ and V_z are the respective velocity components, along the radial, azimuthal and axial directions.

The excitation is assumed to travel in the z-direction and is of the form $I \exp\{i(\omega t - kz)\}$ (taking its real part only), where I is the linear current density of the coil in amps./unit z-direction. Then the total magnetic field comprising the excitation and the induced field is then given by

$$\underline{H} = \left[H_r(r)\hat{i}_r + H_z(r)\hat{i}_z \right] \exp\{i(\omega t - kz)\} \tag{8}$$

where i_r and i_z are the unit vectors along the radial and azimuthal cylindrical polar coordinate systems.

Combining Equations (3), (5) and (6), we obtain the transport equation in magnetic field:

$$\nabla^2 \underline{H} + M_o \sigma \nabla \times (\underline{V} \times \underline{H}) = M_o \sigma \frac{\partial \underline{H}}{\partial t} \tag{9}$$

By using Eq. (8), the transport equation becomes in a cylindrical coordinate system as follows:

$$(\frac{d^2}{dr^2} + \frac{1}{r}\frac{d}{dr} - k^2 - i\omega M_o \sigma)[H_r(r)\hat{i}_r + H_z(r)\hat{i}_z]$$
$$+ M_o \sigma [ik\hat{i}_r + (\frac{d}{dr} + \frac{1}{r})\hat{i}_z] V_z(r) H_r(r) = 0 \tag{10}$$

From Eq. (5) by using the similar Eq. (8) we get the components of the current density as:

$$J_r = 0$$
$$J_z = 0$$
$$J_o = - \frac{dH_z(r)}{dr} - i k H_r(r) \tag{11}$$

Similarly, Eq. (4) is reduced to

$$\frac{dH_r(r)}{dr} + \frac{1}{r} H_r(r) - i k H_z(r) = 0 \tag{12}$$

Again for small flow velocities and/or tube diameters, the convective forces are small compared to viscous, pressure and electromagnetic forces. Hence momentum Eq. (1) finally reduces to the following equation by using Eq. (8):

$$\frac{\partial p}{\partial z} = M_f(\frac{d^2}{dr^2} + \frac{1}{r}\frac{d}{dr})V_z - M_o H_r(r) J_o(r) \tag{13}$$

Making the above mentioned Eqs. dimensionless by introducing the folowing new dimensionless parameters:

$$r^* = \frac{r}{r_o}, \qquad z^* = \frac{z}{r_o}, \qquad k^* = r_o k$$

$$V_z^* = \frac{V_z}{(r_o^2/M_f)\frac{\partial p}{\partial z}}, \qquad H_z^*(r) = \frac{H_z(r)}{r_o^2 \sqrt{\frac{\sigma}{M_f}} \frac{\partial p}{\partial z}} \tag{14}$$

$$J_o^* = \frac{J_o}{r_o\sqrt{\frac{\sigma}{M_f}}\left(\frac{\partial p}{\partial z}\right)}, \qquad H_o^* = \frac{H_o}{r_o^2\sqrt{\frac{\sigma}{M_f}}\left(\frac{\partial p}{\partial z}\right)}$$

$$t^* = \frac{t}{r_o^2\,\rho/M_f}, \qquad \omega^* = \frac{\omega}{M_f/r_o^2\rho}$$

Thus Eq. (13) reduces to:

$$\frac{d^2 V_z^*}{dr^{*2}} + \frac{1}{r^*}\frac{dV_z^*}{dr^*} - G^* H_r^* J_\theta^* - 1 = 0 \tag{15}$$

Similarly the Eq. (10) becomes:

$$\left(\frac{d^2}{dr^{*2}} + \frac{1}{r^*}\frac{d}{dr^*} - k^{*2} - 2\Im\,P_m F_i\right)(H_r^* \hat{i}_r + H_z^* \hat{i}_z)$$

$$+ G^*\left[ik\hat{i}_r + \left(\frac{d}{dr^*} + \frac{1}{r^*}\right)\hat{i}_z\right] V_z^* H_r^* = 0 \tag{16}$$

where the new dimensionless parameters involved in the above equations are:

Magnetic Prandtl number

$$P_m^* = \frac{\mu_o \sigma M_f}{\rho} \tag{17a}$$

$$G^* = \frac{\mu_o G\, r_o^3}{M_f}\frac{\partial p}{\partial z} \tag{17b}$$

and frequency parameter

$$F = \frac{fr_o \rho r}{M_f} \tag{17c}$$

where f is the electrical frequency in c/sec.

Also the equation for current density becomes:

$$J_\theta^* = -\frac{dH_z^*(r)}{dr^*} - ik^* H_r^*(r) \tag{18}$$

and Equation (12) becomes

$$\frac{dH_r^*(r)}{dr^*} + \frac{1}{r^*}H_r^*(r) - ik^* H_z^*(r) = 0 \tag{19}$$

Equation (10) is non-linear in character. We may neglect the non-linear terms for liquids like blood which have small magnetic Reynolds number.

$$R_m = r_o V_o M \sigma$$

Here r_o and V_o are characteristic dimensions of lenght and velocity. In the case of blood R_m is of the order of 10^{-8} for a tube of $r_o = 2$ cm and with blood velocity 100 cm/sec. Thus the equation (16) simply reduces to

$$\left(\frac{d^2}{dr^{*2}} + \frac{1}{r^*}\frac{d}{dr^*} - k^{*2} - 2\pi F P_m i\right)\begin{pmatrix}H_z^*\\H_r^*\end{pmatrix} = 0 \qquad (20)$$

Taking the z-compñent of the magnetic field from Eq.(20), we get

$$\frac{d^2 H_z^*}{dr^{*2}} + \frac{1}{r^*}\frac{dH_z^*}{dr^*} + \alpha^{*2} H_z^* = 0 \qquad (21)$$

where the new parameter

$$\alpha^{*2} = i^3 \, 2\pi F P_m - k^{*2} \qquad (22)$$

and
$$i = \sqrt{-1}$$

The solution of Equation (21) is

$$H_z^* = A J_o(\alpha^* r^*) + B K_o(\alpha^* r^*) \qquad (23)$$

where $J_o(\alpha^* r^*)$ and $K_o(\alpha^* r^*)$ are Bessel functions of 1st kind and 2nd kind with complex argument of zero order. A and B are arbitrary constants to be evaluated by using the following boundary conditions for the magnetic field:

$$H_z^*(1) = H_o^* \text{ at } r^* = 1 \qquad (24a)$$

$$H_z^*(0) \rightarrow \text{finite as } r \rightarrow 0 \qquad (24b)$$

Eq. (24b) yields that $B = 0$, while Eq. (24a) gives

$$A = \frac{H_o^*}{J_o(\alpha^*)}$$

Substituting these constants in Eq.(23), the non-dimensional axially magnetic field component becomes

MAGNETIC PUMPING OF BLOOD IN THE VASCULAR SYSTEM

$$H_z^* = \frac{H_o^* J_o(\alpha^* r^*)}{J_o(\alpha^*)} \tag{25}$$

Similarly the radial component of the magnetic field can be evaluated from Eq. (19) by using Eq. (25) which gives

$$H_r^*(r^*) = \frac{ik^* H_o^*}{\alpha^* J_o(\alpha^*)} J_1(\alpha^* r^*) \tag{26}$$

where $J_1(\alpha^* r^*)$ is the Bessel function of first order. The Eq. for azimuthal current (11) can now easily be solved by using Eq. (25) and (26).

$$J_o^* = \frac{J_1(\alpha^* r^*)}{J_o(\alpha^*)} \frac{(\alpha^{*2} + k^{*2})}{\alpha^*} H_o^* \tag{27}$$

Putting the values for current density and the magnetic field of Eq. (26) and (27), the momentum Equation (15) takes the form:

$$\frac{d^2 V_z^*}{dr^{*2}} + \frac{1}{r^*} \frac{dV_z^*}{dr^*} - 1 - \beta^* J_1^2(\alpha^* r^*) = 0 \tag{28}$$

where for the sake of convenience the new dimensionless parameters β^* is put as:

$$\beta^* = \frac{ik^* H_o^{*2}(\alpha^{*2} + k^{*2})}{\alpha^{*2} J_o^2(\alpha^*)} G^* \tag{29}$$

The solution of Eq. (28) is quite cumbersome. However, when

$$\alpha^* r^* < 1 \tag{30}$$

which implies that

$$k^* < 1 \tag{30b}$$

and

$$w^* r_o^2 M_o G < 1 \tag{30c}$$

then the Bessel functions can be approximated by taking the expansion of the first two terms (see McLachlan/1965/) as:

$$J_o(\alpha^* r^*) \simeq 1 - \frac{\alpha^{*2} r^{*2}}{4} \tag{31a}$$

$$J_1(\alpha^* r^*) = \frac{\alpha^* r^*}{2} - \frac{\alpha^{*3} r^{*3}}{16} \tag{31b}$$

The use of the above Bessel function expansions reduces the Eq. (28) to the form:

$$\frac{d^2 V_z^*}{dr^{*2}} + \frac{1}{r^*} \frac{dV_z^*}{dr^*} = 1 + \frac{\beta^*}{4} \alpha^{*2} r^{*2} \tag{32}$$

Eq.(32) is to be solved using the following no-slip boundary conditions:

$$V_z^* = 0 \quad \text{at} \quad r^* = 1$$
$$V_z^* \to \text{finite as} \quad r^* \to 0 \tag{33}$$

Hence the complete solution for the dimensionless velocity profile is:

$$V_z^* = 1/4 \, (r^{*2} - 1) \left[1 + \frac{\alpha^{*2} \beta^*(r^{*2}+1)}{16} \right] \tag{34}$$

The Eq.(34) contains complex quantities. Performing some algebraic calculations and retaining only the real part, we obtain the equation for the velocity profile as:

$$V_z^* = \tfrac{1}{4}(r^{*2}-1) \left[1 + \frac{k^*(2+k^{*2})\, \omega^* G(r^{*2}+1) H_o^{*2}}{8 \left\{ (2+k^{*2})^2 + \omega^{*2} \, P_m^{*2} \right\}} \right] \tag{35}$$

Inspection of Eq.(35) reveals that the axial velocity depends upon the excitation frequency. Thus there exists an optimum frequency at which the pumping action is maximised. This dimensionless optimum frequency in terms of the liquid properties is obtained:

$$\omega^* = \frac{2+k^{*2}}{P_m^*} \tag{36}$$

which alternatively can be written by substituting the value for P_m (Eq. 17b) as

$$\omega^* = \frac{(2+k^{*2})\varphi}{M_o G\, M_f} \tag{36b}$$

Now transforming the dimensionless velocity profile Eq.(35) and the optimum circular frequency ω^* to the dimensional form by using a set of variables defined by

Eq. (14), we obtain finally the Eqs. for velocity profile $V_z(r)$ and optimum circular frequency of the liquid:

$$V_z(r) = \frac{1}{4}\frac{(r^2-r_o^2)}{M_f}\left[\frac{\partial p}{\partial z} + \frac{\mathcal{T}fkM_o^2(2+r_o^2k^2)(r_o^2+r^2)H_o^2}{(2+r_o^2k^2)^2 + \omega^2 r_o^4}\right] \quad (37)$$

$$\omega^* = \frac{2+r_o^2k^2}{G M_o r_o^2} \quad (38)$$

The above velocity profile equation reduces to the Poissilue flow if the amplitude of the applied magnetic field $H_o = 0$.

The rate of flow Q is given by

$$Q = 2\pi \int_0^{r_o} V_z(r)r\, dr \quad (39)$$

Substituting Eq. (37) in Eq. (39) we obtain

$$Q = -\frac{\mathcal{T}}{24M_f}r_o^4\left[3\frac{\partial p}{\partial z} + 8Fr_o^2\right] \quad (40)$$

where for simplicity we put

$$F = \frac{\mathcal{T}fk^2H_o^2M_o^2G(2+r_o^2k^2)}{4(2+r_o^2k^2)+\omega^2 r_o^4 M_o^2 G^2} \quad (41)$$

By rearranging Eq. (41) we obtain the Eq. for the pressure gradient as:

$$\frac{\partial p}{\partial z} = -\frac{8}{\mathcal{T}r_o^4}M_f Q - \frac{4}{3}Fr_o^2 \quad (42)$$

where the value for F is already given by Eq. (42).

DISCUSSION

In Eq. (42), the first term represents the nonmagnetic part of the pressure gradient, while the second term is the contribution of the magnetic field. Values of the non-magnetic and magnetic pressure gradients have been computed for the case of liquid like blood to magnetic fields of different strengths. For the purpose of the present computation, the folowing values were used. The electrical conductivity of Blood $\sigma = 14 \times 10^{-3}$ mho/cm, dynamic viscosity $u_f = 7.2 \times 10^{-3}$ poise. Optimum excitation frequency for blood as computed from Eq.(37) is 4.6×10^{8} c/s for a tube of diameter 2 cms.

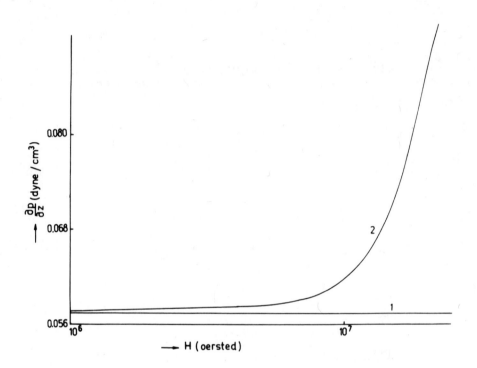

Fig. 2: The figure shows the relationship between the pressure gradient $\partial p/\partial z$ (dyne/cm3) and the magnetic field strength H(Oersted). The electrical frequency f is taken as 50 c/s. The flow rate Q=50 c/sec and the tube radius r_o =2cm. Trace 1 represents the non-magnetic pressure gradient while trace 2 is the magnetic pressure gradient.

Fig. 2 shows the non-magnetic and magnetic pressure gradients for an exitation frequency of 50 c/s for various values of the magnetic field amplitude. Again Fig. 3 plots the pressure gradients at the optimum excitation frequency. It is observed that magnetic pressure gradient increases with increasing magnetic field amplitude; however, the values are relatively large at the

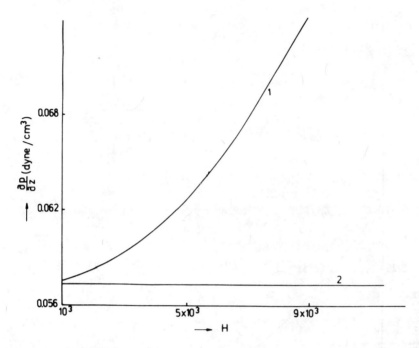

Fig.3: The figure shows the relationship between the pressure gradient $\partial p/\partial z$ (dyne/cm^3) and the magnetic field strength H at the optimum electrical frequency of blood = 4.6×10^8 c/s. Other parameters are the same as those used for platting Figure 3. Trace 2 represents the non-magnetic pressure gradient while trace 1 is the magnetic pressure gradient.

optimum excitation frequency. Thus at the optimum excitation frequency the magnetic effect increases from nearly nil for magnetic field amplitude of 10^3 oersteds to about 26 % to 10^4 oersteds amplitude. Of course, at larger magnetic field amplitudes, this effect would greatly increase.

Figs. 4 and 5 depict the various variations of the non-magnetic pressure gradients when the tube diameter is changed, other parameters being kept constant. It is observed that whereas the non-magnetic part decreases with increase of tube diameter, while the magnetic part increases.

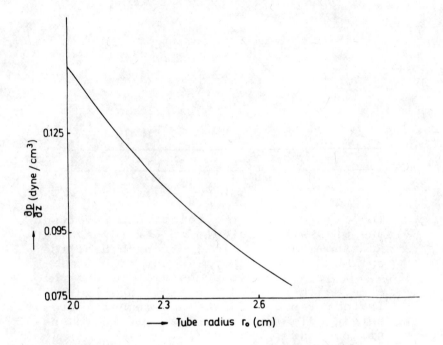

Fig. 4: The figure shows the relationship between the non-magnetic part of the pressure gradient $\partial p/\partial z$ (dyne/cm^3) and the tube radius r_0 (cm) when the electrical frequency is 50 c/s. The value for flow rate is taken as Q = 50 c/sec.

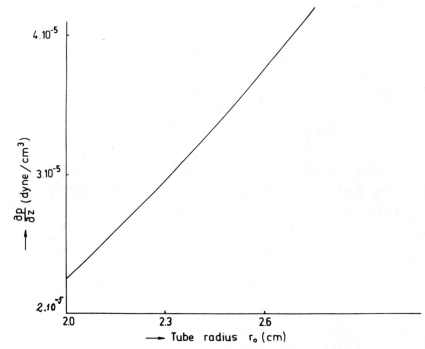

Fig. 5: The figure represents the relationship between the magnetic part of the pressure gradient $\partial p/\partial z$ (dyne/cm^3) and the tube radius r_o (cm) at the electrical frequency f = 50 c/s.

CONCLUSIONS

This paper presents an exact analysis of flow of blood in the presence of moving magnetic field in an idealised artery. In subsequent work the author intends to analyse blood flow when the tube carrying it is elastic or non-ridid.

The computed results suggest that significant acceleration of the flow of blood can be obtained, using even practical levels of magnetic field, however, the excitation frequency has to be much more higher than the usual electric supply frequency 50 c/s or 60 c/s.

ACKNOWLEDGEMENT

The authors wish to thank Dr. G.S. Sekhon, Applied Mechanics Department, Indian Institute of Technology,

New Delhi for his discussions and useful suggestions at varios stages in the preparation of this paper.

REFERENCES

1. Cowling, T.G., 1957, Magnetohydrodynamics - Interscience Publications.

2. Kolin, A., 1936, Rev. Sci. Instum. 16, 109-116.

3. Korchevskii, E.M., and Marochunik, L.S., 1965, Biofizika, 10 No.2, 371-373 (in Russian); Biophysics, 10 No.2, 411-413 (English translation).

4. McLachlan, N.W., 1955, Bessel Functions for Engineers, Clarendon Press, Oxford.

BIOMAGNETIC EFFECT OF THE CONSTANT MAGNETIC FIELD ACTION ON WATER AND PHYSIOLOGICAL ACTIVITY

M.S.Markov, S.I.Todorov, Maria R.Ratcheva

Department of Biophysics, Biological Faculty

Sofia University, Sofia, Bulgaria

The biophysical mechanism of the biological action of the magnetic field and "magnetic" water are connected with structural changes taking place in the external and internal cell water. The magnetobiological problem concerning the possible way of realizing these structural changes have not been made clear by now. No doubt the primary cause of the biomagnetic effect should be sought in the changes of physical and physical-chemical properties of "magnetic" water as a result of the magnetic field influence on water.

With regard to this we have studied the changes of the light absorption, the specific electroconductivity, the magnetic susceptibility as well as the Raman spectrum and physiological activity of water treated in a static state by constant magnetic field with magnetic induction in the range of 450-3500 Gs for 30,60 and 120 minutes.

The magnetic effect was calculated according to the next formula:

$$M_{ef} = \frac{N\ exp. - N\ c}{N\ c} \cdot 100\ \%$$

where N exp. are the corresponding values of the absorption, electroconductivity, etc. in the experimental sample and N and c are the same values in the control sample.

The dependence of the magnetic effect on the magnetic induction makes it possible to conclude that the maximum change of the water absorption property is obser-

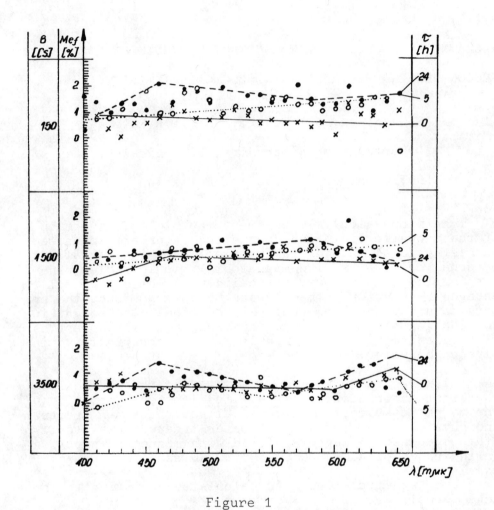

Figure 1

The magnetic effect dependent on magnetic induction and postmagnetic time

ved in the magnetic field with induction B=450 Gs. The maximum change of the "magnetic" absorption is observed in the visuable range 24 hours after the magnetic influence. Considering the dependence of the magnetic effect on the postmagnetic time, the exposition time and the value of the magnetic induction increase of the negative values could be observed as postmagnetic time increases. It is easy to establish that the dependence of the effect at B=450 Gs and t=30 min. greatly differs from the dependence resulting at t=60 and 120 min. Some intensification of the changes in postmagnetic time is observed.

Essential differences between "magnetic" and ordinary water have not been noticed in the Raman spectrum. Probably this is connected with the great half-width of the line, determined by the OH group. Changes in several cm^{-1} at width about a hundred cm^{-1} can not be noticed practically.

Changes of the magnetic susceptibility of distilled and drinking water with different rate of hardness were studied together with Z.Uzunov. It was established that in different cases the character of change was analogous, with an expressed maximum at an induction of 3350 Gs.

On the basis of the data presented about the changes of several physical-chemical characteristics of the magnetic water it is possible to make an attempt to explain the mechanism of magnetic influence. According to the two-structural model, water molecules occupy the knots of the quasicristal water grid. The falling of water molecules or admixture of ions in the interspace of the grid is the reason for the appearence of deffects in the grid structure. Defects are expressed in bending and severing of the hydrogen bonds. These processes depend on the change of the dielectric permeability, electroconductivity and magnetic susceptibility. It is possible to assume the existence of some original resonance mechanism of the magnetic influence. In this way we can explain the maximum effect, resulting at B=450 Gs, t=30 min, and t=24 hours. It is reasonable to suppose that the magnetic field with an induction of B=450 Gs is optimum in the range of 0 - 3500 Gs. At the same time this assumption does not contradict the possible existence of other optimum values at induction more than 3500 Gs.

We got a whole series of experimental data which show the biomagnetic action of the "magnetic" water on physiological processes.

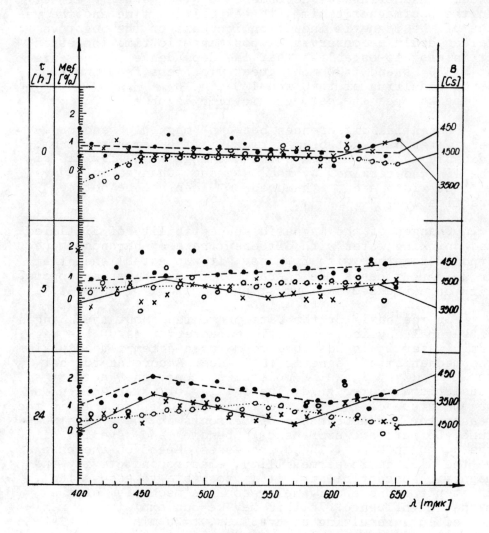

Figure 2
The magnetic effect dependent on postmagnetic time

CONSTANT MAGNETIC FIELD ACTION 445

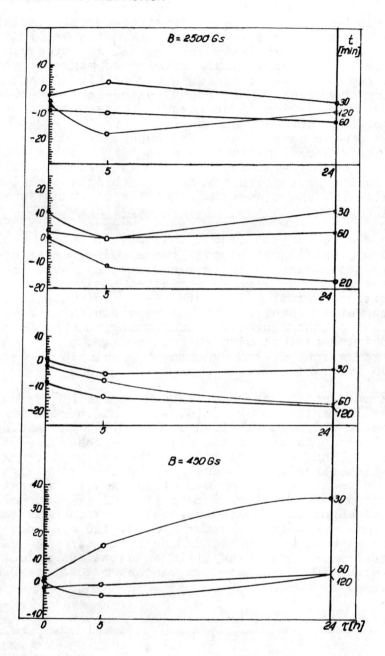

Figure 3

Relative change of the specific electroconductivity dependent on magnetic induction, postmagnetic time, and exposition time.

It has been proved that "magnetic" water stimulates growth of germination of Cucurbita maxima in 38,5%, increases the mitotic activity in the cell meristem root of Vicia faba. The physiological activity of magnetic water is expressed in reducing the quantity of "free" water and increasing the quantity of "bound" water in the Hellianthus annus germination. A decrease of oxygen consumption and the organic acid amount at tissue breathing was established on the same object. The observation of the section of the Zea mays meristem germination shows a growth increase of nearly 42%. The study of the biomagnetic influence on the protoplasm movement in the giant cells of the Nitella sp. shows dependence of the movement speed on the direction of the magnetic power lines.

The changes of the leucocytes, erythrocytes and haemoglobin content in rat blood after a daily radiation of 30 min in a constant magnetic field with an induction of B=450 Gs for 20 days were investigated. Meanwhile the same indices were studied in another group of rats, having drunk "magnetic" water. In both groups a sensitive increase (11-49%) of the leucocytes and haemoglobin content was observed in comparison with the control group. In the same time the weight of the experimental animals increased nearly 30 %. (Table 1).

The changes in the absorption spectrum of Zea mays chlorophyll germination were investigated in relation with the magnetic induction applied and seedling the exposition in the constant magnetic field. The behaviour of maximum λ=637,5 nm, corresponding to the well-known maximum 640 nm was observed. The seeds were exposed to a magnetic field - "direct" action, or were treated with "magnetic" water - "indirect" action. Strong (B38000 Gs) and weak (B=450 Gs) magnetic fields were applied. The seeds were exposed in a magnetic field for t=5,30, 120 min or were irrigated with water treated in the field for the same period of time. In the weak field an anomalous reduction of the intensity of the maximum was observed at t=30 min.

Table 1

index	direct action	indirect action
weight	28%	32%
erythrocytes	1%	5%
leucocytes	35%	13%
haemoglobin	49%	11%

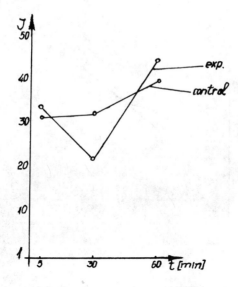

Figure 4

Reduction of the intensity of maximum $\lambda = 637{,}5$ nm in the weak field

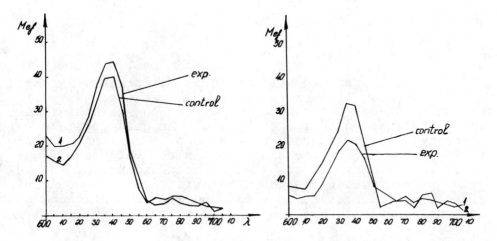

Figure 5

Behaviour of the maximum $\lambda = 637{,}5$ nm in the weak magnetic field

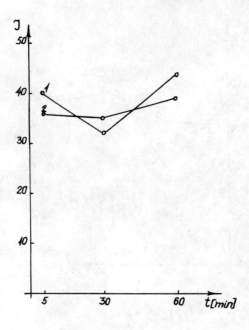

Figure 6

Decrease of the maximum intensity in "magnetic" water

Only 25% decrease of the maximum intensity was observed in cases when magnetic water exposed in a strong field was used. (Fig.6)

It can be supposed that the weak field exerts greater influence as it is more similar to the intensity of the adapted geomagnetic background. The changes observed can be explained with the changes of the quasicrystal chlorophyll associations. The energy of a constant magnetic field can excite the electrons in the chlorophyll molecule and as a result the electron consistence of the system pygment-water changes. The existence of an original resonance effect is possible and some fields cause a biomagnetic effect, while others do not.

The influence of the magnetic field treated water on the production and gustatory guality of the tomato has been proved. As a result of the magnetic treatment of the water, the production increases considerably - 1920 kilogrammes, which equals a 16% increase in comparison with the control plot. The average weight has increased from 59,2 g to 77,2 g, that is 32%. At the same time changes in the dry substance content and the acidity are observed, varying from 9,8% to 14,9%.

These experimental data show that "magnetic" water has a definite physiological activity.

REFERENCES

Aminaev G.A., Sitkin M.I., Butorina N.I. - in Meeting on the Study of the Effect of Magnetic Fields on Biological Objects, Moscow, 1966, p. 5 (in Russian)

Bernal J.D., Fowler R.H. - J.Chem.Phys., 1933, 1, 515.

Cook E. - in: Biological Effects of Magnetic Fields, Plenum Press, New York 1964, v. 1, p. 246.

Gemishev Ts.M., Todorov S.I. - Sofia Univ. Year-book 1971, 63, No.2, 223 (in Bulgarian)

Klassen V.I. in: The Response of Biological Systems to weak Magnetic Fields, Moscow 1971, p.14 (in Russian)

Kohn P.G. - in: The State and Movement of Water in Living Organisms, Cambridge 1965, p. 3.

Markov M.S., Uzunov Ts.D. - Sofia Univ. Year-book 1974, 66, No.1, 241 (in Bulgarian)

Nakhil'nitskaia Z.N. in: The Magnetic Field in Medicine, Frunze 1974, p.100 (in Russian)

Samoilov O.Ia. in: The Structure of Electrolyte Water Solutions and Hydration of Ions, Moscow 1957 (in Russian)

Todorov S.I., Racheva M.R. - Sofia Univ.Year-book 1969, 61, No.2, 225 (in Bulgarian)

Todorov S.I., Goranov A.I., Gemishev Ts.M. - Sofia Univ. Year-book 1969, 61, No. 2, 239 (in Bulgarian)

Todorov S.I., Markov M.S. - Sofia Univ. Year-book 1971, 63, No.1, 239 (in Bulgarian)

Todorov S.I., Markov M.S., Racheva M.R. - Sofia Univ. Year-book 1972, 64, No.1, 189 (in Bulgarian)

Todorov S.I., Racheva M.R., Markov M.S. - Sofia Univ. Year-book 1972, 65, No.2, 265 (in Bulgarian)

LIST OF PARTICIPANTS

Alexander Alexandrov
Central Lab. of Biophysics
Bulgarian Acad. of Sciences
1113 Sofia
Bulgaria

Henry Arnstein
Dept. of Biochemistry
University of London
King's College
Strand, London WC2R 2LS
United Kingdom

Hans-Jochen Bartels
Mathematical Institute
University of Göttingen
Bunsenstrasse 3/5
34 Göttingen
BRD

D. Bellamy
Department of Zoology
University College
Cardiff
United Kingdom

Melvin Blecher
Department of Biochemistry
Georgetown University
   Medical Center
Washington, D.C. 20007 USA

Monique Boumokra
Société Internationale de
   Biologie Mathématique
11 bis av. de la Providence
92-Antony
France

Francis Collot
Société Internationale de
   Biologie Mathématique
11 bis av. de la Providence
92-Antony
France

Dimiter Dimitrov
Central Lab. of Biophysics
Bulgarian Acad. of Sciences
1113 Sofia
Bulgaria

Vladimir Dimov
Institute of Animal Science
Kostinbrod
Bulgaria

James A. Edwardson
Imperial College of Science
   and Technology
Department of Biochemistry
London SW
United Kingdom

Elka Elenkova
Central Lab. of Biophysics
Bulgarian Academy of Sciences
1113 Sofia
Bulgaria

Olga D. Genbacev-Krtolica
Institute for the Application
   of Nuclear Energy in
   Veterinary Agriculture
Zemun
Yugoslavia

LIST OF PARTICIPANTS

Georgi Georgiev
Central Lab. of Biophysics
Bulgarian Acad. of Sciences
1113 Sofia
Bulgaria

Denis Gospodarowicz
The Salk Institute for
  Biological Studies
San Diego, California
USA

Gopal Saran Gupta
Dept. of Biophysics
Panjab University
Chandigarh-160014
India

Nadejda Hadjivanova
Central Lab. of Biophysics
Bulgarian Acad. of Sciences
1113 Sofia
Bulgaria

Michel Hall
University of Southampton
Highfield
Southampton
United Kingdom

Nikola Ivanov
Medical Institute of Varna
Marin Drinov St. 55
Varna
Bulgaria

Elwood Jensen
University of Chicago
950 East 59th St.
Chicago, Illinois
USA

Dobrina Kalitzin
Medical Institute of Varna
Chair of Biochemistry
Marin Drinov St. 55
Varna
Bulgaria

Georgy Kichev
Institute of Biology and
  Pathology of Reproduction
73 Lenin Ave.
1113 Sofia
Bulgaria

Nikolai Koralov
Central Lab. of Biophysics
Bulgarian Acad. of Sciences
1113 Sofia
Bulgaria

Marin Kotarov
Central Lab. of Biophysics
Bulgarian Acad. of Sciences
1113 Sofia
Bulgaria

Stefan Kovachev
Central Lab. of Biophysics
Bulgarian Acad. of Sciences
1113 Sofia
Bulgaria

Pavel Kulev
Central Lab. of Biomech.
Bulgarian Acad. of Sciences
7 November St.
Sofia
Bulgaria

Kamen Kumanov
Institute of Animal Science
Kostinbrod
Bulgaria

Ivan Madjarov
Institute of Animal Science
Kostinbrod
Bulgaria

LIST OF PARTICIPANTS

Alina Maleeva
Institute of Endocrinology
Medical Academy
Boul. Chr. Michailov 6
Sofia
Bulgaria

Dieter Maretzki
Institute of Biochemistry
Humbolt-Universität
Hessische St. 3/4
104 Berlin
GDR

Kaltscho Markov
Center for Biology
Bulgarian Acad. of Sciences
1113 Sofia
Bulgaria

Marko Markov
Biological Faculty
Sofia University
Moskowska St. 49
Sofia
Bulgaria

Stefan Milanov
Institute of Endocrinology
Medical Academy
Boul. Chr. Michailov 6
Sofia
Bulgaria

Ina Mindova
Central Lab. of Biophysics
Bulgarian Acad. of Sciences
1113 Sofia
Bulgaria

Vassil Neitchev
Central Lab. of Biophysics
Bulgarian Acad. of Sciences
1113 Sofia
Bulgaria

Tania Neitcheva
Central Lab. of Biophysics
Bulgarian Acad. of Sciences
1113 Sofia
Bulgaria

Robert Northcutt
Mayo Clinic
Division of Endocrinology
Rochester, Minnesota
USA

Doitchin Panayotov
Institute of Endocrinology
Medical Academy
Boul Chr. Michailov 6
Sofia
Bulgaria

Zlatko Penev
Medical Institute of Varna
Marin Drinov St. 55
Varna
Bulgaria

Temenuzhka Pentcheva
Medical Institute of Varna
Marin Drinov St. 55
Varna
Bulgaria

Alexander Petrov
Institute of Solid State
   Physics
Bulgarian Acad. of Sciences
Boul. Lenin 72
Sofia
Bulgaria

Igor Politaev
Institute of Molecular
   Biology
Academy of Sciences
Vavilova St. 32
Moscow
USSR

Vladimir Prokopov
Central Lab. of Biophysics
Bulgarian Acad. of Sciences
1113 Sofia
Bulgaria

Michail Protich
Institute of Endocrinology
Medical Academy
Boul. Chr. Michailov 6
Sofia
Bulgaria

Robert Rees
Genetics Department
Aberdeen University
2 Tillydrone Ave.
Aberdeen
United Kingdom

Karl-Ernst Reinert
Institute of Microbiology
 and Experimental Therapy
Academy of Sciences of GDR
Beutenberg St. 11
Jena
GDR

Jean Richard
Société Internationale de
 Biologie Mathématique
30 rue des Binelles
92 Serves
France

Brij Saxena
Cornell University
Medical College
1300 York Avenue
New York, N.Y.
USA

Wolfgang Schulze
Central Institute of Heart
 and Circulatory
 Regulation Research
Academy of Science of GDR
Lindenberger Weg 70
1115 Berlin Buch
GDR

Milka Setchenska
Central Lab. of Biophysics
Bulgarian Acad. of Sciences
1113 Sofia
Bulgaria

Radoslav Simeonov
Institute of Microelements
Boul. Lenin 72
Sofia
Bulgaria

Ralitza Staikova
Central Lab. of Biochemistry
Faculty of Medicine
Plovdiv
Bulgaria

Maria Staub
1st Institute of Biochemistry
Semmelweis University Medi-
 cal School
VIII Pecskin u. 9
Budapest
Hungary

Anton Stoimenov
Central Lab. of Biophysics
Bulgarian Acad. of Sciences
1113 Sofia
Bulgaria

Bayard Storey
Dept. of Obstetrics and
 Gynecology
Univ. of Pennsylvania
 School of Medicine
Philadelphia, PA 19174
USA

V. Sud
Dept. of Biophysics
All-India Institute of
 Medical Sciences
New Delhi-110016
India

LIST OF PARTICIPANTS

Gospodin Sveshtarov
Center for Biology
Bulgarian Acad. of Sciences
1113 Sofia
Bulgaria

Ah Ti Tan
Faculty of Medicine
McGill University
3655 Drummond St.
Montreal
Canada

Jamshed Tata
National Institute for
　Medical Research
Mill Hill
London NW7 1AA
United Kingdom

Dintcho Traikov
Repub. Hospital
Boris I St.
Sofia
Bulgaria

Penka Tsocheva-Mitseva
Central Lab. of Biophysics
Bulgarian Acad. of Science
1113 Sofia
Bulgaria

Yacov Varshavsky
Academy of Sciences of the
　USSR
Institute of Molecular
　Biology
Vavilov St. 32
Moscow B-312
USSR

Nicola Vassilev
Central Lab. of Biophysics
Bulgarian Acad. of Sciences
1113 Sofia
Bulgaria

Julia Vassileva-Popova
Central Lab. of Biophysics
Bulgarian Acad. of Sciences
1113 Sofia
Bulgaria

Nelli Visheva
Institute of Endocrinology
Medical Academy
Boul. Chr. Michailov 6
Sofia
Bulgaria

Viktor Vransky
Agricultural Academy
Boul. Dragan Zankov 8
Sofia
Bulgaria

Tony Yanev
Central Lab. of Biophysics
Bulgarian Acad. of Sciences
1113 Sofia
Bulgaria

W. Zainzinger
Beckman Instruments Austria
Stefan-Esders-Platz 4
A-1191 Vienna
Austria

SUBJECT INDEX

Abortion, HPL in, 147
ACTH (adrenocorticotropic hormone), 13, 19, 264, 355, 356
Adenohypophasis, 361-67, 391
Adenosine triphosphate (see ATP)
Adrenaline, 265-66 (see also Epinephrine)
Adrenocortical hormones, 264, 355-56, 401
Aging (ageing)
  adaptation in, 66
  adrenocortical hormones in, 355-56
  central nervous system (CNS) in, 63, 66-68
  hormone dysfunction in, 351
  hormone regulation in, 63-88, 354-55
  hypothalamus in, 254
  information transfer in, 63-88
  process of, 349-51
  sex steroids effects in, 68-82
  stress response in, 356=57
Alcoholism, testicular efficiency and, 371-76
Allosteric interactions, conformational changes and, 127-35

Amino acids
  in age changes, 65-67
  in protein synthesis, 272-73
Anemia, blood cell synthesis in, 285-89, 291-95
Antibody-antigen reactions, 13-19
ATP (adenosine triphosphate)
  activity in sperm, 299
  effects of gonadotropic hormones on, 160
Autoimmunity, 356

Binding, of chorionic gonadotropin (hCG)
  and free mobile cells, 203-08
  to luteal cells, 177-197
  of protein hormones, 203-08
Biochemical research, equipment and technology, 341-45
Bioelectric brain activity, sex hormones and, 385-88
Biomembranes
  elasticity of, 113-14
  flexoelectric effects on, 112-23
  liquid crystalline structure, 111-12

Biopolymer molecules
  informational transfer activities of, 99-109
  investigational perspectives, 141-42
Biotons, 3, 4, 9
Blood flow, magnetic field effects on, 427-40
Blood sugar levels, and age, 352
Bone marrow, erythroid cell development, 285-89

Caffein, effects on lung triglyceride lipases, 423
Calcium ion activity, in cell respiration, 297-304
Cardiovascular system, aging manifestations in, 68
Catecholamine hormones, 391-92
  (see also Epinephrine)
Cells
  binding of IhCG to different types, 188
  blood, 203-08, 285-89
  deep freezing of, 377-83
  enzyme role in division of, 287-89
  erythroid, 285-89
  free moving, 203-08, 391-96
  proliferation of, factors effecting, 297-99, 371-76, 406
  see Sperm cells
  stimulation of growth, 273-75
Central regulatory proteins(CRP), 43-51, 53-55, 63
Centrifuges, 342-43
Chemical mass action, substrate concentration and, 53-62
Chorionic gonadotropin (CG), binding of, 177-97
Chromatin, 65, 267
Collagen, 356, 401, 404
Control systems, for development in time, 35-41
Cortisol, 401-06

Corticosteroids
  in aging research, 68
  autoimmunity and, 356
  thymus changes due to, 401-03
Cyclical AMP (anedosine monophosphate), 297-304
Cyclical recurrence, in hormone regulation, 43-50
Cystol, estradiol-receptor transformation in, 211-23
  DHT binding and, 232, 234
Cytoplasm
  growth response in target, 276-78
  protein synthesis in, 270-76

Dehydroepiandrosterone (DHEA), inhibitory effects of, 279-83
Deuterium-hydrogen exchange, 100-07
alpha-Dihydrotestosterone (DHT), target receptor response of, 231-34
Diabetes mellitus, testicular efficiency in, 371-76
DNA (deoxyribonucleic acid)
  estrogen effects on, 239-41
  informational role of, 99, 109
  synthesis, 267, 286, 297-99, 371-76, 406, 409-17
DNA polymerase, 409-17

Ecdysone, 265, 266, 272
ECG (electrocardiogram), data transmission, 335-39
EEG (electroencephalogram), changes in due to sex hormones, 385-88
Embryoactive substances (ontogenins), 307-08
Endocrinology, of aging, 68
Energetic efficiency, calculations of, 4-7
Energy conversion, molecular, 137-38, 142

INDEX

Environment, as hormonal stimulus, 261-64
Enzymes
  in aging, 65-66
  in blood cell differentiation, 285-89
  gonadotropic hormone effects, 161
  in piezoelectric effects in hormones, 141
  proteolytic, 177, 197
  substrate interactions with, 53-61, 83-84, 103
Epididymal sperm, 297-304
Epinephrine (adrenaline)
  adenohyphasis and, 391
  free moving cells and, 391-96
  lung triglyceride lipases and, 419-25
  response to, 265-66
  sperm cells and, 391-96
Erythroblasts, 285-89, 291-95
Erythrocytes, magnetic field effects on, 446
Erythropoiesis, 267, 285-89, 355
Estriadol-receptor complex, transformations, 211-20
Estrogen, 266, 272 (see also Sex steroids)
  in cytoplasmic protein synthesis, 272-3
  nuclei target cell responses to, 271
  synthetic, in protein synthesis, 237-48
  target receptor transformation and, 211-23
  testosterone conversion to, 231
Estrophilin
  isolation and purification, 220-23
  transformation of, 211-33
External electric fields, in biomembranes, 117-23

Feminization, testicular, 229-34
Flashing-light spectroscopy, 313-332
Flexoeffect, 123

Flexoelectric polarization
  in biomembranes, 111-23
  cybernetic model of, 122
  and piezoelectric effects, 142
Fluctuation equilibrium, in conformational changes, 127, 130-33
Fluid mosaic model, 103, 201
Follicle stimulating hormone (FSH), 187-97, 203-07, 231
Free mobile cells, protein hormone binding to, 203-08

Glycoproteins, 83, 377-78
Gonadotropic hormones, 157-62
Growth hormones, 265, 267, 273-278
Growth of organic bodies, calculations for, 3-11

Hapten-antibody reaction, 13
Heavy water, 100-07
Hemoglobin
  conformational changes in, 106-09
  infra-red spectroscopy, 106
  magnetic field effects, 446
  production of, 285-88
Heterotropic effect, formula for, 76-77
Histone, 267, 288
Homotropic effect, formula for, 75
Hormonal activity, target receptor response, 82-88
Hormonal-antibody reactions, allosteric model, 13-19
Hormonal concentration, periodic regulation and, 82
Hormonal metabolism, 21-33
Hormonal receptor complex, model of, 23
Hormonal receptors (see Target receptors)

Hormonal regulation
  in aging, 63-88, 349-58
  cyclical recurrence of, 45-50
  of growth, mathematical aspects, 3-11
  processes of, 82-88, 261-78
Hormone-hormone complex formation, 251-55
Hormones (see also specific kinds and types)
  evolutionary position of, 263
  growth and development functions, 263-69
  informational transfer functions, 261-62
  protein synthesis stimulation, 275-78
  releasing (RH), 49-51
Hormone substrates, characteristics, 83-84
Human placental lactogen (HPL), synthesis of, 147-55
Hydrogen isotope exchange, 100-6
Hydrophobic interaction of protein and steroid hormones, 254-55, 281
Hyperplasia, 361-67
Hypothalmic-pituitary-adrenal axis, 354-58
Hypothalmic release hormones, 264

Information transfer
  in aging processes, 64-88
  by biopolymer conformation changes, 99-109
  chemical, 229, 234
  by DNA molecules, 99, 109
  in embryos, 307-308
  hormonal functions in, 261-64
  in thymus regeneration, 407-07
  types of processes, 99-100
Insulin, 201-02, 265-66
  effects on lung triglyceride lipases, 419-25
  secretion of, and age, 352
Interstital cell stimulating hormone (ICSH), 63, 70-71

d-Irradiation, gonadal effects from, 361-67
Islets of Langerhans, 352
Isoenzyme inhibition, 279-83

Lactogen, human placental (HPL), synthesis of, 147-55
Lecithin molecule, 114, 116
Leucocytes, magnetic field effects on, 446
Life span, 350-51
Lipid molecules, structural properties, 112-19
Liquid crystals, structural properties, 112-17, 123
Luteal cells and membranes, hCG binding of, 179-99
Luteinizing hormone (LH), 45-48, 178-97, 231, 252
Luteotropic hormone (LTH), 203, 207
Lymphocytes, 401-06
  DNA synthesis in, 409-17
  in thymus grafts, 406
Lysosomes, 288, 355

Magnetic field, physiological effects of, 441-49
Magnetic field strength, and fluid flow pressure, 427-40
Magnetohydrodynamics (MHD), 427-40
"Magnetic" water, physiological effects, 441-49
Mass action, law of, in substrate concentration, 53-62
Membrane changes, in aging, 66
Metabolic changes, hormone regulation of, 264, 275-78
Mitochondria, Calcium ion uptake by, 297-303
Mitosis, estrogen effects on, 237

Neurotransmitters, 67, 86-87, 261
Nucleic acid
  in aging research, 65-66
  conformational transitions of, 103, 109
  in informational transfer, 100, 103
  synthesis of, 237-48, 287
Nucleus, cellular
  genetic transcription and, 268
  growth response in, 276-78

Ontogenines, embryoactive substances, 307-08
Organic body growth, calculations of, 3-11
Oscillation in regulatory hormones, 43-51
Oxytocin effect, in hormone receptor theory, 21-33

Parathyroid hormone, 266
Peptide NH groups, 100-07
Periodicity, in hormonal regulation, 43-51, 82
Pheromones, 261
Piezoelectric effect (PEE)
  anisitropic coefficient of, 138
  in biomembranes, 113-14, 119
  see also Flexoelectric polarization
  in protein crystals, 137-142
Pituitary, anterior (see Adenohypophasis)
Pituitary tissue, prolactin in, 169-75
Placental proteins, synthesis of, 147-55
Plasma membrane (PM), liver, 201
  luteal, 179, 187, 192, 196
Polypeptides, 101-03, 201-02
Prednisone, 355
Prolactin (Pr)
  binding to blood and sperm cells, 203-06
  blood serum and pituitary concentrations, 169-75
  in hormone-hormone interaction, 252
  in male reproductive function, 174
Prostoglandin (PGF-2 alpha), in placental protein synthesis, 147-55
Protein conformation
  allosteric site changes, 127-34
  in biopolymer molecules, 102-108
  in hormone-hormone interaction, 251-55
Protein globules (crystals)
  active transport mechanism, 116-17
  external electric field, 119-21
  flexoelectric orientation, 117
  hydrogen isotope exchange in, 103-05
  piezoelectric effect in, 137-142
Protein hormones (PH)
  in age changes, 63-87
  binding in blood and sperm cells, 203-08
  cyclical regulation in, 43-51
  in sex steroid interaction, 75-80, 251-55
  substrate effects in, 53-61
Protein synthesis
  amino acids in, 272-73
  by bone marrow cells, 286-87
  in cytoplasm, 272-73
  hormonal regulation of, 265-66
  placental, 147-55
  ribosome stimulation and, 273-278
  synthetic estrogen effects on, 237-48

Receptors (see Target receptors)

Releasing hormones (RH), 49-51
Reticulocytes, 292, 294
Ribosomes
    hormone stimulated proliferation of, 273-75
    in protein synthesis, 246-47, 270-71, 286
RNA (ribonucleic acid)
    blood cell content in anemia, 291-95
    concentration effected by synthetic estrogen, 239-48
    messenger (mRNA), 269-72, 288
    see RNA-polymerase
    ribosomal (rRNA), 270-71, 286
    see RNA synthesis
    transcriptional control, 267-268
RNA-polymerase
    in genetic transcription, 267-268
    informational role of, 99
    synthesis capability, 216-20
RNA synthesis
    in bone marrow cells, 296-98
    estrogen effects on, 238
    hormonal stimulation of, 273-278, 285-89
    see also Protein synthesis
    ribosomes in, 272-73

Secretin, 201-02
Semen, deep freezing of, 377-83
Seminal plasma glycoproteins, 377-83
Sex character information, 229-234
Sex factor, in prolactin content, 169-75
Sex steroids (SS)
    in age regulation, 68-80
    EEG responses to, 385-88
    see also Estrogen
    in hormone-hormone interaction, 251-55
    inhibitory effects of, 75, 80, 81, 279
    protein hormone interaction, 75-80, 251-255
    substrate concentration and, 53-61
    target receptor interactions,
Sexual disorders, and spermatogenesis, 371-76
Spermatogenic activity test (SAT), 371-76
Spermatogenesis, factors effecting, 371-76
Sperm cells, human, 203-08
    in epididymis, 297-304
    epinephrine effects on, 391-397
Sperm dilution, 377-83
Substrate
    binding on tetramer enzymes, 129
    concentration and mass action, 53-62
    enzyme interaction, 83-84, 103

Target receptors (TR)
    in age change regulation, 63-74, 83, 86-87
    central signal and, 82
    chorionic gonadotropin binding and, 187, 192
    of epinephrine, 392-93
    estrogen effects on, 211-23
    hepatic, 201-02
    hormonal activation of, 82-86, 285
    protein hormone interaction in, 75, 78-80
    protein synthesis simulation in, 275-78
    in recurrent hormonal regulation, 45-51
    site affinity modulation, in luteal cells, 177, 196
    substrate effects on, 13-61
Template activity, hormonal regulation of, 267
Testicular biogenic activity, 157-62

INDEX

Testicular feminization, 229-34
Testicular function efficiency, 361-67, 371-76
Testosterone, 229-34
Thymus
  corticosteroid effects on, 401-03
  graft implants of, 401-07
  hormones and age changes, 69
  regeneration, 406
Thyreo-stimulating hormone (TSH), 163-67, 178, 264
Thyroid functions, 351-52
Thyroid gland, hormone effects on, 264
Thyroxine, 266
Tonsillar lymphocytes, DNA synthesis in, 409-17
Transcription, genetic, 267-68, 270-72
Translation control, in RNA synthesis, 268-69, 272-73
Tryptophane (Trp), in hormone-hormone interaction, 251-255

Vagus, cervical, EEG response to sex stimuli, 385-88
Vascular systems, magnetic field effects on, 427-40
Vasoactive intestinal polypeptide (VIP), 201-02
Vasopressin, 266
Vitellogenin, 272

Water, magnetic field effects and, 441-49